Global Water Dynamics

BOOKS IN SOILS, PLANTS, AND THE ENVIRONMENT

Editorial Board

Agricultural Engineering	Robert M. Peart, University of Florida, Gainesville
Animal Science	Harold Hafs, Rutgers University, New Brunswick, New Jersey
Crops	Mohammad Pessarakli, University of Arizona, Tucson
Environment	Kenneth G. Cassman, University of Nebraska, Lincoln
Irrigation and Hydrology	Donald R. Nielsen, University of California, Davis
Microbiology	Jan Dirk van Elsas, Research Institute for Plant Protection, Wageningen, The Netherlands
Plants	L. David Kuykendall, U.S. Department of Agriculture, Beltsville, Maryland
	Kenneth B. Marcum, Arizona State University, Mesa, Arizona
Soils	Jean-Marc Bollag, Pennsylvania State University, University Park, Pennsylvania
	Tsuyoshi Miyazaki, University of Tokyo

Soil Biochemistry, Volume 1, edited by A. D. McLaren and G. H. Peterson
Soil Biochemistry, Volume 2, edited by A. D. McLaren and J. Skujiņš
Soil Biochemistry, Volume 3, edited by E. A. Paul and A. D. McLaren
Soil Biochemistry, Volume 4, edited by E. A. Paul and A. D. McLaren
Soil Biochemistry, Volume 5, edited by E. A. Paul and J. N. Ladd
Soil Biochemistry, Volume 6, edited by Jean-Marc Bollag and G. Stotzky
Soil Biochemistry, Volume 7, edited by G. Stotzky and Jean-Marc Bollag
Soil Biochemistry, Volume 8, edited by Jean-Marc Bollag and G. Stotzky
Soil Biochemistry, Volume 9, edited by G. Stotzky and Jean-Marc Bollag
Soil Biochemistry, Volume 10, edited by Jean-Marc Bollag and G. Stotzky

Organic Chemicals in the Soil Environment, Volumes 1 and 2, edited by C. A. I. Goring and J. W. Hamaker
Humic Substances in the Environment, M. Schnitzer and S. U. Khan
Microbial Life in the Soil: An Introduction, T. Hattori
Principles of Soil Chemistry, Kim H. Tan
Soil Analysis: Instrumental Techniques and Related Procedures, edited by Keith A. Smith
Soil Reclamation Processes: Microbiological Analyses and Applications, edited by Robert L. Tate III and Donald A. Klein
Symbiotic Nitrogen Fixation Technology, edited by Gerald H. Elkan

Soil–Water Interactions: Mechanisms and Applications, Shingo Iwata and Toshio Tabuchi with Benno P. Warkentin
Soil Analysis: Modern Instrumental Techniques, Second Edition, edited by Keith A. Smith
Soil Analysis: Physical Methods, edited by Keith A. Smith and Chris E. Mullins
Growth and Mineral Nutrition of Field Crops, N. K. Fageria, V. C. Baligar, and Charles Allan Jones
Semiarid Lands and Deserts: Soil Resource and Reclamation, edited by J. Skujiņš
Plant Roots: The Hidden Half, edited by Yoav Waisel, Amram Eshel, and Uzi Kafkafi
Plant Biochemical Regulators, edited by Harold W. Gausman
Maximizing Crop Yields, N. K. Fageria
Transgenic Plants: Fundamentals and Applications, edited by Andrew Hiatt
Soil Microbial Ecology: Applications in Agricultural and Environmental Management, edited by F. Blaine Metting, Jr.
Principles of Soil Chemistry: Second Edition, Kim H. Tan
Water Flow in Soils, edited by Tsuyoshi Miyazaki
Handbook of Plant and Crop Stress, edited by Mohammad Pessarakli
Genetic Improvement of Field Crops, edited by Gustavo A. Slafer
Agricultural Field Experiments: Design and Analysis, Roger G. Petersen
Environmental Soil Science, Kim H. Tan
Mechanisms of Plant Growth and Improved Productivity: Modern Approaches, edited by Amarjit S. Basra
Selenium in the Environment, edited by W. T. Frankenberger, Jr., and Sally Benson
Plant–Environment Interactions, edited by Robert E. Wilkinson
Handbook of Plant and Crop Physiology, edited by Mohammad Pessarakli
Handbook of Phytoalexin Metabolism and Action, edited by M. Daniel and R. P. Purkayastha
Soil–Water Interactions: Mechanisms and Applications, Second Edition, Revised and Expanded, Shingo Iwata, Toshio Tabuchi, and Benno P. Warkentin
Stored-Grain Ecosystems, edited by Digvir S. Jayas, Noel D. G. White, and William E. Muir
Agrochemicals from Natural Products, edited by C. R. A. Godfrey
Seed Development and Germination, edited by Jaime Kigel and Gad Galili
Nitrogen Fertilization in the Environment, edited by Peter Edward Bacon
Phytohormones in Soils: Microbial Production and Function, William T. Frankenberger, Jr., and Muhammad Arshad
Handbook of Weed Management Systems, edited by Albert E. Smith
Soil Sampling, Preparation, and Analysis, Kim H. Tan
Soil Erosion, Conservation, and Rehabilitation, edited by Menachem Agassi
Plant Roots: The Hidden Half, Second Edition, Revised and Expanded, edited by Yoav Waisel, Amram Eshel, and Uzi Kafkafi
Photoassimilate Distribution in Plants and Crops: Source–Sink Relationships, edited by Eli Zamski and Arthur A. Schaffer

Mass Spectrometry of Soils, edited by Thomas W. Boutton and Shinichi Yamasaki

Handbook of Photosynthesis, edited by Mohammad Pessarakli

Chemical and Isotopic Groundwater Hydrology: The Applied Approach, Second Edition, Revised and Expanded, Emanuel Mazor

Fauna in Soil Ecosystems: Recycling Processes, Nutrient Fluxes, and Agricultural Production, edited by Gero Benckiser

Soil and Plant Analysis in Sustainable Agriculture and Environment, edited by Teresa Hood and J. Benton Jones, Jr.

Seeds Handbook: Biology, Production, Processing, and Storage, B. B. Desai, P. M. Kotecha, and D. K. Salunkhe

Modern Soil Microbiology, edited by J. D. van Elsas, J. T. Trevors, and E. M. H. Wellington

Growth and Mineral Nutrition of Field Crops: Second Edition, N. K. Fageria, V. C. Baligar, and Charles Allan Jones

Fungal Pathogenesis in Plants and Crops: Molecular Biology and Host Defense Mechanisms, P. Vidhyasekaran

Plant Pathogen Detection and Disease Diagnosis, P. Narayanasamy

Agricultural Systems Modeling and Simulation, edited by Robert M. Peart and R. Bruce Curry

Agricultural Biotechnology, edited by Arie Altman

Plant–Microbe Interactions and Biological Control, edited by Greg J. Boland and L. David Kuykendall

Handbook of Soil Conditioners: Substances That Enhance the Physical Properties of Soil, edited by Arthur Wallace and Richard E. Terry

Environmental Chemistry of Selenium, edited by William T. Frankenberger, Jr., and Richard A. Engberg

Principles of Soil Chemistry: Third Edition, Revised and Expanded, Kim H. Tan

Sulfur in the Environment, edited by Douglas G. Maynard

Soil–Machine Interactions: A Finite Element Perspective, edited by Jie Shen and Radhey Lal Kushwaha

Mycotoxins in Agriculture and Food Safety, edited by Kaushal K. Sinha and Deepak Bhatnagar

Plant Amino Acids: Biochemistry and Biotechnology, edited by Bijay K. Singh

Handbook of Functional Plant Ecology, edited by Francisco I. Pugnaire and Fernando Valladares

Handbook of Plant and Crop Stress: Second Edition, Revised and Expanded, edited by Mohammad Pessarakli

Plant Responses to Environmental Stresses: From Phytohormones to Genome Reorganization, edited by H. R. Lerner

Handbook of Pest Management, edited by John R. Ruberson

Environmental Soil Science: Second Edition, Revised and Expanded, Kim H. Tan

Microbial Endophytes, edited by Charles W. Bacon and James F. White, Jr.

Plant–Environment Interactions: Second Edition, edited by Robert E. Wilkinson

Microbial Pest Control, Sushil K. Khetan

Soil and Environmental Analysis: Physical Methods, Second Edition, Revised and Expanded, edited by Keith A. Smith and Chris E. Mullins

The Rhizosphere: Biochemistry and Organic Substances at the Soil–Plant Interface, Roberto Pinton, Zeno Varanini, and Paolo Nannipieri

Woody Plants and Woody Plant Management: Ecology, Safety, and Environmental Impact, Rodney W. Bovey

Metals in the Environment: Analysis by Biodiversity, M. N. V. Prasad

Plant Pathogen Detection and Disease Diagnosis: Second Edition, Revised and Expanded, P. Narayanasamy

Handbook of Plant and Crop Physiology: Second Edition, Revised and Expanded, edited by Mohammad Pessarakli

Environmental Chemistry of Arsenic, edited by William T. Frankenberger, Jr.

Enzymes in the Environment: Activity, Ecology, and Applications, edited by Richard G. Burns and Richard P. Dick

Plant Roots: The Hidden Half, Third Edition, Revised and Expanded, edited by Yoav Waisel, Amram Eshel, and Uzi Kafkafi

Handbook of Plant Growth: pH as the Master Variable, edited by Zdenko Rengel

Biological Control of Crop Diseases, edited by Samuel S. Gnanamanickam

Pesticides in Agriculture and the Environment, edited by Willis B. Wheeler

Mathematical Models of Crop Growth and Yield, Allen R. Overman and Richard V. Scholtz III

Plant Biotechnology and Transgenic Plants, edited by Kirsi-Marja Oksman-Caldentey and Wolfgang H. Barz

Handbook of Postharvest Technology: Cereals, Fruits, Vegetables, Tea, and Spices, edited by Amalendu Chakraverty, Arun S. Mujumdar, G. S. Vijaya Raghavan, and Hosahalli S. Ramaswamy

Handbook of Soil Acidity, edited by Zdenko Rengel

Humic Matter in Soil and the Environment: Principles and Controversies, Kim H. Tan

Molecular Host Resistance to Pests, S. Sadasivam and B. Thayumanavan

Soil and Environmental Analysis: Modern Instrumental Techniques, Third Edition, edited by Keith A. Smith and Malcolm S. Cresser

Chemical and Isotopic Groundwater Hydrology: Third Edition, Emanuel Mazor

Agricultural Systems Management: Optimizing Efficiency and Performance, Robert M. Peart and W. David Shoup

Physiology and Biotechnology Integration for Plant Breeding, edited by Henry T. Nguyen and Abraham Blum

Global Water Dynamics: Shallow and Deep Groundwater, Petroleum Hydrology, Hydrothermal Fluids, and Landscaping, Emanuel Mazor

Additional Volumes in Preparation

Seeds Handbook: Biology, Production, Processing, and Storage, Second Edition, Revised and Expanded, Babasaheb B. Desai

Principles of Soil Physics, Rattan Lal and Manoj Shukla

Field Sampling: Principles and Practices in Environmental Analysis, Alfred R. Conklin Jr.

Sustainable Agriculture and the International Rice–Wheat System, edited by Rattan Lal, Peter R. Hobbs, Norman Uphoff, and David O. Hansen

Plant Toxicology: Fourth Edition, Revised and Expanded, edited by Bertold Hock and Erich F. Elstner

Global Water Dynamics
Shallow and Deep Groundwater, Petroleum Hydrology, Hydrothermal Fluids, and Landscaping

Emanuel Mazor
Weizmann Institute of Science
Rehovot, Israel

MARCEL DEKKER, INC. NEW YORK • BASEL

Although great care has been taken to provide accurate and current information, neither the author(s) nor the publisher, nor anyone else associated with this publication, shall be liable for any loss, damage, or liability directly or indirectly caused or alleged to be caused by this book. The material contained herein is not intended to provide specific advice or recommendations for any specific situation.

Trademark notice: Product or corporate names may be trademarks or registered trademarks and are used only for identification and explanation without intent to infringe.

Library of Congress Cataloging-in-Publication Data
A catalog record for this book is available from the Library of Congress.

ISBN: 0-8247-5322-4

This book is printed on acid-free paper.

Headquarters
Marcel Dekker, Inc., 270 Madison Avenue, New York, NY 10016, U.S.A.
tel: 212-696-9000; fax: 212-685-4540

Distribution and Customer Service
Marcel Dekker, Inc., Cimarron Road, Monticello, New York 12701, U.S.A.
tel: 800-228-1160; fax: 845-796-1772

Eastern Hemisphere Distribution
Marcel Dekker AG, Hutgasse 4, Postfach 812, CH-4001 Basel, Switzerland
tel: 41-61-260-6300; fax: 41-61-260-6333

World Wide Web
http://www.dekker.com

The publisher offers discounts on this book when ordered in bulk quantities. For more information, write to Special Sales/Professional Marketing at the headquarters address above.

Copyright © 2004 by Marcel Dekker, Inc. All Rights Reserved.

Neither this book nor any part may be reproduced or transmitted in any form or by any means, electronic or mechanical, including photocopying, microfilming, and recording, or by any information storage and retrieval system, without permission in writing from the publisher.

Current printing (last digit):

10 9 8 7 6 5 4 3 2 1

PRINTED IN THE UNITED STATES OF AMERICA

PREFACE

Hydrology has been almost taken over by mathematical modeling. However, natural systems are multidimensional and multiparametric, and to understand them real data—observations and measurements obtained in real study areas—are needed. The following are examples of the applied research approach:

The three spatial dimensions are surveyed via a large number of sampling and measuring points, e.g., springs, wells, drillings, geological surveys, and more.

The time dimension is addressed by investigation of the geological and hydrological dynamics of the system, e.g., by dating of the groundwater host rocks, by isotopic dating of the water itself, and on the small scale by seasonally repeated measurements.

Multiparametric observations and measurements include water table positions, water heads, water temperature, and an extended list of chemical and isotopic analyses performed on carefully collected water samples. The final picture of every study case is derived on the basis of a large number of observations and derived conclusions.

Correlations between measured parameters are sought, as they provide indispensable insights into the studied systems, e.g., identification of mixing of water sources; external origins of dissolved ions versus water–rock interactions; temperature-induced processes; evaporation effects; and processes such as dolomitization, de-dolomitization, and absorption.

Spatial distribution of water facies, in depth profiles and between adjacent wells or springs, sheds light on the occurrence of shallow through-flowing groundwater systems, and deeper isolated rock-compartments, that contain fossil formation waters as well as oil and gas.

The book addresses topics related to groundwater exploitation and preservation, petroleum genesis and exploration; thermal water recreation and energy production; nuclear waste repositories; and the educational aspects of these topics.

Emanuel Mazor

CONTENTS

Preface   iii

Part I. The Geohydroderm and Its Major Groundwater-Containing Geosystems

1. Water Propelled Geological Processes and Shaped the Landscapes of Our Planet   2
   - 1.1 Water—Earth's Sculptor   2
   - 1.2 Water—The Unique Fluid on Our Planet   5
   - 1.3 The Special Properties of Water that Are the Base of All the Phenomena Dealt with in this Book   7
   - 1.4 Key Roles of the Oceans in the Dynamics of the Global Water Cycle   10
   - 1.5 Fresh Water Erodes Mountains but Exists Thanks to Them   11
   - 1.6 Formation Water, Entrapped in Isolated Rock-Compartments, Has a Meteoric Isotopic Composition and an Imprint of Evaporitic Brines   11
   - 1.7 Location of Land and Sea Changed Constantly   11
   - 1.8 Petroleum Hydrology   12
   - 1.9 Earth Exhibits Rocks that Are Unique Resources of this Planet—Products of Water-Induced Processes   13
   - 1.10 The Dynamics of the Global Water Cycle Propelled Biological Evolution   13
   - 1.11 Summary Exercises   15

2. Exploring and Understanding the Geohydroderm by Sequences of Observations and Conclusions ... 16
   2.1 Global Groundwater Research Within the Geohydroderm ... 16
   2.2 The Active Cycle of Fresh Surface Water and Unconfined Groundwater ... 18
   2.3 Interstitial Water Entrapped in Rocks Beneath the Vast Oceans ... 24
   2.4 Fossil Formation Waters Entrapped Within Sedimentary Basins and Rift Valleys ... 27
   2.5 Halite and Gypsum ... 31
   2.6 Shallow and Deep Groundwaters Are Indispensable Geological Records ... 31
   2.7 Brine-Spray-Tagged Meteoric Formation Water Is Also Common Within Crystalline Shields ... 33
   2.8 Petroleum Occurrence and Genesis ... 35
   2.9 Warm and Boiling Groundwaters ... 37
   2.10 Summary Exercises ... 39

3. Basic Research Concepts, Aims and Queries, Tools, and Strategies ... 41
   3.1 Basic Research Concepts and Terms ... 41
   3.2 Research Aims and Queries ... 44
   3.3 The Research Tools ... 53
   3.4 Research Strategies ... 61
   3.5 Summary Exercises ... 63

Part II. Shifting of Water and Salts Between Oceans and Continents

4. Shallow Cycling Groundwater, Its Tagging by Sea Spray, and the Underlying Zone of Static Groundwater ... 66
   4.1 Groundwater Facies of the Geohydroderm ... 66
   4.2 Sea Spray Salts Concentrated Along a Large River System—The Murray River Basin, Australia ... 68
   4.3 Sea Spray Salts Concentrated in a Closed Lake System Within an Arid Zone—Yalgorup National Park, Australia ... 75
   4.4 Sea Spray Salts Concentrated in Unconfined Groundwater—Campaspe River Basin, Australia ... 76
   4.5 Sea Spray-Tagged Fresh and Saline Groundwaters in the Unconfined Groundwater System at the Crystalline Shield of the Wheatbelt, Australia ... 78

Contents

	4.6 Sea Spray Versus Brine-Spray Tagging	82
	4.7 Sea-Derived Ions Serve as Benchmarks Identifying Water–Rock Interactions	82
	4.8 Gravitational Flow in the Unconfined Groundwater System and Static Water Storage Beneath	83
	4.9 Summary Exercises	91
5.	Interstitial Waters in Rock Strata Beneath the Oceans	92
	5.1 Extending Our Hydrological Curiosity to Beneath the Oceans	92
	5.2 The Deep Sea Drilling Project	93
	5.3 Water Content in Suboceanic Sediments	93
	5.4 The Widespread Marine Facies of Interstitial Water (Cl ~ 19 g/L, Cl/Br ~ 300, Diagenetic Changes Are Common)	94
	5.5 Continental Brine-Tagged Facies: Salinity Higher than Seawater Cl/Br 200 or Lower, Ca–Cl Present	98
	5.6 Information Retrievable from Below-Ocean Interstitial Waters	108
	5.7 Interstitial Water is Connate Water, Entrapped in Its Host Rocks Since the Initial Stage of Sedimentation	111
	5.8 Interstitial Waters Tagged by Brine-Spray Disclose that the History of the Mediterranean Sea Basin Included a Continental Stage	111
	5.9 Geological Evidence Proves that the Mediterranean Sea Underwent a Phase of Drying Up	112
	5.10 Summary Exercises	113
6.	Salt, Gypsum, and Clay Strata Within Sedimentary Basins Disclose Large-Scale Evaporitic Paleo-Landscapes	115
	6.1 Minerals Formed Along the Continuous Evaporation Path of Seawater and Notes on the Composition of the Residual Brines	115
	6.2 Formation of Halite and Gypsum Deposits Necessitated Evaporation of Tremendous Amounts of Seawater During Extended Time Intervals	116
	6.3 Evaporitic Paleo-Facies: Information Recorded by Associated Formation Waters	117
	6.4 The Permian "Saline Giant" of the Salado Formation—An Ancient Evaporitic Megasystem	118
	6.5 Evaporite Deposits Are Common in Sedimentary Basins	119

6.6	Silurian Salt Deposits Were Not Dissolved by the Nearby Formation Water	120
6.7	Recent Lowering of the Dead Sea Lowered the Coastal Groundwater Base Flow and Initiated Rapid Dissolution of a Buried 10,000-Year-Old Halite Bed	121
6.8	The Many Preserved Salt Beds Manifest the Preservation of Connate Groundwaters	122
6.9	Limestone–Clay Alterations Reflect Alternating Sea Transgressions and Regressions	123
6.10	Summary Exercises	123

Part III. Deep Groundwater Systems—Fossil Formation Waters

7. The Geosystem of the Fossil Brine-Tagged Meteoric Formation Waters — 126
    - 7.1 Formation Waters Within Sedimentary Basins — 126
    - 7.2 Formation Waters Within Rift Valleys — 140
    - 7.3 Fossil Nonsaline Groundwaters Tagged by $CaCl_2$, Formed During the Messinian, at the Land Bordering the Dried-Up Mediterranean Sea — 149
    - 7.4 Some Physical Aspects of Formation Waters — 151
    - 7.5 The Fruitcake Structure of the Formation Waters and Petroleum-Containing Geosystem — 153
    - 7.6 A Brief History of the Basic Concept of Connate Groundwater — 154
    - 7.7 The Bottom Line: Brine-Spray-Tagged Formation Waters Provide Markers of Paleo-Landscapes, Water Age, and Paleoclimate — 155
    - 7.8 Solving a Great Puzzle: Why Are Recent Groundwaters Sea Tagged and Commonly Rather Fresh, Whereas Formation Waters Are By and Large Saline and Brine Tagged? — 155
    - 7.9 Summary Exercises — 157

8. Fossil Formation Waters Range in Age from Tens of Thousands to Hundreds of Millions of Years — 158
    - 8.1 Confinement Ages of Connate Waters and Criteria to Check Them — 158
    - 8.2 Isotopic Dating of Fossil Groundwaters — 158

	8.3	Hydraulic Age Calculations—An Erroneous Approach to Confined Goundwaters, Which Are Static	161
	8.4	Radiogenic ^{40}Ar Dating	163
	8.5	Mixed Water Samples Are Commonly Encountered	164
	8.6	Isotopic Dating of Very Old Groundwaters	166
	8.7	Conclusions and Management Implications	169
	8.8	Summary Exercises	170
9.	Brine-Tagged Meteoric Formation Waters Are Also Common in Crystalline Shields: Geological Conclusions and Relevance to Nuclear Waste Repositories		171
	9.1	The Special Nature of Data Retrieved from Boreholes in Crystalline Rocks	171
	9.2	Observations Based on Data from the Fennoscandian and Canadian Shields and Deduced Boundary Conditions	176
	9.3	What Typifies Formation Waters Within Crystalline Rocks?	190
	9.4	Results from the KTB Deep Research Boreholes	194
	9.5	Isotopic Dating of the Fossil Groundwaters Within Shields	196
	9.6	Working Hypothesis: Tectonic "Fracture Pumps" Introduced Meteoric Groundwater to Great Depths	200
	9.7	The Saline Waters in Shields Serve as a Geological Record	200
	9.8	Nuclear Waste Disposal Implications	201
	9.9	Summary Exercises	202

Part IV. Petroleum Hydrology

10.	Anatomy of Sedimentary Basins and Petroleum Fields Highlighted by Formation Waters		205
	10.1	Petroleum and Associated Formation Waters Are Complementing Sources of Information	205
	10.2	Petroleum-Associated Formation Waters in the Western Canada Sedimentary Basin	206
	10.3	Petroleum-Associated Formation Waters Within Ordovician Host Rocks, Ontario, Canada	212
	10.4	Kettleman Dome Formation Waters Associated with Petroleum—Key Observations and Concluded Boundary Conditions	213

10.5	Shallow Formation Water and Petroleum in Devonian Rocks, Eastern Margin of the Michigan Basin	217
10.6	Petroleum-Associated Brines in Paleozoic Sandstone, Eastern Ohio	222
10.7	Formation Waters of the Mississippi Salt Dome Basin Disclose Detailed Stages of Petroleum Formation	228
10.8	Norwegian Shelf: Petroleum-Associated Formation Waters, Upper Triassic to Upper Cretaceous	235
10.9	Lithostratigraphic Controls of Compartmentalization Were Effective from the Initial Stage of Subsidence and Further Evolved Under Subsidence-Induced Compaction	238
10.10	Summary Exercises	239

11. Evolution of Sedimentary Basins and Petroleum Highlighted by the Facies of the Host Rocks and Coal — 240
    11.1 Sediments Formed in Large-Scale Sea–Land Contact Zones — 240
    11.2 Lithological Evidence of Subaerial Exposure Phases — 247
    11.3 The Lithological Record of Inland Basins and Rift Valleys — 249
    11.4 Rock-Compartment Structures and Their Evolution — 252
    11.5 Compartmentalization Was Effective from the Initial Stage of Subsidence and Further Evolved Under Compaction — 253
    11.6 Summary Exercises — 253

12. Petroleum and Coal Formation in Closed Compartments—The Pressure-Cooker Model — 255
    12.1 Did Petroleum Migrate Tens and Even Hundreds of Kilometers? — 255
    12.2 Coal—A Fossil Fuel Formed with No Migration Being Involved — 259
    12.3 Boundary Conditions Set by Formation Waters and Petroleum and Coal Deposits — 262
    12.4 The Pressure-Cooker Model of Petroleum Formation and Concentration Within Closed Compartments — 263

12.5	Another Case Study Supporting the Pressure-Cooker Model	267
12.6	Pressure-Regulating Mechanisms Within Rock Sequences Discussed in Light of the Fruitcake Structure of Isolated Rock-Compartments	268
12.7	Summary Exercises	269

Part V. Hydrology of Warm Groundwater and Superheated Volcanic Systems

13. Mineral and Warm Waters: Genesis, Recreation Facilities, and Bottling — 271
    - 13.1 The Anatomy of Warm Springs — 271
    - 13.2 Medicinal and Healing Aspects of Warm and Mineral Waters — 286
    - 13.3 Developing the Resource—The Hydrochemist's Tasks — 288
    - 13.4 Local Exhibitions Disclosing the Anatomy of Warm and Mineral Water Sources and Their Properties — 289
    - 13.5 Bottled "Mineral Water" — 290
    - 13.6 Summary Exercises — 290

14. Water in Hydrothermal and Volcanic Systems — 292
    - 14.1 Hydrothermal Systems — 292
    - 14.2 Yellowstone National Park, Western United States — 293
    - 14.3 Cerro Prieto, Northern Mexico — 304
    - 14.4 The Wairakei, Tauhara, and Mokai Hydrothermal Region, New Zealand — 310
    - 14.5 Noble Gases in a Section Across the Hydrothermal Field of Larderello, Italy — 314
    - 14.6 Fumaroles of Vulcano, Aeolian Island, Italy — 319
    - 14.7 The Hydrology and Geochemistry of Superheated Water in Hydrothermal and Volcanic Systems — 326
    - 14.8 Summary Exercises — 327

Part VI. Implementation, Research, and Education

15. Data Acquisition, Processing, Monitoring, and Banking — 329
    - 15.1 Sample Collection and In Situ Measurements — 329
    - 15.2 Checking the Laboratories' and Data Quality — 330
    - 15.3 Types of Wells — 331
    - 15.4 Multiparameter Studies — 332

	15.5	Multisampling	334
	15.6	Monitoring Networks	336
	15.7	Effective Data Banks	339
	15.8	Summary Exercises	340
16.	Conclusions and Research Avenues	341
	16.1	Criteria to Check Working Hypotheses Related to Global Water Occurrences	341
	16.2	Geosystems that Host Fluid Water—Research Topics	348
	16.3	Geological Records—Research Avenues	350
	16.4	Summary Exercises	352
17.	Educational Aspects of Water, the Unique Fluid of Planet Earth	354
	17.1	List of Educational Topics	355
	17.2	National Water and Man Museums	362
	17.3	Local Exhibitions and Water and Man Demonstration Centers	363
	17.4	Educational Water Recreation Parks	364
	17.5	Spas	364
	17.6	Teaching at School and Student Mini-Research Projects	364
	17.7	Teaching at Universities	365

Epilogue: Three Energy Sources and One Transporter—The Geo-Quartet Unique to Planet Earth *367*

Answers and Discussion of the Exercise Questions *369*

References *381*

Index *391*

Global Water Dynamics

PART I
THE GEOHYDRODERM AND ITS MAJOR GROUNDWATER-CONTAINING GEOSYSTEMS

~~~~~~~~~~~~~~~~~~~~

The geohydroderm
is the
water-containing skin
of
Planet Earth

~~~~~~~~~~~~~~~~~~~~

1
WATER PROPELLED GEOLOGICAL PROCESSES AND SHAPED THE LANDSCAPES OF OUR PLANET

The study of the domain of water touches its distribution in the three dimensions of space and along the time dimension that provides insight into its dynamic processes. Water, in one form or another, is found all over our planet and is encountered at depths of thousands of meters within the rocky crust. Water has been around since the early days of Earth, it has a history of around 4 billion years.

As liquid water in large amounts is absolutely unique to Earth, so are the outcomes and products of the water-involved geological processes. If Earth is to us a friendly home, it is thanks to all that water has created.

1.1 Water—Earth's Sculptor

The surfaces of two of our neighbors in the solar system, Mercury and Mars, are well exposed, and so is the face of our moon. On these planetary bodies we see the little-changed primordial landscape of meteoritic impact craters of all sizes, disclosing the last stages of accretion (Figs. 1.1, 1.3, and 1.4). In strong contrast, Earth has lost that primordial landscape due to continuous dynamic processes that go on to this day. Venus is covered by an atmosphere a hundred times denser than ours (Fig. 1.2) and is rich in CO_2. The surface temperature is around 400°C, so it is no place for water.

Water had a dominant role in the transformation of Earth's landscape, playing the part of the terrestrial sculptor, propelled by two giant sources of energy: (1) the internal energy active in pushing the crustal plates and building of mountains and (2) the sun that pumps the water from the oceans

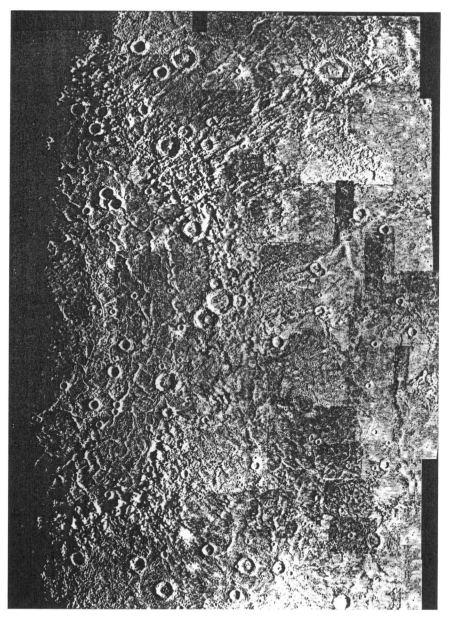

Fig. 1.1 A typical landscape of Mercury, disclosing uncountable impact craters. The length of the picture is around 1000 km. (Courtesy of NASA.)

Fig. 1.2 The dense atmosphere of Venus. Surface temperature is on the order of 400 °C; the pressure of the atmosphere is about hundred times that of Earth. (Courtesy of NASA.)

placing it on the landscape. As a result, water flows down and erodes the scenery, but the internal terrestrial forces keep creating new reliefs. During this endless cycle of erosive activity, sediments have accumulated in subsided bains and beneath the oceans, and in them are entrapped connate waters that we encounter in deep wells, often in close association with petroleum deposits.

All these processes are parts of the global water cycle, and all this history has direct bearing on what we study in the frame of chemical and isotopic hydrology of groundwater, petroleum, and the oceans.

Fig. 1.3 The landscape of Mars.

1.2 Water—The Unique Fluid on Our Planet

Before we dive into the fascinating topic of the four-dimensional global water cycle, let us marvel at our terrestrial treasure. The solar system has nine planets that revolve around the sun. Mercury, closest to the sun, is very hot and has a very thin atmosphere, mainly of helium and argon. Venus is second closest to the sun; the surface temperature is around 470°C, and it has a dense CO_2 atmosphere. Then comes Earth. Mars is the fourth planet, it is small and has a thin atmosphere, mainly of CO_2. Jupiter, the fifth planet, is the largest; it is very cold and has a helium and hydrogen atmosphere. The other planets—

Fig. 1.4 The landscape of our moon; the region is about 150 km across. (Courtesy of NASA.)

Saturn, Uranus, Neptune, and Pluto—are all far away from the sun, are very cold, and are made mainly of frozen hydrogen and helium.

Earth is the "blue planet," so called because of the oceans that cover two-thirds of the surface (Fig. 1.5). The size and distance from the sun are such that the physical conditions on this unique planet are suitable for water to exist in its three phases—ice, liquid water, and vapor. Furthermore, the size and density of Earth are such that the gravity pull is effective enough to keep the water vapor.

There are slight signs of some ice as part of the polar caps of Mars, which contain mainly frozen CO_2, and from time to time theories are put forth that some water existed on Mars in the past. The clear fact is that Earth is the only planet in the solar system that has *liquid water* and plenty of it. So far there has been no observation of liquid water in any body outside the solar system.

So let us respect our planet and treasure its skin of water. After all, water is the base of life—another uniqueness of Earth!

1.3 The Special Properties of Water that Are the Base of All the Phenomena Dealt with in this Book

As human beings we take water for granted, but as scientists let us devote some time to marvel at the extraordinary properties of this substance, which occupies such a key role in our planet. The list of special properties of water and their bearing on our environment and our life is amazingly extended, and every explorer can make up a different list. The following are just a few examples:

Water has three phases—fluid, gas, and solid—and it readily passes from one phase to another within the temperature range prevailing on the Earth's surface. On the other planets and the moon the prevailing temperature is in the range of freezing alone or vapor alone. The crucial outcome: The global water cycle is based on the three physical phases of water.

The solid phase of water is lighter than the fluid phase. This is anomalous; the solid phase of other materials is heavy and sinks in the fluid. The crucial outcome: If ice were heavier than water, we would have only tiny oceans! The water at the polar sections freezes seasonally, but the floating ice melts during the warm seasons, or floats to warmer regions and melts there. If ice were heavy, it would sink to the bottom of the ocean and there it would be isolated from the seasonal sun. Thus, huge parts of the oceans would be masses of ice, and only at the upper few meters would there be water during the warm seasons. This, in turn, would severely limit evaporation and cloud formation, and the oceanic streams would

Fig. 1.5 Earth as seen from space featuring a medium dense atmosphere; through the clouds are seen the vast oceans that give Earth the name "the blue planet." (Courtesy of NASA.)

be extremely limited in extent. The same holds true for lakes in cold climates.

Water can dissolve salts and gases but be readily separated as pure vapor. The sun evaporates vapor from the ocean, but the salts remain in the sea. The crucial result: Pure vapor clouds are driven by the wind (another process propelled by the sun), and the vapor condenses into rain and snow, which are the source of terrestrial fresh water that is eventually drained back to the ocean. In this mode the same water molecules have taken part uncountable times in the endless water cycle.

Water readily dissolves CO_2 and turns into an effective acid. The main source of CO_2 is provided to water on its passage through the root zone of soils. A very visual outcome of rocks attacked by this acid are the karstic features seen in many limestone terrains. But, actually, all rocks get dissolved by CO_2-induced water–rock interactions. This is a major mode of erosion by which material of the continental landscapes is transported into the sea.

Water has low viscosity and is not sticky. The basic outcomes: (1) Water flows efficiently as surface water. (2) Water infiltrates into the soil cover. (3) It flows through permeable rocks. (4) Thanks to the low viscosity the oceans get well mixed.

Water is heavy enough to be moved by the gravitational field. The crucial outcome: Runoff and underground flow are possible.

Water can split into subparcels and reunite. The fundamental result: Infiltration through little voids between soil and rock particles is feasible, uniting to a larger scale flow in the seaward base flow and then joining the huge water body of the ocean.

Water has a specific gravity that is just right. The crucial outcome: Water vapor does not escape from Earth.

The vapor and fluid phases are transparent. The valuable result: Sunlight can pass through the atmosphere and through the upper few hundred meters of seas and lakes—a prerequisite of life.

The water phase is a greenhouse gas. The crucial outcome: The temperature on Earth is kept balanced.

The 4×10^9-year-old terrestrial catalyst. Water exists on Earth for some 4 billion years and it is still around as it was on its first day. It is involved, for example, in photosynthesis; it is involved in the formation of sedimentary rocks; and it mobilizes and deposits ores; but it always comes back as pure water.

Oil is immiscible with water, a property that is intrinsic in the formation of clean petroleum deposits.

Oil floats on water, a property that is intrinsic in the migration of oil to traps (within the original rock-compartment; Chapter 10).

Water has a very high dissolution capacity of chlorides. The crucial outcome is efficient flushing of salts from the continent and their concentration in the ocean—operations that are a major part of the salt cycle.

Water is an effective lubricant as we know from wet floors that are slippery, landslides that occur following rainy seasons, and the hydrolubrication and hydrofracturing that are involved in seismic processes.

Last but not least, Water is the base of life.

1.4 Key Roles of the Oceans in the Dynamics of the Global Water Cycle

The ocean surface is the base of drainage of runoff and of erosion. Observations reveal that rivers flow to the ocean and drain to its surface and that rivers immediately respond to changes in the sea level, which have happened throughout geological history.

The sea surface is the base of erosion, as is revealed by the observation that beds of active rivers are never dug beneath the respective sea level. They were substantially deeper in the geological past, whenever the sea level dropped, but their beds were filled up by erosion products the moment the sea level rose.

The ocean surface is the base of drainage of the groundwater base flow, as is manifested by the observation that the active through-flow of groundwater is toward the sea surface. The water table in shallow coastal wells fluctuates daily with the high and low tides and responds to seasonal fluctuations of sea level. Similarly, the groundwater level followed the pronounced sea level drop during the glacial periods.

The ocean is the source of fresh water, driven landward by clouds. Water that stays on the continent's surface gets gradually saline by two major processes: dissolution of salts and other water–rock interactions and evaporation. If these were the only processes, fresh water would disappear. However, the water of the continents is drained back to the ocean, along with the salts, and the sun evaporates pure water that is wind driven to the continent, supplying new fresh water. Thus, the ocean is the source of fresh water found on the continents.

The ocean is the major reservoir of chlorides and other soluble salts. The ocean is undersaturated with regard to chloride compounds, and thus acts as a very efficient sink of these compounds. Small amounts of Cl and accompanying ions are driven landward by the wind, but they are cycled back with the seaward draining water.

The ocean is the receptor and storage place of continental erosion products. Water pumped by the sun and precipitated on the land relief runs back to the ocean, carrying large amounts of eroded rock materials that are sedimented on the ocean floor. During sea regressions such rocks get exposed to a new phase of erosion and transport into the sea. Thus, rocky material is shifted between marine and continental environments.

1.5 Fresh Water Erodes Mountains but Exists Thanks to Them

Ocean water, evaporated by the sun, is constantly lifted to the mountains, where it becomes runoff that fills up the fresh water lakes and rivers, recharges the groundwater, and creates springs. The fresh water runoff erodes the landscape by transporting the rocky material into the sea. By this process Earth would acquire a low flat landscape, and seawater would eventually invade the continents and cover them. At such a stage fresh water could not be formed any more, and Earth would be covered all over with saline seawater. This situation reminds us of the saying "Don't cut the branch you are sitting on." Water does it, and rough estimates reveal that this outcome would be reached in a period on the order of 20 million years.

To our luck, this is not what we observe—the landscape is rich in high mountains, the product of processes propelled by the terrestrial internal energy. Orogenic activity creates new mountains, and our landscape is kept high above sea level, continuing the process of rain and snow formation. *Thus, we have fresh water thanks to the mountain-building activity.* This is one of the many amazing aspects of the global water cycle.

1.6 Formation Water, Entrapped in Isolated Rock-Compartments, Has a Meteoric Isotopic Composition and an Imprint of Evaporitic Brines

Observations reveal that formation water encountered in neighboring drill holes often varies in its chemical and isotopic composition, disclosing entrapment in hydraulically separated rock-compartments (Chapter 7). The confinement ages and isotopically determined ages indicate these are indigenous connate waters. Almost as a rule, formation waters have light isotopic δD values, reflecting an origin as continental meteoric water, and the chemical composition reflects tagging by evaporitic brine-spray.

1.7 Location of Land and Sea Changed Constantly

Many of the sedimentary rocks that host the continental meteoric formation waters are of a marine origin. This observation-based conclusion reveals that in many parts of the globe sea transgressions and regressions alternated frequently (Chapter 7). Thus, the continental land we see exposed today is a brief episode in an ever-changing playground between sea and land. Marine fossils found in rock covering the continents testify that at their time the respective piece of land was covered by the ocean and alternations of regression and transgression occurred many times.

Other mechanisms that constantly shifted the location of sea and land were plate tectonics and continental drift, which must have had a strong influence on the global water cycle, i.e., the dispersion of water between oceans and continents.

1.8 Petroleum Hydrology

Oil and gas are fluids that are found in close association with formation waters (Chapter 10). The chemical and isotopic characteristics of the latter reveal boundary conditions that are relevant to the understanding of the formation of petroleum deposits:

Entrapment within hydraulically isolated rock-compartments. The composition and pressure of petroleum-associated formation waters vary between neighboring petroleum exploration drill holes and producing wells, revealing entrapment within separated rock-compartments. Hence, the associated oil or gas is entrapped in hydraulically isolated compartments as well. The same picture is outlined by the composition of the petroleum in producing fields—different compositions and pressures disclose the existence of numerous neighboring separated traps.

The ingredients needed for petroleum formation were contained in the rock-compartments since their initial formation. The isolated rock-compartments were formed at the moment their rocks subsided and were confined by overlying rocks (Chapter 11). Certain rock-compartments contained from the beginning all the ingredients needed for the formation of petroleum deposits: *organic raw material, permeable rocks*, and *formation water*.

Water prevented collapse of rocks as a result of compaction by the weight of the overlying rocks. This maintained the hydraulic connectivity that was essential in petroleum concentration in traps that make up exploitable deposits.

The petroleum-associated formation water subsided into the oil formation temperature window. We have seen that the isotopic composition of the hydrogen of formation waters reveals δD values of meteoric water, but the oxygen has often a relatively heavy $\delta 18O$ value, indicating temperature-induced isotopic exchange with the host rocks (Chapter 10). This, in turn, indicates that the petroleum-associated formation water subsided to the depth of the petroleum-formation temperature window.

Petroleum migration was limited to the original rock-compartments. As stated, formation waters reveal confinement within rock-compartments that are separated by clay-rich rock barriers (Chapters 9 and 11). Petroleum migration necessitates hydraulic connectivity, and therefore migration was limited to the original host compartments. This leads to the pressure cooker model of oil formation and concentration (Chapter 12).

1.9 Earth Exhibits Rocks that Are Unique Resources of this Planet—Products of Water-Induced Processes

Our daily life is anchored in the use of a long list of rocks that we take for granted, but it is amazing that all are products of water-induced processes, and they exist over the whole world on our planet alone. A few examples follow:

Building rocks like limestone, sandstone, pebbles, and the raw materials of lime and cement—all are sedimentary rocks.

Clays, needed for ceramics, cement, and other uses, are marine and lacustrine sediments.

Rock salt is precipitated from seawater evaporating in lagoons and sabkhas.

Gypsum an important ingredient of cement, is another rock precipitated from evaporating seawater.

Phosphorite serves as a major fertilizer and in a long list of industrial products and was formed in specific sections of the oceans.

Iron ore deposits are all sedimentary and connected to water-related concentration processes.

Bauxite, a major raw material for aluminum is a sort of paleo-soil formed in tropical climates.

Copper is produced from a variety of ores that have been concentrated by hydrothermal processes.

Soil, the traditional medium of agriculture, is a water-induced weathering product formed on the continents' land surface.

These are just a few examples. Literally, almost all the materials used by us are unique to Earth and their formation was water dependent.

1.10 The Dynamics of the Global Water Cycle Propelled Biological Evolution

Since the voyage of Charles Darwin we have been aware that within a terrain that is steady for a long enough time, the flora and fauna reach an evolutionary stability at which given species dominate. Individual offsprings are constantly born with a variety of mutations, but they do not fit the given stable environment and do not evolve. In contrast, in a changing environment mutational offsprings get a chance to thrive and eventually replace previously dominant species. A prominent arena of life-important dynamics was the constantly changing landscape that thrived as a result of the alternating sea transgressions and regressions, and the rising mountains that were intensively eroded by the water cycle.

The remarkable biological evolution advanced as a result of the geological dynamics, and the ever-changing global water cycle played a crucial role in this process.

The fossil record, preserved in sedimentary rock systems, is incomplete, and the research of the connate waters, addressed in the present book, complements it substantially in disclosing, for example, that uncountable times the sea covered continents and then retreated, only to return again and retreat once more. This dynamic cycle caused constant changes in the environment, providing the arena for the biological evolution. The salt content, relative ion abundances, and isotopic composition of fossil groundwaters reflect the enormous variability of the paleo-environments, and the water dating methods provide the relevant time scale.

We can go one step back and wonder what was the role of water in the origin of life. It seems that it was essential in several ways; examples follow:

Cleaning up a dense CO_2 primordial atmosphere. A large amount of sedimentary carbonate rocks cover the earth. A simple calculation shows that if we go back to the early history of Earth, before the marine carbonates were precipitated, all the involved CO_2 was in the atmosphere, which was as dense as the CO_2 atmosphere of our neighbor Venus. When our planet cooled enough, this CO_2 was gradually cleaned away and stored in the limestone and dolomite rocks. This was a prerequisite for the origin of life.

Providing the initial free atmospheric oxygen. At the early terrestrial history water vapor was photochemically decomposed, the hydrogen escaped into space, and the oxygen was consumed by a list of terrestrial materials. Eventually free oxygen accumulated, the ozone layer could be formed, and the remaining water got shielded. Thus, the initial dose of free oxygen was introduced into the ancient atmosphere, making room for advanced forms of life to evolve—thanks to the liquid water and the vapor.

Creating a countless assortment of landscapes and ecological niches that life could select for its coming into being. Scientists are still wondering at which environmental setup happened the miracle of the origin of life, but surely the activity of water prepared an enormous assortment of candidate locations.

Liquid water was an essential ingredient needed to sustain living forms, as it is to this day.

1.11 Summary Exercises

Exercise 1.1: Comparing the landscape of the Moon and that of Earth, we realize the great impact of water in the latter case. What is it?

Exercise 1.2: What benefit do we have from the fact that ice floats on fluid water?

Exercise 1.3: Water is an everlasting material—on which observations is this conclusion based?

Exercise 1.4: "Don't cut the branch you sit on" is an old wise saying. Is water almost violating it by eroding down the mountains? Discuss.

Exercise 1.5: Do we have a good reason to respect the water resource? What is the basic step we have to take in order to minimize anthropogenic negative intervention?

2
EXPLORING AND UNDERSTANDING THE GEOHYDRODERM BY SEQUENCES OF OBSERVATIONS AND CONCLUSIONS

Two-thirds of the Earth's surface is covered by water, the largest reservoir being the oceans, accompanied by fresh surface water bodies and polar ice and snow caps. But water is present also in the entire upper part of the lithosphere. The term *geohydroderm* has been suggested to address this global terrestrial water domain (Mazor, 1986). The geohydroderm is composed of a number of large-scale global subsystems for which the term *geosystems* is suggested; these are further addressed in section 2.1.2.

2.1 Global Groundwater Research Within the Geohydroderm

The geohydroderm term provides an umbrella concept for a wide look at the unique terrestrial fluid, from the time of its formation some 4 billion years ago and in its spatial distribution down to a depth of thousands of meters within the rocks.

2.1.1 The Challenge of Studying the Concealed Resource

The scientific understanding of the systems of nature is tricky. Let us have a look at the domain of geology: The geological setup around Paris is entirely different from the setup around New York or the formations encountered in the central Sahara. Furthermore, drilling to a depth of 200 m will in each location bring up certain sets of geological data, but deepening the drilling to

500 m will bring up substantially different observations. And additional complexity is experienced in regard to the research resolution. We will get a nice set of results studying an entire folded mountain system, but another researcher who looks in detail at a single outcrop will end up with entirely different data sets.

The question comes up how do we learn about the general rules of nature? How can we see the forest beyond the trees? In the field of groundwater hydrology the situation is similarly complicated, as the subject of our investigation is concealed underground, and we have access to it only at very limited points—springs, wells, and drillings. At each of these points we study the nature of the rocks, the level of the water table, the water temperature, a long list of dissolved ions, the concentration of isotopic age indicators, the isotopic composition of the water phase, and so on. But we are confronted with a major unknown; what happens to the water systems between the points of our studies? A most fruitful approach is to become familiar with as many case studies as possible, in each conduct observations and draw conclusions, and from these build up working hypotheses. The latter are sound only if they include criteria to check their validity in new study areas, thus leading to more sets of observations and conclusions, serving to support the working hypothesis or to rephrase it.

The subjects addressed in the present book are of a wide scope, having in common the strong relation to water stored underground. Part I outlines the logical order of sets, or bundles, of observations and derived conclusions and the overall picture. The tools of research are presented in Part II and a parade of case studies is presented in Parts III–V organized by geosystems. In each chapter a series of key conclusions is reached, and these overlap in many cases, thus demonstrating the wide extent of their significance. Finally, Part VI addresses issues of implementation, further research avenues, and aspects of education.

2.1.2 The Geohydroderm and Its Major Groundwater-Containing Geosystems

Water is encountered within the voids of rocks all over Earth's crust. The major large-scale systems, for which the term *geosystem* is suggested, are as follows:

The oceans that cover two-thirds of the planet's surface.
The sedimentary rocks underlying the vast oceans are all found to contain interstitial water (Chapter 5).
The unconfined continental shallow groundwater flow zone, operating within the rocks of the continental landscape reliefs and draining to the ocean surface (Chapter 4).

Formation water entrapped within the sedimentary basins and rift valleys beneath the shallow groundwater flow zone, occasionally associated with petroleum (Chapters 7, 8, 10, and 11).

Formation water entrapped within crystalline shields, known from drill holes thousands of meters deep (Chapter 9).

Water associated with magma chambers and active volcanos (Chapters 13 and 14).

Thus, two-thirds of the Earth's surface is covered with water, and all the rocks of the upper part of Earth's crust also contain water. The term *geohydroderm* has been suggested for this global water-containing rock system (Mazor, 1986).

2.1.3 Information Encoded into the Shallow Active Water Cycle Is a Prerequisite to Understand the Deeper Geosystems

The term *meteoric water* pertains to water of an atmospheric origin, and it is used to define groundwater formed in continental setups. As we will see later on, all waters entrapped within the continental part of the geohydroderm, even at a depth of thousands of meters, had their origin at the Earth's surface. The deeper fossil formation waters originated as a part of ancient shallow active water systems that subsequently got entrapped and subsided.

The processes going on in the shallow active water cycle encode into groundwater a remarkable amount of information. Acquaintance with this information is also the key to understanding fossil formation waters, the related petroleum deposits, and high-temperature water systems. The fossil formation waters turn out to be by and large meteoric and tagged by brine-spray, thus disclosing an origin at ancient low flatlands that were the arena of frequently alternating sea transgressions and regressions. Mapping of formation water ages and facies provides a tool for the spatial mapping of the paleo-flatlands. This, in turn, leads to the understanding of petroleum deposits.

A separate branch of research focuses on the dynamics of groundwater in hydrothermal and volcanic systems, applying for example, the atmospheric noble gases (ANG) as a useful tool to identify boiling and condensation processes.

2.2 The Active Cycle of Fresh Surface Water and Unconfined Groundwater

The basic laws of gravity-induced water flow are derived from the following set of key observations and conclusions.

2.2.1 Surface Water Gravity Flow

Gravitational flow.   *Observation*: Runoff water on the landscape always flows downward. *Conclusion*: The flow is induced by the terrestrial field of gravity.

The zone of hydraulic potential difference.   *Overall conclusion*: The previous observation and conclusion leads to the generalization that runoff water flows within the *zone of hydraulic potential difference*, which extends from the highest landscape point to the lowest one (Fig. 4.11d).

The sea surface is the plain of zero hydraulic potential.   *Observation*: All rivers that flow to the sea reach its surface; no river stops above sea level, and no river digs its bed beneath sea level. *Conclusion*: The sea surface is the terminal base of drainage of runoff water; it is the *plain of zero hydraulic potential*.

Gravity flow is conditioned on the availability of free space—empty of solids and fluids.   *Observation*: Runoff flows along the steepest paths available on the landscape. *Conclusion and discussion*: Gravitation pulls vertically down, but water cannot flow through solids, so it flows at the free space above them. This is similar to a ball rolling on a land slope; it rolls along the steepest path available (Fig. 2.1a).

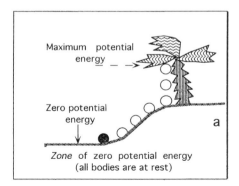 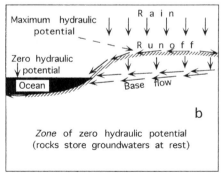

Fig. 2.1   Motion of a ball determined by the gravitational pull vertically down and the availability of free space for motion: (a) freefall and rest at the ground; rolling in free space available on top of the sloping ground and rest upon reaching flat ground. (b) Below ground surface no free space for downward motion is available and all bodies are at rest. Runoff and rivers flow in this mode.

2.2.2 Underground Flow Patterns

The following observations can best be made in regions that are not disturbed by human activity:

Aerated and saturated zones and the water table. *Observation*: In all wells there is an upper part empty of water, and beneath which there is water. *Conclusion and nomenclature*: There is a zone at which the voids between the soil and rock fragments contain air—the *aerated zone*; and beneath it there is a zone at which the voids of the rocks are filled with water—the *saturated zone*. The upper surface of the saturated zone is called the *water table* (Fig. 4.9).

Discharge. *Observation*: The depth of the water table in shallow wells drops following dry seasons. *Conclusion*: The groundwater in the saturated zone flows away; it is drained to some place.

Recharge. *Observation*: Following rainy seasons and snow melt seasons the water table in shallow wells rises. *Conclusion*: A part of the precipitation water that reaches the land surface infiltrates and recharges the groundwater system.

Drainage. *Observation*: Measurements in a large number of wells in a study area always reveal that the water table is slightly inclined. *Conclusion*: This is another indication that the shallow groundwater system flows toward a base of drainage.

The sea surface is the terminal base of drainage of groundwater. *Observation*: Shallow wells reach fresh groundwater even tens of meters from the seashore. *Second observation*: The regional water table is observed everywhere to be inclined toward the sea surface—not above it nor beneath. *Conclusion*: The sea surface acts as the terminal base of groundwater drainage. *Discussion*: This is similar to the sea surface as has been shown above serving as the terminal base of drainage of rivers and runoff water.

The water table fluctuates with the changes in the sea level. *Observation*: Measurements in coastal observation wells reveal rises and falls that follow the tidal and seasonal rises and falls of the sea level. *Conclusion*: This is an independent demonstration that the sea surface is the base of drainage.

The rare case of coastal submarine springs issuing through paleo-karst channels. *Observation*: In a few sites divers identified small springs issuing on the sea floor close to the shore. *Discussion*: This feature has been observed in limestone terrains and is explained by groundwater flow in remnants of ancient karstic conduits that originally drained to a lower paleo–sea surface (e.g., during the ice ages). When the sea level rose, the lower conduits were soon filled up in most places.

Exploring and Understanding the Geohydroderm

The gravitationally flowing groundwater is young. *Observation*: Tritium and ^{14}C dating reveals ages in the range of recent to less than 10^4 years. *Practical conclusion*: This provides us with an independent tool to identify recent through-flowing groundwater.

Fresh water occasionally encountered in near-shore sub-marine drillings is ancient. *Observation*: Fresh groundwater is occasionally encountered in off-shore drillings and is found to contain no measurable tritium and low to zero concentration of ^{14}C. *Conclusion*: This is fossil groundwater, entrapped when the sea level rose following the termination of the Ice Age.

Groundwater flow is restricted to the zone of hydraulic potential difference. *Discussion*: The listed observations and conclusions reveal that, like the surface water, groundwater flows within the zone of hydraulic potential difference. The highest potential is at the highest points of the landscape, and zero hydraulic potential prevails at sea level. *The sea surface is the plain of zero hydraulic potential* (Fig. 4.10).

2.2.3 The Empty Space Requirement for Flow of Water

First principles discussion: Water cannot flow through a solid, nor through a space already occupied by water. The only possible mode of flow is through empty space (air is pushed aside). For this reason, the water flowing in rivers stops the moment the sea is reached; the arriving water spreads along the empty space above the sea surface. The same is true for the infiltrating groundwater; once it reaches the saturated zone it cannot penetrate it, and hence it spreads along the free voids of the aerated zone above the water table. As a result, the water heaps up and flows laterally to the base of drainage.

The direction and extent of water flow is the intermediate between the downward pull of terrestrial gravity and the availability of free space to flow through.

2.2.4 Water Chemistry: Sea Spray Tagging and Water–Rock Interactions

Possible sources of Cl dissolved in flowing groundwater. *Observation*: The only Cl-containing rock is halite (NaCl), and all other common rocks are devoid of Cl. Water in contact with halite dissolves it readily, attaining a Cl saturation concentration of about 180,000 mg/L. *Conclusion*: Any groundwater that contains significantly less Cl discloses that its host rocks are devoid of halite. *Second conclusion*: Low concentrations of Cl must have an external origin, outside the rock system.

An external source common to Cl and correlated ions. *Observation*: In different samples collected from surface water systems in a region, it is a common observation that the concentrations of ions such as Br, Na,

Sr, Li, and often also Ca, Mg, K, and even SO_4 and others are well correlated with the concentration of Cl (e.g., Figs. 4.2; 4.4; 4.6; and 4.7). *Conclusion*: This indicates that these ions all arrived from a common external source.

Sea spray. *Observation*: The seawater values plot directly on the mentioned ionic correlation lines (e.g., Figs. 4.4; 4.6; 4.7; and 4.8). *Conclusion*: The external source is sea spray, which is transported landward by the wind in clouds and as dry dust. These salts are washed into the runoff and down into the groundwater systems (Mazor and George, 1992).

Such tagging of surface water and the shallow groundwater by sea spray has been observed all over the globe, even hundreds and thousands of kilometers inland (Mazor, 2003).

Sea spray tagging is specific to the young, ongoing water cycle. *Observation*: The active flowing water is in most cases fresh, i.e., of a salinity that is significantly lower than that of seawater, but as a rule it is tagged by sea spray, recognizable by a Cl/Br weight ratio of around 293, and CaCl, being negligible. *Second observation*: These waters are young—zero to 10^4 years old. *Conclusion*: Tagging by sea spray typifies practically all the young water systems occurring on the continents.

We will latter on see that a different tagging—by evaporitic brine-spray—typifies all the deeper fossil formation waters. This difference provides key information on the terrestrial geological history, as we will see.

External sources of ions, introduced from outside the rock systems. The sea spray discussed in the previews section is a source of ions that reach groundwater from outside the rocks with which the water comes into contact. These are *external*, or *allochthonous*, ions. They can be recognized by one or more of the following patterns:

1. Chloride contained in concentrations that are significantly below saturation with regard to halite, indicating that it originates from outside the rock system.
2. All the ions that are correlated with the Cl concentration are external as well.
3. Seawater ion/Cl ratios indicate an origin from airborne sea spray.
4. The external ions occur in the same relative abundances, irrespective of the nature of the host rocks.
5. The external ions are washed in by the recharge water.

Internal sources of ions, introduced by water–rock interactions. Ions that reach groundwater via interaction with the host rocks are termed *internal*, or *autochthonous*, ions. The following are some examples:

1. Water in contact with soil dissolves biogenic CO_2 turns into an acid, and interacts with rocks, contributing to the water the HCO_3 anion

and to a lesser extent the CO_3 anion. These anions are accompanied by various cations, according to the involved rock types. The solubility of the named anions is low and, hence, groundwater does not get saline by the CO_2-induced interactions with rocks.
2. Direct dissolution of halite results in a very high Cl content of around 180 g/L, balanced by Na. Dissolution of gypsum results in SO_4 concentration of around 2700 mg/L, balanced by Ca. Exposure of groundwater to these two rock types is extremely rare.
3. Dolomitization and de-dolomitization are processes that exchange Ca for Mg or vice versa. Positive Ca–Mg correlations, such as seen in Figs. 4.4, 4.5, and 4.7, rule out these processes.
4. Various ion exchange reactions with clay have been observed, but seem rarely to be quantitative.

The external ions serve as benchmarks identifying water–rock interactions. The ions found in groundwater are the sum of the contributions from the external and the internal sources. Once the external ions are identified, via their above listed patterns, their concentration can be subtracted from the observed ion inventory, and only the remaining ions may be of an internal water–rock interaction. Applying this procedure, it turns out that in most cases the contributions by water–rock interactions are rather limited.

Conservative and reactive ions. Conservative ions stay in the groundwater once they reach it from an external source. Prominent examples are Cl and Br; their seawater ratio is well maintained in surface waters and in shallow groundwaters, indicating lack of losses or gains by water–rock interactions, i.e., they are conservative. All ions that reveal linear correlation lines and seawater values plot on them are conservative in that setup as well (Fig. 4.7). These often include Na, Mg, Sr, Li, and occasionally also SO_4 and Ca.

In contrast, HCO_3 is a reactive ion, often deviating from the seawater values (Fig. 4.4), and often so is SO_4.

Water–rock interactions are fast, and thereafter the water–rock systems are chemically stable. *Observation*: Water–rock interactions are observed in young groundwaters to be fast, and equilibrium is reached in a matter of a few months to at most a few years. *Second observation*: Through-flowing groundwater of even a few thousand years age clearly reveals the tagging by sea spray. *Conclusion*: Once the rapid water–rock interactions reach equilibrium, the water–rock system is commonly stable.

Formation in a dry climate environment is disclosed by elevated groundwater salinity. *Observation*: Groundwater formed at present in arid regions is saline (Fig. 4.7). *Conclusion*: By this observation it is interpreted that saline formation waters reflect prevalence of a dry paleoclimate.

2.2.5 Rivers and Shallow Cycling Groundwater Clean the Continents from Soluble Compounds

Observations: Rain contains 5 to 10 mg/L Cl; rivers contain tens to hundreds mg/L of Cl; and shallow flowing groundwater contains tens to several thousands of mg/L Cl. A similar order of increasing concentrations is observed for a long list of chloride-accompanying ions. *Conclusion*: All these dissolved ions are washed into the oceans. *Discussion*: The continents have two main sources of soluble compounds: (1) the sea spray and (2) the products of water–rock interactions. The bulk of these soluble products is washed into the sea, a function we often take for granted. However, in certain arid landscapes we can see salts covering the ground, e.g., salinas or salt pans. These are regions of very poor drainage, demonstrating the important job commonly done by the active water cycle—rinsing the continents.

2.2.6 Isotopic Tagging of Meteoric Water

Isotopic tagging of continental groundwaters—the global meteoric water line.
Observation: The stable isotope compositions of waters from all over the world, expressed by the δD and $\delta^{18}O$ values, plot on a well-defined line on δD versus $\delta^{18}O$ diagrams—the *global meteoric water line GMWL* (Figs. 4.5 and 4.7). In certain cases groundwater values plot on a slightly different "local meteoric water line." *Practical conclusion*: This isotopic pattern typifies meteoric waters and serves as a most useful indicator to identify the continental origin.

2.2.7 Tagging of Meteoric Water by Atmospheric Noble Gases

Observation: Water exposed to air contains the atmospheric noble gases He, Ne, Ar, Kr, and Xe in well-known concentrations that portray a characteristic relative abundance pattern (Figs. 13.2 and 13.9). *Second observation*: The atmospheric noble gases are observed in formation waters in concentrations that indicate "reasonable" intake temperatures and display the characteristic atmospheric abundance pattern. *Practical conclusion*: This noble gas pattern typifies meteoric waters, and serves as a most useful indicator to identify the continental origin.

2.3 Interstitial Water Entrapped in Rocks Beneath the Vast Oceans

An intriguing question is which type of water do the rocks underlying the oceans contain? Drillings of the international Deep Sea Drilling Project penetrated the rocks lying beneath a few thousand meters of seawater (see Chapter 5). Rock cores from a depth of hundreds to over one thousand meters were obtained from all the ocean basins, and the water they contained was extracted and analyzed. Water was found to fill voids that occupy 25 to 80% volume of the host rocks (section 5.3).

Interstitial waters in the rocks beneath the vast oceans are also confined in hydraulically isolated rock-compartments. *Observation*: The chemical and isotopic composition of the interstitial waters is observed to vary substantially, often abruptly, at the drilled depth profiles (e.g. Figs. 5.1–5.4). *Conclusion*: These waters are entrapped in hydraulically separated rock-compartments.

Seawater did not replace the interstitial water contained within the underlying rock strata. *First principles discussion*: The rocks beneath the ocean floor are deep below sea level, i.e., they are in the zone of zero hydraulic potential. Hence, seawater cannot penetrate into the underlying rocks and cannot flush their original interstitial water. *Observation*: The interstitial waters differ from seawater compositions in various ways, disintegration of the sulfate being a common case. *Second observation*: In a number of cases continental water is encountered disclosing stages of subaerial exposure, e.g., during the Messinian event of the drying up of the Mediterranean Basin (section 5.5). *Conclusion*: Seawater did not replace the water contained in the underlying rocks, in agreement with the first principles conclusion.

Most suboceanic interstitial waters can be traced to have originated from seawater that underwent some diagenetic changes. *Observations*: Most interstitial water samples contain 19 ± 1 g/L Cl; their weight ratio of Cl/Br is around 293; and the stable isotopes of the water have δD and $\delta^{18}O$ values close to the seawater value, i.e., $0 \pm 2\%$. A common deviation from seawater composition is a relatively low SO_4 content (Figs. 5.1 and 5.4). *Conclusion*: The suboceanic interstitial waters reveal a wide range of compositions, but in most cases an origin from entrapped seawater is recognizable. The best benchmarks provide the concentrations of conservative ions like Cl and Br.

At certain locations suboceanic interstitial waters of a continental origin have been encountered. *Observation*: Certain core sections, hundreds of meters beneath the sea floor, revealed interstitial water with a concentration of Cl and other ions of up to four times the concentration in seawater, and Cl/Br ratios of 200 or less (section 5.5.1). *Discussion*: In order to reach a Cl concentration of four times the concentration in seawater, original seawater had to evaporate substantially (a water loss of at least 75%). Such intensive evaporation could be attained only in coastal flatlands that were occasionally invaded by seawater. A second way such water could be formed is by groundwater that was tagged by brine-spray. In both cases the described interstitial water originated on land. *Conclusion*: The transgressing seawater covered the continental paleo-landscape, without penetrating into the underlying rocks and without replacing their included continental groundwater.

Seawater–continental groundwater contact. *Observations*: (1) Nonpumped coastal drill holes and observation wells reveal fresh water down to the sea level depth, and various water types beneath. The latter vary from semisaline to higher than seawater salinity; (2) detailed chemical analyses reveal time and again that in nonpumped wells the deeper saline water has a composition that differs from the composition of seawater; (3) the deeper saline waters have in various cases been dated and found to be with no tritium and low ^{14}C, i.e., several thousands of years old. *Conclusion*: The below sea level groundwater in nonpumped coastal wells is fossil water that was entrapped in the rocks during a phase of lower sea level, i.e., when the base of drainage was lower. Seawater does not flow laterally into the continental sediments.

Second observation: Coastal wells that were overpumped, until their water table descended several meters beneath sea level, in many cases became saline, and the chemical analyses revealed seawater intermixing. *Conclusion*: In these cases seawater flowed in the aerated zone downward to the "pumping cone" created in the water table in the area of the overpumped wells (Fig. 2.2).

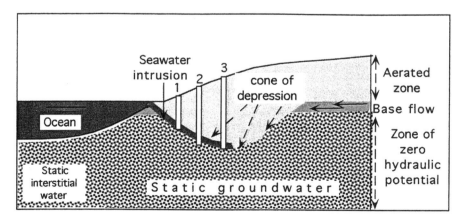

Fig. 2.2 Seawater flowing into the cone of depression formed in the water table by overpumping.

2.4 Fossil Formation Waters Entrapped Within Sedimentary Basins and Rift Valleys

The term *formation water* is applied to groundwater encountered in drillings beneath the ongoing active groundwater flow. Formation waters occur at depths beneath sea level, i.e., in the *zone of zero hydraulic potential* (Fig. 2.1). They have been encountered in wells penetrating down to a depth of a few thousand meters within subsided basins and rift valleys (Chapter 7), often associated with petroleum deposits (Chapter 10). Formation waters are common also within crystalline shields (Chapter 9), and they are often involved in ongoing hydrothermal processes and volcanic manifestations (Chapter 14).

Systematic multiparameter study of formation waters, all over the globe, reveals that a remarkable amount of essential information is encoded by them.

Formation waters are fossil, saline, and tagged by brine-spray. *Observations*: Formation waters are saline, for example, the Cl concentration is commonly within the range of 10^3 to 10^5 mg/L, the Cl/Br weight ratios are in the range of 80 to 200, and $CaCl_2$ is prominent besides the NaCl (Figs. 7.1, 7.6, 7.13). *Discussion and conclusion*: The salinity and ionic ratios of formation waters differ substantially from the composition of the local recent groundwater, evidence that one is dealing with fossil groundwater disconnected from the land surface. *Second observation*: The formation waters reveal high isotopic ages, in the range of 10^5 to 10^8 years (Chapter 8). *Conclusion*: This is independent evidence that these are fossil waters.

The δD values of formation waters indicate these are meteoric waters, formed on exposed continents. *Observation*: The isotopic composition of the hydrogen of formation waters is light; the δD values are in the range of $-160‰$ to $-20‰$. *Conclusion*: These values are well within the range of meteoric, i.e., continental, groundwater (Figs. 10.3, 10.5, 10.14). This implies that by and large formation waters were formed on paleo-continents during repeated stages of sea regressions. The continental groundwater was preserved within the host rocks that were covered by the alternating sea transgressions, which laid down new shallow marine sediments that confined the underlying rock strata (Fig. 2.3).

A remarkable spatial variability of formation water characteristics indicates storage within hydraulically separated rock-compartments. *Observation*: Formation waters encountered in adjacent wells and at different depths in the same drill holes display a vast variability of salinity and chemical composition (e.g., Figs. 7.2; 7.4; 7.5). *Conclusion*: The formation waters are entrapped in hydraulically isolated rock-compartments.

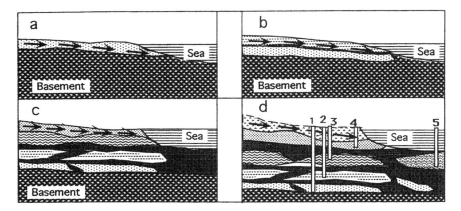

Fig. 2.3 Successions of meteoric water introduced into exposed marine sediments, reflecting successions of sea retreats and invasions.

The thickness of these rock-compartments, mappable by formation water composition similarity or diversity, is often of tens to hundreds of meters, and the aerial distribution is up to a few square kilometers. *Second observation*: Significant pressure differences of formation water are encountered in adjacent wells and at different depths of the same drill holes. *Conclusion*: This is an independent indication of formation water entrapment within hydraulically isolated rock-compartments.

A fruitcake structure is revealed by the fossil formation waters. *Observation*: Entrapment within closed rock-compartments characterizes fossil groundwaters in all the geosystems, namely, sedimentary basins, crystalline shields, and the extensive rock systems located beneath the oceans. *Conclusion*: A three-dimensional view of the formation water–containing geosystems reveals rock pockets of different sizes filled with different types of groundwater and, occasionally, oil and gas. In a picturesque way, we deal with a *fruitcake structure*.

A dense three-dimensional network of impermeable barriers is an integral element of the fruitcake structure of the sedimentary basins. *Observation*: Sedimentary rock sequences are rich in permeable rocks like limestone or sandstone that host water, as well as petroleum, in hydraulically interconnected voids. These water-hosting rock units alternate with impermeable clay-rich beds that act as hydraulic seals, preventing water through-flow. *Second key observation*: The sedimentary rock strata filling the sedimentary basins reveal lateral facies changes, often over rather short distances of hundreds of meters—a structure well seen in outcrops of sedimentary rocks. Thus, the permeable rock strata are laterally noncontinuous; they are frequently *interrupted* by clay-rich beds, of a thickness ranging from a few centimeters to many meters. *Conclusion*: These clay-rich beds act as the hydraulic seals that engulf the rock-compartments that host the different types of formation water (and petroleum) encountered in the sedimentary basins. *Second conclusion*: Looking at the sealing beds that engulf the water-hosting rock-compartments, we see that they make up a *three-dimensional network of the impermeable barriers*. This network is an integral part of the fruitcake structure.

Fossil formation water originated as paleo-groundwater that was present in its host rocks while the latter were confined by overlaying sediments. The raisins and other fruits are placed in a cake while it is formed; they can not be "shot in" afterward. The same must be true for the different types of formation waters; they were each enclosed in its hosting rock-compartment when the latter was covered and confined by new overlying sediments (Fig. 2.3).

All waters entrapped within the continental sections of the geohydroderm had their origin at the Earth's surface. This is borne out by several of the above listed conclusions:

Meteoric δD signature, indicating an origin from precipitation formed at the continent surface

Prominence of external ions, brought in from the land surface, along with the infiltrated water phase

Atmospheric noble gases signature, indicating an origin from water that equilibrated with air

Entrapment within closed compartments of sedimentary rocks, indicating confinement upon coverage by accumulating sediments

Lack of observable diffusion effects indicates lack of hydraulic connectivity between the rock-compartments. *Observation*: There are ample examples of different water types stored in adjacent very old rock-compartments (Chapters 7 and 10). The different water salinities and ionic compositions were well preserved over extended geological time periods. *Conclusion*: Diffusion between such rock-compartments was negligible. *Discussion and second conclusion*: Diffusion through solids is known to be negligible, but diffusion between different solutions that are hydraulically interconnected can be substantial. Hence, the preservation of different fluids in neighboring rock units indicates they are hydraulically disconnected, e.g., by clay-rich rocks.

Permeable and impermeable rocks, aquifers and aquicludes, and artesian wells. *Observation*: Self-flowing high-yield artesian wells in many places produce large volumes of water from deep rock units, some even over a thousand meters deep. *Conclusions*: (1) The producing rock unit is *permeable*—it includes voids that are interconnected so they can accommodate the water and let it flow through into the free space of the drill hole to reach the land surface (that acts under this setup as a base of drainage), and (2) the pressure is *lithostatic*, i.e., the weight of the overlying rocks causes compaction that is forceful enough to lift the water.

Additional observations: (1) Adjacent wells in artesian fields, and wells reaching different depths, often produce water of different compositions; (2) each of the different wells produces continuously the same type of water; (3) these are fossil connate waters, of high confinement ages (Chapter 8), at places even of a Paleozoic age. *Conclusions*: (1) The different water types are entrapped in hydraulically separated rock-compartments and (2) the compartments are engulfed by *impermeable* rocks, e.g., clay and shale.

(*Nomenclature*: Each producing rock unit is an *aquifer*, and each sealing rock layer is an *aquiclude*.)

The hydraulic seals must have been effective from the early stage of confinement of the individual rock-compartments. *Discussion*: The observation that formation waters stored in sedimentary rocks have a fruitcake structure implies that the hydraulic sealing between the rock cells was formed right away when the host rocks were buried and confined by succeeding sediments. Clay layers of tens of centimeters or more are common in the sedimentary rock sections, and they provided the observed hydraulic isolation. Subsidence-induced compaction by the overlying rocks and a gradual temperature increase are two mechanisms that increased the sealing property of the clay units.

Basin-wide groundwater through-flow is ruled out by the fruitcake structure. *Discussion*: Basin-wide through-flow would (1) require basin-wide hydraulic connectivity and (2) would have replaced the original groundwaters by a uniform type of new recharged water. The fruitcake structure clearly discloses that this is not the case. This is in good agreement with the first principle conclusion that the deeper parts of the sedimentary basins are in the zone of zero hydraulic potential, in which no gravitational flow occurs.

Connate groundwater. *Nomenclature*: Lane (1908) suggested the term *connate water* for water entrapped within its host rocks since it was often confined by overlying rocks as a result of a subsidence process. *Observation*: A major part of the encountered formation waters meets the specifications of connate water. *Conclusion*: The age of connate water is between the age of the host rock and the age of the directly overlying confining rocks. Connate waters as old as the Cambrian are known (Chapter 8).

2.5 Halite and Gypsum

Observation: Salt rock and gypsum are formed mainly by intensive evaporation of seawater in coastal lagoons and sabkhas, common in evaporitic low flatlands. *Second observation*: Halite and gypsum deposits are readily washed away by invading seawater or runoff water, and are easily washed by flowing groundwater that comes into direct contact with them. *Third observation*: Halite and gypsum beds are rather common among the rocks filling sedimentary basins and rift valleys. *Fourth observation*: Halite and gypsum beds are known from all geological ages, going back to the Cambrian (Chapter 6).

Conclusions: (1) The halite and gypsum deposits encountered in the sedimentary basins represent a much larger scale of paleoformation of the evaporites; (2) evaporitic environments were common at the sedimentary basins during the respective geological periods; (3) the salt and gypsum beds are stored within hydraulically isolated rock-compartments that are engulfed by impermeable rocks, e.g., clays and shales; (4) the halite and gypsum deposits underwent effective coverage and confinement soon after their formation, e.g., by marine sediments deposited by a transgressing sea.

2.6 Shallow and Deep Groundwaters Are Indispensable Geological Records

Far-reaching paleo-geographical and tectonic conclusions can be derived from the assortment of compositions of shallow and deep groundwaters.

High-relief landscapes characterized by recent sea spray–tagged meteoric groundwater. *Observations*: (1) Currently formed meteoric groundwater is typically tagged by sea spray (Chapter 4); (2) the present configuration of the major part of the continents is of a high relief, and coastal low flatlands are scarce. *Conclusion*: In elevated landscapes water-carrying clouds as well as sea spray are transported landward, and the formed groundwater is tagged by sea spray.

Additional observation: No currently formed groundwater gets entrapped and preserved; all the recent groundwater is readily drained back into the oceans. *Conclusion*: This is a direct outcome of the elevated landscape relief. This explains why no sea spray–tagged fossil formation water has been found.

Ancient flat evaporitic lowlands disclosed by brine-tagged meteoric formation water. *Observations typifying formation waters*: (1) Continental (meteoric) origin (light δD values); (2) elevated salinity; (3) tagging by Cl/Br weight ratios that are less than 200; (4) contents of $CaCl_2$ besides the dominating NaCl. *Conclusion*: These are meteoric groundwaters tagged by brine-spray. The light δD values disclose formation on the continent, and the evaporitic tagging indicates formation in vast evaporitic landscapes.

Discussion: The required evaporitic landscapes must have been the arena of frequent invasions by the sea, followed by regression and land exposure. The exposed land was vast, judging by the large-scale extension of the sedimentary basins that contain the brine-tagged meteoric formation waters. The regressing sea left behind vast areas of lagoons and sabkhas that supplied the salt-spray that was dissolved in the shallow groundwaters formed on the continent. The following sea transgression deposited new marine sediments that confined the former ones, including their contained meteoric brine-tagged groundwater.

Orogenic tranquility phases displayed by the brine-tagged meteoric formation water. *An additional conclusion, based on the mentioned observations*: The prevalence of the vast ancient land stretches that were evaporitic, flat, and low discloses long geological periods at which mountain building was negligible; the landscape reached erosive maturation, and the leading tectonic activity was large-scale subsidence of sedimentary basins.

Prevalence of dry paleoclimate phases displayed by the saline meteoric formation water. *Observations*: (1) Currently formed groundwater (sea spray tagged) is commonly fresh, but in arid zones it is often saline—even more than seawater (section 4.5); (2) in contrast, formation waters are all saline; (3) formation waters are brine-spray tagged, demonstrating ancient evaporitic facies. *Conclusion*: Arid climate prevailed at the evaporitic flat low lands that were the environment in which the formation waters were formed and entrapped in subsiding rock-compartments. *Discussion*: These arid paleoclimate conditions might well have been the result of (1) the lack of elevated landscapes limiting the rain yield of the clouds and/or (2) during the sea regressive stages large sections of the exposed continents may have been far inland.

Frequent sea regressions recorded by the meteoric formation waters. *Observation*: The formation waters are on the one hand continental/meteoric (light δD values); on the other hand they are entrapped within rock-compartments that are mainly built of shallow marine facies. *Conclu-*

sion: The host rocks disclose mainly sea trangression phases, during which the rocks were formed, whereas the sequence of different types of formation waters record the continental exposure phases, during which the formation water was formed.

Discussion: Looking at the thick sequences of marine rock strata, the conclusion may be reached that they disclose continuous, long-lasting transgressions. It is in this relation that the record of the meteoric formation waters is crucial—they indicate frequent alternating phases of sea regressions.

Subsidence into the oil formation temperature range, disclosed by ^{18}O-enriched formation water. *Observation*: All the formation waters have light δD values, well in the range of continental/meteoric groundwaters, yet the $\delta^{18}O$ values are often heavy compared to meteoric water values (plotting to the right on δD–$\delta^{18}O$ diagrams). *Conclusion*: These formation waters underwent an ^{18}O isotopic enrichment by temperature-induced exchange with rocks. This indicates, in turn, that the respective formation waters subsided, along with their hosting rock-compartments, deep enough. *Second conclusion*: This indicates, in turn, that formation water associated petroleum all the way through, from the surface-related formation of sediments that contained organic raw material and groundwater down to the petroleum-formation temperature window.

Unlifting and erosion evidenced by shallow ^{18}O-enriched formation waters. *Observation*: Oxygen-18 enriched meteoric formation waters, at times associated with petroleum, are also encountered occasionally at shallow depths. *Conclusion*: This is an indication that the respective region underwent erosion and uplifting that resulted in the present shallow occurrence of the formation water and associated petroleum.

Discussion: The association of petroleum with ^{18}O-enriched meteoric formation waters indicates in such cases that we deal with erosion and uplifting, and not with upward migration of petroleum.

2.7 Brine-Spray-Tagged Meteoric Formation Water Is Also Common Within Crystalline Shields

The crystalline shields, e.g., of north Canada, Fenoscandinavia, or Western Australia, are built of igneous and metamorphic rocks. Drillings connected to ore exploration and mines revealed the occurrence of saline water to a depth of over 1000 m, and the deep KTB drilling in Germany and the research drilling in the Cola Peninsula encountered such waters down to a depth of over 9 km (Chapter 9). Processing of published data reveals that these waters are indistinguishable from the discussed formation waters encountered within the sedimentary basins. The origin of these waters and a list of practical management conclusions are fascinating.

An upper, highly fractured zone hosts at present the groundwater through-flow zone, drained to the sea surface. *Observations*: (1) The land relief of the crystalline rocks, protruding above sea-level, is relatively rich in open fractures; (2) runoff infiltrates into these fractures; (3) shallow wells encounter fresh groundwater; (4) this groundwater is tagged by sea spray; (5) its δD and $\delta^{18}O$ values are light; and (6) it reveals a young isotopic age (Chapter 9). *Conclusion*: A shallow active groundwater system is developed in all crystalline rock exposures, the base of drainage being the sea surface. *Discussion*: The rather intensive shallow fracturing is the result of (1) stress release following removal of the overlying rocks by erosion and (2) water–rock interactions that dissolve the rocks and wash away the products.

At greater depth the rocks are distinctly less fractured, often containing formation water that is saline, meteoric, tagged by brine-spray, and old. *Observations in mine galleries and exploration drill holes in crystalline shields reveal* (1) the fracture density is relatively low; (2) part of the fractures are filled with water; (3) this water is saline and tagged by brine-spray; (4) the δD and $\delta^{18}O$ values are light; and (5) isotopic dating indicates ages of more than 10^7 years (Chapter 8). *Conclusions*: (1) The composition differs significantly from the recent local groundwater, indicating that it is fossil formation water; (2) the isotopic composition indicates a meteoric origin, i.e., ancient groundwater descended into the deep fractures; and (3) the low fracture density indicates a one-time entrapment of the water with no through-flow going on.

The crystalline rocks were exposed as part of the flat paleo-landscapes of evaporitic conditions. *Working hypothesis*: The fossil meteoric water, encountered at a depth within the crystalline shields, was formed in the vast evaporitic paleo-landscape that prevailed at the geological periods of the formation and filling of the neighboring sedimentary basins.

The formation waters record a phase of exposure with no coverage by sedimentary rocks. *Another conclusion*: The brine-spray-tagged meteoric water could enter the crystalline rock systems only through exposed fractures, and hence at that geological period the respective shields were exposed and had no sediment cover.

Spatial diversity of the composition of the entrapped waters reveals storage within hydraulically isolated fracture-compartments. *Observations*: (1) The fossil waters encountered in adjacent drillings, and at different depths within single drillings or mines, reveal a significant spatial variability, and (2) the water yield in pumped wells is low, as compared to wells in sedimentary rocks. *Conclusion*: The formation waters in the crystalline rocks are stored in small sets of interconnected fractures that make up *fracture-compartments* that are hydraulically isolated from each other.

The deep meteoric formation water indicates fracturing started at the land surface and propagated downward. *Further conclusions drawn on the basis of the above-listed observations*: (1) As the encountered formation water is meteoric, it was originally formed on the land surface. Subsequently, a stage of dilatation and tectonic activity formed fissures that gradually penetrated to greater depths, and the formation water gradually trickled down (as outlined in Fig. 9.1).

Practical conclusions. *Observation*: The reported chemical and isotopic dating data reveal that the majority of the collected water samples were mixtures of recent and fossil waters (Chapter 9). *Conclusions*: (1) The boreholes short-circuited different fracture-compartments, in many cases also letting in recent groundwater or mixing water from different compartments, and (2) reports on observed water at certain depths seem to be erroneous as a result of water migration in the man made mining channels and boreholes (Chapter 9).

Nuclear waste repositories in crystalline shields—lessons to be learned. (1) The extent of the active through-flowing groundwater system has to be established; (2) the deeper zone of hydraulically isolated fracture-compartments within crystalline shields seems suitable for repositories of nuclear wastes, as they preserved even very old fossil groundwater in hydraulically well-isolated fracture-compartments; and (3) in such repositories all open boreholes and exposed fractures have to be well sealed in order to prevent short-circuiting of fluids (section 9.8).

2.8 Petroleum Occurrence and Genesis

The discussed information encoded in formation waters is of far-reaching relevance to petroleum occurrence and its formation environment.

Formation water is an integral part of the inventory of petroleum-containing rock-compartments. *Observations*: (1) Formation water is frequently observed and separated at the heads of oil and gas wells and (2) formation waters are frequently encountered in oil exploration and production wells, entrapped within hydraulically separated rock-compartments (Chapters 7 and 10). *Conclusion*: Petroleum is associated with formation water since the initial stage of confinement of their joint hosting rock-compartments.

Pronounced spatial variability of the characteristics of water and petroleum indicates confinement in hydraulically isolated rock-compartments. *Observation*: The formation waters encountered in oil and gas fields reveal a clear spatial variability, different fluids being encountered within drillings placed hundreds of meters to a few kilometers apart, and within depth differences of tens to hundreds of meters (e.g., Figs. 10.4; 10.6; 10.19). *Conclusion*: Oil and gas deposits are entrapped within efficiently isolated rock-compartments. *Additional observations*: Oil and gas deposits, and the highly soluble salt rock, are preserved in "traps" over long geological periods. *Conclusion*: This provides independent evidence that the discussed fluids are enclosed within efficiently isolated rock-compartments. *Thus, gas and oil deposits are part of the fruitcake structure of sedimentary basins.*

Petroleum raw materials and other ingredients needed for its formation often originated at evaporitic low flatlands, frequented by sea transgressions and regressions. *Observations*: (1) Halite beds are often found in close association with petroleum deposits, and (2) as concluded above, the associated formation waters were formed in large-scale evaporitic flat lowlands, frequented by sea invasions and retreats. *Conclusion*: Petroleum formation can often be traced back to evaporitic flat lowlands in which shallow marine and continental sediments made up the required organic raw material, host rocks, and water.

The elevated temperature requirement for petroleum formation is evidenced by the ^{18}O values of the associated formation water. *Observation*: The formation water directly associated with petroleum deposits is commonly observed to be enriched in ^{18}O isotopes. *Conclusion*: This is the result of temperature-induced isotope exchange with the country rocks. Thus, the formation waters provide an independent indication that petroleum formation occurred due to temperature increase that resulted from the increase in subsidence depth.

Discussion: The common notion that petroleum formation is temperature induced is based on laboratory experiments and the observation that many petroleum deposits are encountered at a depth of several kilometers, i.e., in elevated temperature environments. The associated fossil formation waters disclose independently that high temperature is indeed effective.

Petroleum migration is limited to the boundaries of the hosting rock-compartments. *A far-reaching conclusion, derived from the fruitcake structure* and the network of impermeable barriers, is that petroleum migration remains within the individual rock-compartments (e.g., Fig. 12.1). *First principles discussion*: Petroleum migration is envisaged as flotation on associated water bodies present in the rocks. However, migration is conditioned on hydraulic connectivity, but the fruitcake structure of closely packed isolated rock-compartments indicates that hydraulic connectivity exists only within the compartments, and the compartments themselves are each engulfed by impermeable rocks. Hence, petroleum migration could occur only at limited scales within the hosting rock-compartment.

Another first principles discussion. A common notion, prevailing in the literature and in textbooks, is that petroleum compounds migrated from "source rocks" to "reservoir rocks," that the migration was both vertically up and lateral, and migration distances of several kilometers and even hundreds of kilometers have been noted in many case studies (section 12.1). The dense three-dimensional network of impermeable barriers, which is an integral element of the fruitcake structure of the sedimentary basins, rules out such large-scale migration of petroleum compounds.

Saline water was most probably an essential ingredient (catalyst?) of petroleum formation. *Research question*: The observed affiliation of formation water with petroleum compounds raises the question whether water was an essential ingredient in the formation of the petroleum products. For example, did the water and the contained salts act as a catalyst that, for example, lowered the temperature threshold of petroleum formation?

2.9 Warm and Boiling Groundwaters

The information encoded into hydrothermal waters resembles all that we have discussed for the regular formation waters, with the addition of boiling and condensation effects.

The terrestrial heat gradient warms up groundwater that is buried deep enough. *Observations*: (1) In mines and in drillings, in nonvolcanic regions, the temperature increases with depth, revealing a terrestrial heat gradient that near the surface is 25 to 30°C/km; and (2) groundwater encountered in deep boreholes has a temperature that fits the depth and local heat gradient. *Conclusion*: The Earth is an internally hot planet. *Practical conclusion*: The temperature of the water of a spring or a well provides a tool to estimate the depth of groundwater storage. *Practical query*: Are the warm groundwaters part of the renewed through-flowing system, or are these fossil formation waters? *Common findings*: Waters heated to around 50°C, i.e., at least 25°C above the local average ambient temperature, are found to be fossil (Chapter 13).

How does fossil warm groundwater ascend in springs and wells in rift valleys—the tectonic pumps. *Observation*: In rift valleys, and along major fault lines, fossil (saline, brine tagged, and enriched in ^{4}He) thermal (40 to 80°C) water often ascends as springs and in boreholes. *Discussion and conclusions*: The elevated water temperature, of 40 to 80°C, indicates the water was stored at a depth of several hundred meters to over a thousand meters, which is beneath the active groundwater through-flow system. The question comes up; how did the water reach the deep storage sites? There are two different mechanisms: (1) We deal with formation water that subsided in the rift valley, along with the host rock, or (2) we deal with paleo-groundwater that penetrated underground through temporarily opened fault plains. Both these mechanisms are generated by tectonic processes. A second question is how does this deep fossil water ascend? The mechanism seems once more to be tectonic—compression, as part of a temporal tectonic activity, and ascent along dilated fault planes.

As the discussed water was brought down by tectonic activity and pushed up by another mode of this mechanism, the term *tectonic pumping* is suggested for the motion of fossil thermal groundwaters in rift valleys and along major fault systems.

Hydrothermal groundwater is superheated by shallow magma chambers. *Observations*: (1) In certain locations there are manifestations of superheated groundwater issuing as geysers; (2) temperatures of superheated groundwater have been observed in research boreholes to reach up to 350°C; (3) in all such locations shallow magma chambers have been detected (Chapter 14); and (4) in all these cases a fossil hot water end member has been identified. *Conclusion*: The water has been stored in the rocks before the magma intruded, and the lithostatic pressure was high enough to allow superheating of the local formation water. Some water ascends along currently open flow paths.

Noble gases reveal the meteoric origin and physical processes operating in hydrothermal systems. *Observation*: An atmospheric relative abundance pattern of Ne, Ar, Kr, and Xe is typically encountered in hydrothermal water phases. *Conclusion*: These waters are of meteoric, i.e., of continental origin, like the formation waters. *Second observation*: The atmospheric noble gases are enriched, relative to the equilibrium concentrations with air, in sampled steam phases and depleted in boiling water phases. *Conclusion*: The ascending steam is of an initial stage of boiling, and the hot water is the residual phase (Chapter 14).

Physical processes operating in hydrothermal systems revealed by the dissolved ions. *Observation*: The concentration of dissolved Cl, Na, Br, Li, and occasionally other ions, measured in different sources of the same hydrothermal field, plot on positive linear correlation lines that extrapolate the zero values (e.g., Figs. 14.2 and 14.8). *Conclusions*: (1) The composition differs from recent local groundwater, disclosing an origin under different environmental conditions, and hence this is fossil water; (2) these ions are of an external origin, brought in with the fossil groundwater, and (3) the wide range of ion concentrations and the extrapolation to the zero values disclose both *dilution* (by condensed steam and/or shallow groundwater) and *steam removal* (during ascent and as a result of intensive exploitation).

Radiogenic He and Ar reveal the water is fossil. *Observation*: The concentrations of ^{4}He and of ^{40}Ar are higher than the atmospheric concentrations, calculable via the accompanying ^{36}Ar, Kr, and Xe (Chapter 14). *Conclusion*: The excess ^{4}He and ^{40}Ar are nonatmospheric; they are radiogenic, disclosing that these are fossil waters that were entrapped in rocks before the event of the magmatic intrusion.

Informative exploitation effects. *Observations*: Long-operating steam wells in production fields reveal (1) a gradual decrease in the concentration of CO_2 and (2) a continuous decrease in the concentration of the atmospheric and radiogenic noble gases (Figs. 14.16 to 14.18). *Conclusion*: These are indications that production is mainly of the steam phase, which is continuously reproduced underground, resulting in gradual depletion of the gases in the residual water phase. *Second observation*: In old steam wells sometimes tritium starts to appear in the produced steam. *Conclusion*: This is an indication that recent shallow water penetrated the upper part of the system, reflecting a drop in the pressure of the complex.

Siting of new exploitation wells. *Practical discussion and conclusion*: New wells are preferentially placed near the power station and, thus, near the existing wells. On the other hand, the radius of exploitation of the new well may overlap with that of an existing well, thus reducing its production. Hence, in new boreholes the steam has to be analyzed—if the results resemble the original composition in nonexploited wells, the borehole may be equipped as a production well, but if the concentration of gases is low, tritium is seen and other exploitation effects are noticed—the borehole is better abandoned (and plugged).

2.10 Summary Exercises

Exercise 2.1: The present chapter is organized by sequences of observations and data, followed by conclusions and other logical sections. Please reorganize along these lines the following text and identify which additional data are required: "The African plate was pushed to Spain at the Messinian period, closing the Mediterranean Sea, and as a result the rivers of the surrounding lands deepened their beds, thick salt beds were laid down on the bottom of the drying basin, and the water table of the groundwater, formed at that stage in surrounding countries, was deeper."

Exercise 2.2: Is gravity-driven flow of groundwater to be expected at depths that are below sea level? Discuss.

Exercise 2.3: How can sea spray tagging be identified? And how will we recognize brine-spray tagging?

Exercise 2.4: Is the chlorine observed in groundwater of an external source or of an internal one? What are these terms? Discuss.

Exercise 2.5: Can fossil groundwater provide information on the climate that prevailed at the time and location of its origin?

Exercise 2.6: What is "meteoric water"? Which markers are at our disposal to identify meteoric water?

Exercise 2.7: From which observations have the hydraulically isolated rock-compartments and the fruitcake structure been deduced?

Exercise 2.8: What is meant by the term "formation water"? By which observations can fossil formation water be identified?

Exercise 2.9: Are connate groundwaters a rare feature? Discuss.

Exercise 2.10: Did seawater replace the interstitial water stored in the underlying rocks? Explain.

Exercise 2.11: Why is knowledge of the $\delta^{18}O$ value of a studied groundwater useful?

Exercise 2.12: Steam ascends in a newly drilled exploration borehole. What should be analyzed? How will you know whether the borehole should be equipped to function as a producing well?

3
BASIC RESEARCH CONCEPTS, AIMS AND QUERIES, TOOLS, AND STRATEGIES

Practically every parameter that is measurable in water- and petroleum-related systems is of interest and worth our attention. A major task of the researcher is to identify the aims of every planned study. The second stage is to apply the proper research tools, and finally the strategies have to be selected.

The present chapter is, in a way, a concise summary of the methodological aspects of the published case studies that are presented and discussed in the following chapters. It may be read again after reading through the book as a sort of methodological summation.

Research tools and modes of data interpretation are discussed in detail in a book that, in a way, is a companion to the present book: "*Chemical and Isotopic Groundwater Hydrology: The Applied Approach*" (Mazor, 2003).

3.1 Basic Research Concepts and Terms

3.1.1 Basic Concepts Related to the Unconfined Through-Flowing Groundwater

Permeability is the property of a rock allowing it to contain water and let it flow through interconnected voids.

Impermeability is the property of a rock preventing water from flowing through it.

Aerated zone is the upper zone that contains air in the free voids of the soil and rocks, manifested by the upper dry section of wells.

Saturated zone lies beneath the aerated zone, where all the voids are filled with water. From the saturated zone water flows into boreholes and wells.

Aquifer is the term for a permeable rock unit that contains groundwater in exploitable amounts. Examples of good aquifer rocks are noncemented conglomerates, noncemented sandstones, fractured limestone, and dolomites.

Aquiclude is an impermeable rock unit that does not let groundwater flow through. Most common aquiclude rocks are clay rich, e.g., clay, mudstone, marl, and shale.

Unconfined groundwater system depicts the shallow through-flowing groundwater system. The water table can freely ascend or descend on account of the overlying aerated zone, hence it is unconfined.

Water table is the plain that tops the saturated zone of the unconfined groundwater system. The water table, measured in adjacent wells, turns out to be nearly horizontal, with a slight inclination toward the base of drainage.

Base of drainage is the place to which runoff, rivers, and the unconfined groundwater system drain. There may be local bases of drainage, e.g., lakes, but by and large the base of drainage comprise the ocean surfaces.

Perched aquifer and perched water table are terms related to a groundwater-containing permeable rock stratum that exists within a regional aerated zone that is underlain by a local impermeable aquiclude. These are local sources of groundwater that occassionally sustain low yield wells or issue as springs.

Through-flow zone is another term for the unconfined groundwater system, emphasizing that the system receives recharge, which flows to the base of drainage.

Base flow. Runoff partially infiltrates and flows vertically down in the aerated zone until it reaches the saturated zone at the water table. There, it piles up and flows laterally toward the base of drainage, constituting the base flow.

Zone of hydraulic potential difference is the part of the landscape relief that protrudes above the base of drainage, which commonly is the sea surface. Surface water and groundwater flow within this zone due to gravitation. The flow operates from points of high potential, i.e., high topography, to the plain of zero hydraulic potential, which is the sea surface (see Fig. 2.1).

Meteoric water is water that originated on the continent. It is recognizable by the light isotopic composition of the water (e.g., Figs. 4.5; 4.7).

Sea spray tagging describes that a studied meteoric groundwater contains Cl, Br, Na, and other ions in relative abundances to seawater, e.g., a weight ratio of Cl/Br around 293. The source of the tagging salts is airborne sea spray (Chapter 4). Recent groundwater through-flowing in the unconfined zone is typically tagged by sea spray.

Brine-spray tagging describes that a studied meteoric groundwater contains Cl, Br, Na, Ca, and other ions in relative abundances to those that are known from evaporitic water systems, e.g., Cl/Br weight ratio in the range of 200 to 80 and prominent contents of $CaCl_2$, besides the common NaCl (Chapter 7). The source of the tagging salts is airborne brine-spray and dust. Fossil formation waters are observed to be as a rule tagged by brine-spray.

3.1.2 Basic Concepts Related to the Confined Static Groundwater

Zone of zero hydraulic potential/zone of no-flow extends beneath the base of drainage, i.e., beneath sea level. Groundwater within this zone is static, entrapped in rock-compartments (Fig. 2.1b).

Confined groundwater system is a term for groundwater setups that are located beneath the unconfined through-flow zone, i.e., in the zone of zero hydraulic potential. Confined groundwater has no access to the aerated zone and is confined by an overlying aquiclude.

Artesian water system is the term for groundwater that is entrapped in isolated rock-compartments and ascends in boreholes that penetrate the latter. The pressure is due to compaction by the overlying rocks.

Fossil water is groundwater that was formed during past geological periods, and it is not recharged at present, i.e., it is disconnected from the land surface.

Connate water is groundwater that has been entrapped in its host rocks since the latter subsided and were confined by overlying rocks.

Rock-compartment. A permeable rock unit that contains water and/or petroleum and is engulfed by impermeable rocks that act as hydraulic barriers.

Confinement and confinement age of fossil water. The sediments accumulated in a sedimentary basin were deposited mainly in shallow seawater, and during sea retreats they were exposed on the land. There they got filled with meteoric groundwater. During the following sea invasion new sediments were deposited, on top of the previous ones, thus confining them along with their contained paleo-groundwater.

Thus, the age of the confined connate water is between the age of the host rock and the age of the rocks in the overlying rock-compartment.

Petroleum migration is a term describing the process by which petroleum compounds, dissipated in host rocks, moved after their formation to rock traps at which they accumulated (within the hosting rock-compartment) (Chapter 12).

3.1.3 Common Errors Made with the Term Aquifer

Aquifers erroneously defined by stratigraphic units. Stratigraphic terms such as, for example, "Jurassic Aquifer," "Bunter Sandstone Aquifer," or "Murray Group groundwaters," have no hydrological meaning, and are defined without even the slightest relation to groundwater occurrences. The stratigraphic units contain a large number of alternating permeable and impermeable rock strata, constituting a succession of potential aquifers and aquicludes; and in addition there are frequent lateral facies changes.

In the shallow unconfined groundwater system flow to the base of drainage at the sea opens fractures and dissolution channels, and a sequence of rock strata are hydraulically interconnected, constituting the *unconfined groundwater system*, which in coastal plains may have a thickness of a few meters and in mountainous terrains may occupy over 1000 m.

In the deeper confined groundwater systems fossil water is confined in rock-compartments, defined by their permeability/impermeability configurations, varying in thickness from even a few meters to tens and up to hundreds of meters, and varying much in the lateral extension. The stratigraphic terms do not relate at all to these configurations. Extreme misuse of stratigraphic terms has been made in basin-wide through-flow models.

Aquifers erroneously defined by geographic locations. Geographic terms such as "Upper Silesian Coal Basin Aquifer," "Coastal Aquifer," "Florida Limestone Aquifer," or the "Blumau Aquifer" are misleading, as they lump together bundles of aquifers and aquicludes, and the territorial extension is undefined as well.

3.1.4 Groundwater Systems

Groundwater systems is the proper term for regional groundwater setups, e.g., "Long Island groundwater system," "Cleveland Coastal Plain groundwater system," or "East Pyrenean groundwater system." The term *system* indicates that one deals with an assemblage of a multitude of hydrological units that for a closer acquaintance have to be further described.

3.2 Research Aims and Queries

There is no limit to the quantity of possible measurements, if we consider the variety of parameters, the possible frequency of sampling and measuring, and

3.2.1 Determination of Water Quality

One major purposes water research is to find usable water resources. Thus, every water source has to be analyzed to determine its quality in order to define the most profitable usage. Examples of water usage, roughly in the order of quality, are as follows:

Bottling water, defined by very low ion concentrations, no smell, good taste, and of course no contaminants and biological sterility. Precise specifications have to be obtained from the local authorities.

Drinking water. Similar to bottling water, but with slightly more liberal specifications. Detailed specifications have to be obtained from local authorities.

Irrigation water. Slightly more saline than drinking water; specifications vary according the type of crops, climate, drainage quality, and irrigation technique applied.

Water for therapeutic and recreational requirements. Practically every warm and mineral water has a potential to serve for therapeutic and/or recreational purposes. Detailed analyses are needed in order to identify optimal uses (Chapter 13).

Water suitable to specific industrial applications. The water requirements of industry include washing and flotation processes, for which various types of saline water and treated water are suitable.

Water for desalination. Most desalination is foreseen to be conducted on seawater at coastal locations, but desalination of slightly saline groundwater may in certain cases be profitable (the concentrates have to be drained to the sea!).

3.2.2 Distinction Between Recent Flowing Groundwater and Static Fossil Formation Water

For practical reasons, discussed below, it is important to find out for every groundwater encountered whether it is part of the ongoing flowing system or whether it is fossil entrapped groundwater. Two major methods are at hand: (1) determining the age of the water; roughly up to 10^4 years is recent, and water older than 10^5 years is fossil; (2) sea spray tagging indicates the water belongs to a recent flowing groundwater system, and brine-spray tagging indicates the groundwater is fossil. Both methods have to be applied in order to cross-check the diagnosis.

This distinction is needed in order to get answers to the following questions:

Is the studied groundwater a renewed resource? Recent water is renewed and, hence, a reliable resource, whereas fossil water has the disadvantage of being a one-time resource.

Vulnerability to anthropogenic contamination. Recently recharged groundwater has been hydraulically connected to the land surface and pollutants can be washed in. On the other hand, fossil entrapped groundwater has the advantage of being disconnected from the land surface, and therefore immune to pollution (but care has to be taken in the installation of boreholes and wells).

Identifying sources of saline water intrusions. Intensive pumping may result in a gradual increase of the salinity, eventually causing a deterioration that will necessitate closing of wells. The source of the salination has to be identified; near the seashore it may be intrusion of seawater, but it may also be encroachment of deeper fossil groundwater. To identify the saline source, detailed chemical and isotopic measurements are needed, from the first moment of salinity rise and follow-up. Seawater composition will identify seawater encroachment, and dating will tell whether one deals with direct connection to the open sea or with fossil seawater, locally entrapped during the geological past. In the first case pumping has to be severely reduced so that the local water table will recover and the sea intrusion will be completely stopped. In the other case, the fossil saline water may reside in a small rock-compartment that will soon be emptied, and the well will recover. A third possible outcome may be salination by fossil groundwater that is less saline than seawater, and hence it may turn out that a new salinity value will be reached that is still tolerable.

3.2.3 Identification of Hydraulic Continuities and Hydraulic Barriers

This major task of hydrological studies is of prime importance both in the shallow through-flow zone and in the deeper zone of confined formation waters and petroleum deposits.

Adjacent wells and/or springs that have different compositions tap different groundwater units. Actually it is sufficient that significant differences are established by a single parameter, but a check with additional ones is recommended. Useful parameters include salinity differences, the ionic abundance ratios, and water temperature.

Adjacent wells and/or springs of similar composition seem to tap the same hydrological unit. Water sources may appear similar by salinity, Cl concentration, or temperature, but data of additional parameters are needed, including age indicators.

Mixing of two groundwater types indicates existence of two distinct groundwater units. In this case the end-members have to be identified as well as the mixing mechanism.

Identification of hydraulically separated shallow flowing groundwater systems. For example, adjacent coastal drainage basins or mountainous terrains with adjacent basins draining to separate subdrainage directions must be distinguished. The information is essential in estimating the outcome of overpumping or coping with pollution hazards.

Identification of the lithology, stratigraphic units, and extension of hydraulic barriers is a needed component in the understanding of a studied groundwater system, as well as the identification of petroleum-hosting compartments within a larger field. In the practical work the first clue of the existence of separated fluid-hosting compartments may come from the water or petroleum properties or from knowledge of the rock sequence.

3.2.4 *Determination of the Origin of the Dissolved Ions*

Groundwater systems are composed of a water phase, dissolved ions, and hosting rocks. Understanding the interplay of these three factors is the key to understanding groundwater and associated petroleum deposits.

External ions that were washed in and their identification. External ions originated from outside the host rocks and were washed in by the recharged water phase. Chlorine is a most conspicuous external ion. Ions that reveal positive correlation with the concentration of Cl, or among each other, originated from outside the rock system. The list of these ions varies from one case study to another. An independent check of the external origin of correlated ions is their similar composition within different rock types.

Sea spray versus brine-spray. Major sources of external ions are the sea spray, common in the currently active through-flow system, and the brine-spray that typifies fossil formation waters.

Internal ions that originated from water–rock interactions and their identification. Internal ions originated from inside the hosting rocks, as a result of water–rock interactions. A major role in this respect is played by the CO_2-induced water–rock interactions, the CO_2 being picked up by the recharged water from the soil. The resulting anion is HCO_3, and the balancing cations are mainly Ca in the case of interaction with limestone, Ca and Mg in the case of interaction with dolomite, and Na in certain igneous rocks. In rather rare cases ion exchanges take place and dolomitization or de-dolomitization.

A convenient way to spot the internal ions in studied groundwater is to first identify the external ions and then reduce them from the water composition; the rest are likely to be internal.

3.2.5 Understanding Artesian Systems

Artesian groundwater is specified by its pressure that sustains self-flowing wells. The question raised in case studies is the source of the pressure: Is it *hydrostatic* i.e., exerted by groundwater placed at a higher point in a semi-confined setup? Or is it *lithostatic* pressure within a confined system that is pressurized by the overlying rocks? In the first case we deal with a renewed resource, which is vulnerable to pollution; and in the second case it is a one-time resource, which is immune to pollution. Hydrostatic artesian systems are rare, whereas lithostatic artesian systems are common. The first is identifiable by recent water, and the latter is disclosed by fossil groundwater.

3.2.6 Identification of Geological and Tectonic Stages

Traditionally, rock facies were the main record that served to reconstruct the geological and tectonic history of a studied area. It turns out that formation water is a vast source of additional information, complementing the record documented by the rocks. The following are some examples.

Mode of sediment accumulation in sedimentary basins. The facies of rock sequences seen in the sedimentary basins has been envisaged as marginal marine on the basis of the rock record alone. The meteoric formation waters, entrapped in the rocks, disclose frequent subaerial land exposure, indicating part of the sediments are terrestrial mud or sand by origin.

Paleo stages of sea regressions. The rock-compartments that host the connate formation waters are mainly built by shallow marine rocks. In contrast, the contained connate groundwaters are meteoric water that penetrated when the rocks were subaerially exposed during a sea transgression phase. Detailed three-dimensional mapping of the rock-compartments in a study area provides an estimate of the frequency of the respective paleo-regressions.

What is behind frequent changes of limestone and clay in rock profiles? Sedimentary rock sequences frequently contain beds of clay-rich rocks, their thickness varying from a few centimeters to several meters. The frequent land exposures discussed by the connate meteoric formation waters draw attention to the open possibility that the clay-rich beds were formed, in part, from mud that evolved on the evaporitic paleo-landscapes. Thus, the alternation of limestone and clay beds may be the result of sea transgressions and regressions.

An elevated periphery existed at the margin of the ancient evaporitic low flatlands. The brine-tagged fossil formation waters reflect formation on flat, evaporitic lowlands. At the same time, the quantitative accumulation of sediments in the subsided basins discloses the existence of a mountainous periphery that supplied the raw materials, washed down by erosion. According to the observable extension of the respective sedimentary rocks and entrapped formation waters, the paleo-extension of the flat low landscape relief and the elevated periphery can be reconstructed.

Identification of paleoclimates is feasible, e.g., fossil saline groundwater reflects an arid paleoclimate, whereas fossil fresh groundwater may reflect a humid paleoclimate (interestingly, this case is not observed by fossil groundwaters).

Identification of deep subsidence. Traditionally, oil has been taken as an indication that the original organic raw material–hosting rocks subsided, after their formation, to a depth of several kilometers, until they reached the oil-formation temperature window of over 120°C. The ^{18}O isotopic enrichment, observed in most formation waters, serves as an independent indicator that the hosting rocks subsided to the depth where temperatures around 120°C prevailed. The advantage of the formation waters as temperature/depth indicators is that they are widespread, often in regions that do not contain oil.

Identification of subsequent uplift and removal by erosion. Formation waters enriched in ^{18}O are occasionally observed at shallow depth, of a few hundreds of meters. In these cases the interpretation is that subsequent to the deep subsidence the hosting rock-compartments were uplifted, and overlying strata were eroded away. This paleo-tectonic diagnostic tool has been calibrated by cases in which shallow ^{18}O enriched formation water has been associated with shallow oil deposits.

3.2.7 Application of Formation Water to Explore Petroleum

Formation water is a constant companion of petroleum shows and deposits, thus providing useful related information:

Identification of formation waters that are possibly associated with nearby petroleum deposits. Formation waters that are associated with petroleum differ from regular formation waters by their detectable concentrations of petroleum compounds. This provides a tool for petroleum exploration. The potential of this method is high as formation waters are encountered along all deep drilling operations. High concentrations of petroleum compounds raise the prospect of finding associated petroleum deposits.

Understanding and mapping the structure of individual petroleum fields. Oil and gas fields are heterogeneous, composed of separated petroleum-containing rock-compartments interwoven with only water-containing compartments. The detailed structure of every field can be mapped by the properties of the water as well as the petroleum occurrences. Information of this kind is essential in placing central production plants and siting of new wells.

Proper reinjection of petroleum brines. Brines separated from petroleum have to be safely reinjected, and for optimal solutions the structure of each producing field has to be well known.

3.2.8 Hydrology and Physical Processes Operating in Hydrothermal and Volcanic Systems

The properties of thermal waters disclose the hydrological and volcanic setups and the nature of ongoing processes.

Did the water enter the rock system before or after the magma intrusion? First principles consideration tells us that as the magma chambers and adjacent rock-compartments are hot and pressurized, recharged water cannot flow into them. Thus, we are left with two likely mechanisms: (1) The rocks into which the magma intruded contained fossil formation water that, thus, predated the magma intrusion; and (2) volcanic eruptions are connected to pressure release (demonstrated in extreme cases by the formation of collapse calderas), and when an eruptive phase is terminated, groundwater from the surrounding rocks can rush in, and in oceanic volcanoes seawater can be pumped in as well (Chapter 14). The intrusion of new water may have a part in the renewal of eruptions.

The isotopic dating methods are an essential part of the study of these aspects of volcanic system activity.

Is mixing of different water types taking place in nondisturbed hydrothermal study sites? Various water manifestations sampled in hydrothermal systems reveal different salinities—is there essentially one end-member that is diluted to varying degrees? Or are there several water types that intermix? Detailed chemical analyses can supply the answer at each study site. Published case studies revealed one type of groundwater that is concentrated at certain locations by steam separation, and in other points dilution by condensed steam (Chapter 14).

Which ions are external and which are internal? Chemical data provide the answer in the same way as discussed above in context of the nonthermal formation waters.

How old are the thermal fluids? The answer to this query comes from dating of the water with isotopic age indicators.

Depth of boiling of ascending superheated water is key information required to understand a studied system. Beneath this depth all the water is of one phase—a superheated fluid, and above this depth separation into steam and residual water takes place.

Depth of original penetration of the water is greater than the depth of boiling and equal or deeper than the depth of the relevant magma chamber.

Boiling forming an early steam phase takes place at which location? The early separated steam can be recognized by enrichment in the active and noble gases.

Identification of residual water—where does it issue? In certain boiling springs? Is it produced by certain wells? Low concentration of the atmospheric noble gases typifies residual water.

Intermixing with condensed vapor may be traced, for example, by a relative lowering of the salinity, yet maintaining the high temperature.

Production-induced encroachment of shallow cold groundwater is recognizable by monitoring the water and steam properties of the producing wells. Appearance of tritium and other anthropogenic contaminants is an example how arrival of recent shallow water can be noticed. The results provide an insight into the production-induced dynamics within the studied system.

Siting new steam production wells is done by various considerations, e.g., as close as possible to existing or planned central production facilities, but not within the radius of production of an operating well. If in a new exploration drill hole the initial steam composition is encountered, the borehole may be equipped as a production well, but if the already changed composition is observed, then an adjacent well is producing from that niche and the borehole has to be plugged.

Which ions originate from high temperature water–rock interactions? For example, the often observed HCl and HBr, where were they formed?

Which ions are from a deep "magmatic" origin? Compounds that are in excess of what could come from external sources, or from water–rock interaction alone, are likely to originate from magma, for example, CO_2 or SO_4 and H_2S.

How are fumaroles formed? Are they a product of steam separation? Or is the whole deep-seated fluid bursting out? The atmospheric noble gases are informative in these cases. In the first case they are distinctly enriched, and in the second they are as in air-saturated water.

3.2.9 Selecting Optimal Sites for Nuclear Waste Repositories

The hydrology of planned repositories is crucial in selecting optimal sites. The following are examples of some aspects:

Has the recharged through-flow groundwater access to the planned repository? In other words, is the locally encountered groundwater recent or fossil? What is its age?

Is locally entrapped groundwater corrosive? What can be done? The materials of the waste containers have to be resistant to potential corrosion by the local groundwater, so their composition has to be well studied, taking into account the local heating by the wastes. If the original entrapped water is limited in volume, can it be pumped away?

What is the nature of the hydraulic connectivity of the local rocks? In principle, minimal hydraulic connectivity is looked for. The nature of the selected rock unit has to be understood in terms of fracture connectivity or isolation in the case of crystalline rocks and permeability or conductivity in the case of sedimentary rocks.

Might local heating by the wastes open undesired flow paths? For example, is fracturing of the surrounding rocks anticipated, and to what distance; might rock dissolution be locally intensified?

Can existing boreholes and wells cause damaging short-circuiting? The case studies discussed in Chapter 9 reveal ample examples of groundwater shifting through boreholes, for example, in crystalline shields. This aspect has to be addressed by the final construction of a repository.

3.2.10 Identification of Therapeutic Qualities of Mineral and Warm Groundwaters

In order to turn a warm and mineral water spring or well into a therapeutic or recreational complex, the groundwater qualities and available quantities have to be well known and understood. A list of questions and considerations is discussed in section 13.2.

3.2.11 Earthquake Hydrology

The water table of unconfined groundwater systems and the pressure head in confined systems respond to earthquake events. The responses before an event, during it, and thereafter have good potential to provide insight into the distribution of compression–extension regions and other details of the earthquake anatomy. Careful processing of collected data and observations can lead to the establishment of useful earthquake prediction and research hydrological monitoring networks.

3.2.12 What do the Rocks Beneath the Oceans Contain?

The oceans are a huge body of saline water; do the underlying rocks host seawater as well? Are there cases of interstitial meteoric water formed during sea regression events? Hundreds of research drillings retrieved rock logs from all oceans, up to a depth of 1000 m. The enclosed interstitial water was in many cases analyzed, and a wealth of geological information was found to be encoded in these waters, with much bearing to sedimentary basins and to petroleum geology. This subject is presented in Chapter 5.

3.3 The Research Tools

The list of research tools is vast and involves specialized laboratories to which the field researcher submits collected samples for analyses. The hydrochemist, hydrologist, and petroleum specialist are facing the inspiring challenge to link the ends. They have to be familiar with the sample collection methods, the parameters to be measured, and the processing of the resulting data. Close connection to the laboratory staff experts is recommended, exposing them to the research aims and getting their instructions for the sample collections.

3.3.1 Physical Parameters

The physical parameters are unique in the sense that they all have to be measured in the field in real time:

Water temperature is an efficient indicator of the depth of storage of a studied groundwater.

Unconfined groundwater system. A spring or well that has a temperature that practically equals the local annual average surface temperature belongs to the unconfined groundwater system.

Medium depth through-flowing groundwater. Water at several degrees Celsius above the average annual ambient temperature reveals a depth of flow that can be estimated, knowing the local terrestrial heat gradient.

Deep confined fossil groundwater. Hot waters, tens of degrees warmer than the average land surface temperature, indicate fossil groundwater stored at a depth of thousand meters or more.

Steaming and superheated hydrothermal groundwaters are connected to shallow magma chambers.

Water table and water pressure measurements provide insight into the geometry and dynamics of groundwater systems.

Static and dynamic water table. There is a principal difference between the depth and altitude of a water table monitored at a nonpumped observation well, the *static water table*, and the water table measured in a pumping well, the *dynamic water table*. Both types of measurements are informative and essential. The common use of the term "water table" relates to the static water table. The dynamic water table is a measure that specifies an individual well.

The regional water table of the unconfined groundwater system is best reflected in observation wells that are placed far enough from any operating well so that in a pumping test of any of the adjacent wells no water table drop is noticeable in the observation well. Observation wells are commonly part of a regional monitoring network.

The dynamic water table is relevant to operational "red lines." Each well has its optimal pumping yield—as high as possible to supply all demands, but always within a safe margin so that the water table is not lowered too much, causing seawater or deep saline groundwater to flow in.

Inclination of a regional unconfined groundwater table reveals the boundaries of an unconfined system and identifies its terminal base of drainage.

Perched groundwater tables can be identified by their being above the regional unconfined water table.

Daily fluctuations of the water table in coastal observation wells disclose that the groundwater base flow is to the sea surface, and hence it respond to the tidal events.

Seasonal fluctuations. High water tables are conforting as they reflect the degree of renewal of the resource, and low water tables are alarming as they reflect overpumping or the effect of droughts.

Response to pumping tests in adjacent wells. Drop of water table in wells as a response to pumping tests in neighboring well establishes hydraulic interconnection patterns.

Response to exploitation in artesian wells is one of the guides in siting new production wells.

Response to exploitation in neighboring petroleum wells reveals they tap the same rock-compartment, information needed in sitting new wells.

Semi-artesian and artesian pressure measured during drilling of new boreholes. Special care is needed to sense the encountering of new groundwater bodies as the drilling proceeds. Of prime importance is to stop the drilling operation and measure to what altitude the new water ascends in order to define the separated water compartments and their confined pressure or rise to the surface. This information is acute for siting new wells.

Discharge of springs and yields of wells are important parameters, needed for management strategies as well as for understanding the exploited systems.

Marked seasonal variations of spring discharge help to identify karstic systems as well as springs fed by local perched groundwater systems.

Constant decrease in spring discharge calls for an investigation. Pumping in a nearby well may be the cause, or changes in the recharge intake region, e.g., new highways, dams, or urbanization.

Gradual decrease of the yield of wells is alarming as it may well indicate man-made intervention, e.g., overpumping or operation of new adjacent wells.

Safe yield is a term of optimization of a well operation—the highest yield that can be produced for a long time while maintaining the water quality.

A time lag between the rainy season or snowmelt and water table rise indicates the local recharge is mainly by infiltration through interconnected voids.

Rapid rise of the water table following the rainy season or snowmelt discloses effective recharge via preferred flow paths.

3.3.2 Chemical Parameters

Some of the chemical parameters are preferably measured in the field, as they may change during sample transport to the laboratory, but most chemical parameters have to be analyzed in well-equipped laboratories. Ionic concentrations are expressed in weight per volume, e.g. mg/L; or in equivalents per volume, e.g. meq/L. The equivalent is expressed in amount of molecules, and is derived by observed weight divided by the molecular weight. For example, seawater contains around 19,000 mg/L of Cl; the atomic weight of Cl is 35.5; hence, seawater contains $19,000/35.5 = 535$ meq/L Cl.

Correct sampling is a key issue in getting meaningful laboratory analyses. Examples are avoiding mixing with residual drilling fluids and avoiding mixing of waters from different aquifers as a result of short-circuiting in a well or drill hole.

Common dissolved ions are applied in the following modes: *Water quality* is largely defined by the concentration of about eight major ions, present in gram or milligram per liter concentrations, that determine the water quality with respect to usage, e.g. drinking quality, irrigation, or industrial. *Groundwater facies* is defined by the salinity as well as relative ionic abundances. Examples of water facies are continental sea tagged, continental tagged by brine-spray, fresh water, seawater, and residual brine (bitterns). *Mapping the three-dimensional distribution of rock-compartments* that host formation water and/or petroleum is reliably done by the dissolved ions data.

Natural mixing of different waters is best identified by the pattern of composition diagrams of the common ions.

Trace elements encompass about a dozen elements of importance, which are present in parts per million or per billion concentrations, each highlighting a specific aspect of the origin and anatomy of studied groundwater systems. Every ion contributes information—the more that are analyzed, the better is our understanding of the studied system.

The drinking water standard includes a long list of the maximum tolerable concentrations of dissolved ions. Every well that supplies drinking water must from time to time be analyzed for the entire list.

Pollutants constitute a long list of potentially endangering compounds, and the hydrochemist needs access to laboratories that are experienced in such inspections.

3.3.3 Stable Isotopes of the Water

The δD and $\delta^{18}O$ values are expressed in ‰ deviation from a seawater standard. The light isotopes of water evaporate more efficiently than the heavier ones. As a result, water that evaporates from the ocean is isotopically distinctly lighter than seawater, and this signature is preserved in rain, snowmelt, and groundwater. The following are common patterns:

Global and local meteoric water lines. A large portion of the world's recent and fossil groundwater samples plot on a δD versus $\delta^{18}O$ diagram along a line of

$$\delta D = 8\, \delta^{18}O + 10$$

called the *global meteoric water line (GMWL)*. In certain study areas the water samples plot on somewhat different lines, called *local meteoric water lines (LMWL)*. This global pattern is a key parameter in the identification of recent and fossil meteoric groundwaters, i.e., groundwaters formed on the continent.

Enrichment in the heavy isotopes identifies partially evaporated water. Wells in unconfined water systems in arid zones reveal isotopic values that are relatively heavy, plotting along typical evaporation lines.

Little evaporated fresh groundwater has distinctly negative values of δD and $\delta^{18}O$. Seawater has $\delta D = 0$, and $\delta^{18}O = 0$, except for closed sea sections that experience somewhat higher evaporation losses, identifiable by slightly heavier isotopic values. Evaporitic residual brines have δD and $\delta^{18}O$ values that are distinctly heavier than the seawater values.

Temperature-induced ^{18}O enrichment is noticeable in most formation waters and in thermal waters. The respective samples plot to the right (heavy values) of the δD versus $\delta^{18}O$ diagram. Such isotopic enrichment discloses a deep subsidence history.

3.3.4 Atmospheric Noble Gases

The atmospheric noble gases (ANG) are He, Ne, Ar, Kr, and Xe, which are dissolved in water that equilibrates with air. The solubility is temperature dependent. The ANG are identifiable by their isotopic composition (Mazor, 1972).

Identification of the meteoric origin of water by the ANG contents. Practically all water types that have been analyzed for their noble gas concentrations revealed values that resembled air-saturated water, at common ambient temperatures (Mazor, 1972). This is an independent tool to identify meteoric waters, in parallel with the stable isotopes.

Groundwater is a closed system with regard to the dissolved ANG as the atmospheric input is well preserved in cold and warm groundwaters, recent and old (Mazor, 1972; Herzberg and Mazor, 1979).

Reconstruction of ambient paleo-temperatures. The concentrations of the dissolved atmospheric Ar, Kr, and Xe serve to calculate the ambient paleotemperature that prevailed at the time a studied groundwater was recharged and confined. Case studies revealed paleo-temperatures in the range of 6 to 30°C (Mazor, 1972; Stute et al., 1992).

Physical processes in superheated water systems are disclosed by the noble gases. Comparison to measured concentrations of the ANG with expected values in air-saturated water identify steam phases (enriched in ANG) or residual water (depleted in ANG).

Meteoric origin of the thermal waters has been established by the relative abundances of the ANG (Mazor, 1972, 1975).

Deep storage of the water is disclosed by positive correlations between the atmospheric noble gases and radiogenic ones, indicating the water stayed in the host rocks for a long time and steam removal takes place during ascent.

3.3.5 Radioactive and Radiogenic Dating Methods

Isotopic dating methods include radioactive isotopes like tritium, ^{14}C, and ^{36}Cl and radiogenic stable isotopes like ^{4}He and ^{40}Ar. The first group is the basis of decomposition dating methods and the second group contributes to cumulative dating methods.

The age of groundwaters in the geosystems of the geohydroderm extends over the vast range of 10^1 to 10^8 years. The dating methods cover this range in the following order (Mazor, 2003):

Tritium. Tritium(T) is the heavy rare radioactive isotope of hydrogen, also written ^{3}H. It decays into ^{3}He, with a half-life of 12.3 years. Tritium is formed naturally by interaction of cosmic rays with the atmosphere. Tritium content in water is expressed in tritium units (TU), defined by the ratio to the stable hydrogen isotope, i.e., T/H. 1 TU is defined as T/H = 10^{-13}. The natural content in rainwater is around 5 TU.

Tritium was introduced into the atmosphere in significant amounts by nuclear bomb tests, which were stopped by an international treaty at 1963. In the year 2000 the tritium content in new groundwater was practically back to the natural value. Tritium serves to identify recently recharged groundwater, i.e., its dating range is in the last few decades. The applicability of the dating method is somewhat extended by measurement of the formed stable isotope ^{3}He (Schlosser et al., 1988).

Carbon-14. The heavy rare radioactive isotope of carbon. It decays into nitrogen-14, with a half-life of 5730 years. Carbon-14 is formed by the interaction of cossmic rays with the atmosphere. The content in water is expressed relative to the ^{14}C/^{12}C ratio in an international standard of oxalic acid. The concentration of ^{14}C in the bulk carbon of the standard is defined as 100% modern carbon (100 pmc).

Carbon-14 has been introduced into the atmosphere by nuclear bomb tests, raising the content to around 160 pmc. There is a list of details that have to be addressed in the application of ^{14}C for groundwater dating, and the field hydrologist will get the aid of the laboratory in both instructions for sample collection and age calculation. The range of the dating method is 10^3 to 2×10^4 years. The method is semiquantitative, but the orders of magnitude of the age are reliable and of utmost value (Mazor, 2003).

Chlorine-36. This is the heavy rare radioactive isotope of chlorine. The half-life is 301,000 years. Chlorine-36 is formed by cosmic rays interacting with the atmosphere, and it is incorporated into the hydrological cycle. The content is expressed in units of 10^7 atoms/L of water. In addition, ^{36}Cl is formed within the rock system, its production rate being proportional to the content of uranium, thorium, and chlorine in the rocks. This in situ production amounts to a few percent of the atmospheric contribution. A minor ^{36}Cl contribution was also made by the nuclear bomb tests.

The dating method is semiquantitative but highly informative, in the age range of 3×10^5 to 10^6 years (Mazor, 1992).

Helium-4. This isotope of the noble gas helium is produced as alpha particles emitted in the radioactive disintegration of ^{238}U, ^{235}U, and ^{232}Th. The age, t, is calculable from the following equation of ^{4}He production (Zartman et al., 1961):

$$He = 10^{-7} t\xi \, He(1.21U + 0.28Th)d(\text{rock : water})$$

where U and Th are the concentrations of uranium and thorium in the host rocks, expressed in g/g (the respective coefficients incorporate the half-lives and number of helium atoms formed along each disintegration chain; ξ He is the emanation efficiency of He from the rocks (nearly 1), t is the storage time, or age, of the water in years; and d is the rock density. The rock water ratio in common rocks ranges between 4 to 20 and is calculable from the effective porosity.

The produced ^{4}He is to a large degree dissolved in the formation water that is in contact with the rocks. Thus, the older the water is, the higher the amount of accumulated ^{4}He. Recent water contains atmospheric ^{4}He in a concentration of about 4×10^{-8} cc STP/L, old groundwaters contain orders of magnitude higher concentrations.

The method is semiquantitative, but the orders of magnitude of the age are dependable, within the age range of 10^4 to 10^8 years (Mazor and Bosch, 1990, 1992).

Argon-40. This stable isotope of the noble gas argon is produced by the radioactive decay of ^{40}K, a rare radioactive isotope of potassium, present in rocks in the parts per million level. The disintegration half-life is 1.3×10^9 years. An equation similar to that presented above for the ^{4}He dating can be structured for ^{40}Ar dating.

Recent water contains a small amount of ^{40}Ar that is part of the dissolved atmospheric noble gases, and the addition of radiogenic ^{40}Ar in old formation waters can be several orders of magnitude higher. The effective groundwater dating range is 10^5 to 10^8 years.

The water age is a function of the measured concentration of radiogenic ^{40}Ar, the concentration of ^{40}K in the reservoir rocks (a few ppm) and the water/rock ratio. This semiquantitative dating method is further discussed in section 8.4.

Comparison to confinement ages. The confinement age, defined above, is the age of connate water stored in a hydraulically isolated rock-compartment. The age is between the age of the host rock and the age of the overlying confining rock unit. A good agreement between the confinement age and the isotopic dating age provides a valuable confirmation that the order of magnitude of the water age is correct.

A methodological note. Each of these water dating methods has its own shortcomings, but the dating is reliable in terms of orders of magnitude, and this is what counts and what is of enormous value in the understanding of the formation and structure of studied groundwater systems, as is demonstrated by case studies in the following chapters.

The methodology and practical use of the listed isotopic groundwater dating methods is addressed in detail in the companion book *Chemical and Isotopic Groundwater Hydrology: The Applied Approach* (Mazor, 2003).

Mixing of groundwaters disclosed by contradicting concentrations of age-indicating isotopes. In a rather large number of reported case studies water samples contained a significant concentration of a short-range age indicator, e.g., tritium, and also significant concentrations of a long-range age indicator, e.g., ^{4}He. A straightforward interpretation of such results will lead to the contradicting conclusion that such a water is young and old at the same time. The simple explanation is that we deal in these cases with a mixture of young and old water, for example, by short-circuiting a shallow and a deep aquifer that are tapped by a well. Examples of such contradicting concentrations of age indicators are given in Chapters 8 and 9.

The age of the mixed end-members can be estimated (Mazor, 1992b; Mazor et al., 1986, 1992; Vuataz et al., 1983). Once a mixing case is identified, precaution has to be taken in the treatment of all the other measured parameters—in a case of mixing even the temperature or depth of the water table are mixed values; hence the great importance of analyses of the age indicators.

Mixing is identifiable from chemical analyses made of several water sources collected in the same area, but the age indicators can provide the information from analyses of a single sample.

Age indicators are useful to distinguish between flowing and entrapped groundwaters. Age of a studied water is a simple and straight forward parameter to identify through-flowing groundwater from stagnant water that is entrapped in the zero hydraulic potential zone of closed rock-compartments. The first has a young age, e.g., up to 10^3 years, and the latter is old, e.g., 10^5 years or more.

3.3.6 Petroleum-Related Organic Compounds Dissolved in Formation Waters

These are a promising tool to search for petroleum deposits. The concept is simple: Formation water horizons passed by drillings, made for many purposes, have to be sampled and analyzed for traces of dissolved petroleum compounds. It is expected that formation water closely associated with a petroleum deposit will be enriched in such compounds. Three-dimensional mapping of the results may point toward zones of potential petroleum deposits.

3.3.7 Anthropogenic Pollutants

These have to be analyzed in order to identify over time cases of deterioration of the produced water. The list of potential pollutants is long and includes, for example, pesticides, fertilizers, industrial pollution, and bacteria.

3.4 Research Strategies

There are several modes in which to conduct field research, as demonstrated by the following examples.

3.4.1 Multiparameter Studies

Concentration of a single ion, e.g., just Cl, provides limited information on a studied groundwater. Each parameter provides information and boundary conditions for the understanding of the studied system. The great benefit from multiparameter studies is the modes of correlation, or lack of correlation, that are observable.

The added coast of the multiparameter analyses are small compared to the whole drilling or monitoring operation, but the breadth and scope of the gained information is enormous (Mazor, 1976). This includes the physical, chemical, and isotopic parameters.

3.4.2 Multisampling Studies and Study Area Surveys

Data obtained for a whole group of samples, simultaneously collected at springs and wells of a study area, provide the following kinds of information:

Number and identity of occurring groundwater types and their spatial distribution
Cases of mixing, natural or manmade, and identification of the end-members
Identification of the type of tagging e.g., sea spray or brine-spray.

3.4.3 Time Series

Repeated in situ–measured data and laboratory multiparametric analyses of samples collected at the same source disclose the dynamics of a studied groundwater or petroleum system. For example:

Single observation well. The time series can reveal whether one deals with a single large-scale groundwater body (constant values); importance of seasonal recharge (periodic changes), encroachment of another type of groundwater (directional change in properties); etc. This kind of deduced information pertains to the regional system.

Single pumped well. The time series reveals whether pumping is within the frame of safe exploitation (constant or periodic variations) versus overpumping (directional change, commonly deteriorating in quality).

Single spring. The time series indicates a karstic source (significant rapid seasonal variations, mainly in discharge) or influence of adjacent pumping wells (directional lowering of discharge and/or salinity and change of composition).

Single oil or gas well. The time series provides the picture of the exploitation induced changes occurring within the petroleum-hosting rock-compartment. Such information provides a follow-up of the productivity of the well and helps to site new wells in nonexploited niches.

Single boiling spring. The time series provides clues to whether shallow water is added following rainy seasons and whether nearby human activity threatens the spring.

Single hydrothermal production well. The time series reveals gradual changes caused by the well operation, providing indicators needed for the siting of operational wells in nonexploited niches.

Single fumarole. Here the time series is of special interest as it may reflect a rhythm in the low-scale activity of the semidormant volcano or provide an indication of increased activity that may lead to a renewed eruption.

3.4.4 Importance of Historical Data

To understand ongoing processes, especially in regions of intense human activity, references of the situation before the activity began, or before it was quantitative, are needed as benchmarks. This is true for springs and wells, artesian wells, petroleum wells, and hydrothermal manifestations. Hence, researchers are encouraged to seek available historical data from local archives, regional data centers, and scientific publications.

3.4.5 Sampling and Measurements During Drilling Operations

These are essential to gain knowledge of the spatial distribution of groundwater bodies at the site of study. Every drill hole has a specified aim, but let us always remember that with proper planning of the drilling operation, valuable information on the nature of the local setup can be gained. This strategy is well known with relevance to the lithology—rock cuttings are often systematically collected and studied. The same holds true for encountered water units; the driller has to notice every occurrence of a new water body, stop the drilling, lower a pump and clean up the well, measure the water head and temperature, and collect samples for extensive laboratory analyses.

3.4.6 Sampling and Measurements During Pumping Tests

Pumping tests are geared to establish the safe yield at which the future well can be operated for a long period of time. The interpretation of the results obtained during a pumping test is made on the assumption that a single groundwater unit is involved, but this point has to be checked. Before the test is started, temperature and water table have to be measured, and samples for laboratory measurements have to be collected. During the test the collection of about 10 samples is recommended, along with measurements of temperature, electrical conductivity, and the dynamic water table. If no changes are recorded, only one groundwater body was tapped; but in certain cases sudden changes in the properties will be noticed, indicating inflow of another water body, often of an inferior quality. In such cases the recommended production pumping yield has to be low enough to ensure the water quality.

3.4.7 Cross Sections

Geological and hydrological cross sections are a convenient mode to study a terrain for both its water resources and petroleum deposits. The lines of such cross sections can be selected along directions that include a maximum of wells and drill holes for which data are available. New drill holes can be preferentially located along such cross section lines.

3.4.8 Depth Profiles

Data obtainable from boreholes in a study area can be plotted as depth profiles, providing a readable depiction of the vertical distribution of identifiable groundwater and/or petroleum compartments.

3.5 Summary Exercises

Exercise 3.1: Which term relates to the property of a rock to contain water and let it flow through interconnected voids?

Exercise 3.2: Which of the following terms is correct: (1) Mississippian Aquifer, (2) Long Island groundwater system. (3) Colorado Aquifers, (4) the Tel Aviv Coastal Plain groundwater system.

Exercise 3.3: The 21-year-operating Sweet Well suddenly became saline. What can be done to understand what happened and determine what to do?

Exercise 3.4: The water pumped from a 250-km-inland shallow well revealed a salinity that is half that of seawater and a Cl/Br ratio of 293. How is this possible?

Exercise 3.5: How can the study of formation waters shed light on the origin and occurrence of petroleum deposits?

Exercise 3.6: When dealing with deep fossil groundwater, is it enough to send samples for the long-range dating methods alone in order to save the costs of short-range dating, analyses? Discuss.

PART II
SHIFTING OF WATER AND SALTS BETWEEN OCEANS AND CONTINENTS

*The saline oceans
are the source of the marvelous*
fresh continental water,
and this water
feeds the oceans with salts

4
SHALLOW CYCLING GROUNDWATER, ITS TAGGING BY SEA SPRAY, AND THE UNDERLYING ZONE OF STATIC GROUNDWATER

The unconfined groundwater geosystem extends over all the continents, covering around one-third of the Earth's surface. The number of different groundwater chemical and isotopic compositions, temperature values, ages, and modes of occurrence is countless. Yet the principles of flow in this geosystem are simple; surface water and groundwater flow are propelled by the vertically downward pull of terrestrial gravity, restricted by the availability of free space to flow in. Surface water and through-flowing groundwater are tagged by sea spray, whereas the main water–rock interactions are CO_2 induced.

A large amount of environmental and geological information is recorded in the unconfined groundwater, discussed in the present chapter in light of observations and conclusions derived from a number of case studies.

4.1 Groundwater Facies of the Geohydroderm

Groundwaters occur in a countless number of different salinities, ratios of dissolved ions, isotopic compositions of the water phase, and ages. The specific composition of each groundwater system reflects the environmental conditions at which the groundwater was formed. Thus there are different groundwater facies that can be categorized into major groups as in Table 4.1.

A survey of the occurrences of groundwater facies brings up a number of key observations:

   1. Practically all groundwaters are meteoric, as reflected by the negative values of δD and $\delta^{18}O$. Entrapped seawater, or residual

Table 4.1 Facies and Modes of Formation of Recent and Fossil Groundwaters and Identifying Criteria

Groundwater facies	Chemical and isotopic markers
Meteoric, fed by sea spray	Cl/Br ~290; light δD and $\delta^{18}O$ values that plot near the GMWL
Meteoric, fed by brine-spray	Cl/Br <200; a Ca–Cl component; light (negative) δD and $\delta^{18}O$ values that plot near the GMWL
Entrapped seawater	Cl ~19 g/L; Cl/Br ~290; chemical composition as seawater, δD ~0 and $\delta^{18}O$ ~0
Entrapped residual evaporitic brine, before halite precipitation stage	Cl <190 g/L; Cl/Br ~290; positive (heavy) δD and $\delta^{18}O$ values
Entrapped residual evaporitic brine, beyond halite precipitation stage	Cl ~190 g/L; Cl/Br <200; positive (heavy) δD and $\delta^{18}O$ values
Entrapped residual evaporitic brine that dolomitized limestone	Cl ~190 g/L; Cl/Br <200; Ca–Cl predominating, positive (heavy) δD and $\delta^{18}O$ values
Groundwater that dissolved halite	Cl ~190 g/L, mainly balanced by Na; Cl/Br > 3000; light δD and $\delta^{18}O$ values that plot near GMWL
Seawater that dissolved halite	Cl ~190 g/L, largely balanced by Na; Cl/Br > 3000; δD ~0 and $\delta^{18}O$ ~0
Groundwater that dissolved gypsum	SO_4 ~2700 mg/L, balanced by Ca; light δD and $\delta^{18}O$ values that plot near GMWL
Groundwater, heated to >120°C, e.g., during deep subsidence	Negative δD values, but relatively positive $\delta^{18}O$ values (plot to the right of GMWL)

evaporation brines (bitterns) are not encountered, as is evidenced by the light isotopic composition of practically all groundwaters encountered in wells drilled on the continents.
2. Two groundwater facies are by far the most common: sea-tagged meteoric groundwater and brine-tagged meteoric groundwater.
3. Practically all recent shallow cycling groundwaters belong to the sea-tagged facies; whereas practically all ancient formation waters, stored in closed rock-compartments, belong to the brine-tagged facies.

The facies of rocks is a common concept in geology and provides a major tool to learn about the paleo-conditions under which a studied rock type has been formed. The same is true for a groundwater facies—once it is identified, it sheds light on the ongoing water cycle, active in the zone of rocks situated on the continents above sea level and the genesis of fossil connate groundwaters entrapped in rock-compartments at depths that are below sea level.

The following sections deal with the various types of groundwater facies, and the far-reaching geohydrological conclusions that may be drawn from them.

4.2 Sea Spray Salts Concentrated Along a Large River System—The Murray River Basin, Australia

Salinity increase has been observed at the course of the 2500-km-long Murray River in southeast Australia. Data of samples collected at 17 locations along the river (Fig. 4.1) have been reported by Herczeg et al. (1993). The following patterns are observable:

Steady increase of concentrations disclosing the importance of evaporation. Concentrations of Cl, HCO_3, SO_4, Br, Na, Mg, Ca, and K steadily increase along the flow direction (Table 4.2 and Fig. 4.2). This trend reflects the important role of evapotranspiration that returns water into the atmosphere, leaving the dissolved salts behind. The latter accumulate during dry seasons on top and within the soil system, and they are redissolved by strong rain events that produce runoff that washes the salts into the river system.

Relative abundance of dissolved ions varies along the river flow path revealing tagging by sea spray. The fingerprint diagram of Fig. 4.3a portrays the water composition in the studied 17 sites along the river. Can we see a logical direction? In Fig. 4.3b the seawater line has been added, and it can be seen that the river water differs from seawater mainly by containing relatively too much Ca and CO_3. So in Fig. 4.3c Ca and CO_3 have been left out. At once, the picture is clear: the more saline the water is (in the downstream direction), the clearer is the similarity to the composition of seawater. Thus, K, Mg, Na, Cl, and SO_4 in the river water stem from sea spray, and the Ca and HCO_3 stem from water–rock interactions.

The sea spray origin is disclosed in composition diagrams. The composition diagrams of the Murray River, plotted in Fig. 4.4, reveal the following patterns:

1. Positive correlations are seen between the concentration of Cl and the concentration of the other ions, a direct indication of a common origin.
2. The Cl, Na, and Br data reveal a linear correlation and relative abundances as in seawater (Fig. 4.4a). This confirms that these ions

Shallow Cycling Groundwater

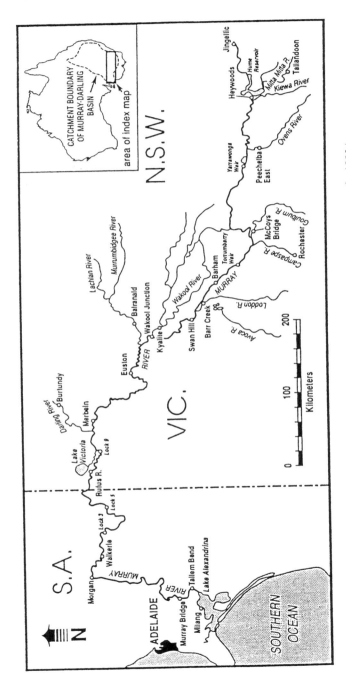

Fig. 4.1 The huge River Murray basin, with 17 stations surveyed along it. (From Herczeg et al., 1993.)

Table 4.2 Average Chemical Composition (mg/L) for the Years 1978–1986 at 17 Points Along the River Murray (Fig. 4.1), Arranged from the Water Head (2322 km flow length from the ocean) to Close to the Outlet (87 km from the ocean)

Location	KP	EC	Cl	SO$_4$	HCO$_3$	Na	K	Mg	Ca
Jingellic	2322	41	1.9	1.0	20	3.2	0.7	1.3	2.7
Hume Dam	2194	54	4.3	2.3	21	4.3	1.0	1.7	2.7
Heywoods	2193	56	4.3	2.2	23	4.3	1.0	1.8	2.8
Yarrawonga	1960	62	6.0	2.0	24	5.6	1.4	2.0	2.8
Torrumbarry	1611	101	14	3.5	29	11	1.8	3.1	3.9
Barham	1505	103	14	3.4	32	11	1.8	3.3	3.9
Swan Hill	1391	250	50	10	34	30	2.2	7.0	6.7
Euston	1099	267	51	9.5	43	29	2.4	7.3	7.8
Merbein	864	393	84	15	48	43	2.7	9.5	9.2
Lock 9	765	383	71	18	68	47	4.1	11	14
Rufus R. Junc.	696	438	78	21	82	50	4.8	12	15
Lock 5	562	494	94	23	79	58	4.8	13	15
Lock 3	431	633	125	31	91	77	5.3	15	18
Waikerie	383	718	147	34	82	92	5.9	15	19
Morgan	322	734	158	36	94	96	5.8	17	19
Murray Br.	115	758	166	38	95	105	5.8	17	21
Tailem Bend	87	778	171	38	94	101	6.0	17	21

Source: Data from Mackay et al. (1988).

originate from sea spray and are practically unmodified by water–rock interactions in the studied system.

3. The correlation lines of Cl with Mg, K, and SO$_4$ reveal relative enrichment of the latter in comparison with the respective seawater ratios (Fig. 4.4b), and this tendency is even more pronounced in the Ca and HCO$_3$ correlation lines (Fig. 4.4c). These deviations from seawater composition reveal contributions by water–rock interactions, predominantly by CO$_2$-enriched water that interacts mainly with carbonates and some silicates.

The sea spray cycle. Based on the listed observations and conclusions the following picture emerges:

1. In the fresh water, at the river head, the ions that originate from CO$_2$-induced water–rock interactions dominate. HCO$_3$ is the dominant anion, and Ca is the dominant cation in limestone terrains, and it is matched by Na and K in crystalline rock terrains. The sea spray ions are present, but are overmasked by the first group of ions.

Shallow Cycling Groundwater

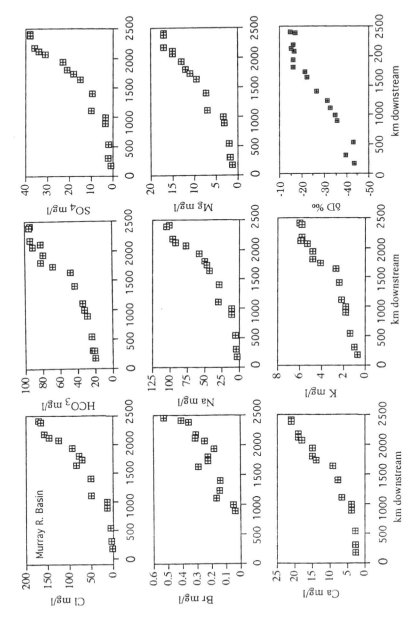

Fig. 4.2 Ion concentrations from the 17 sampling points along the axis of River Murray (see Table 4.2). The concentration of the dissolved ions increases downstream. (From Herczeg et al., 1993.)

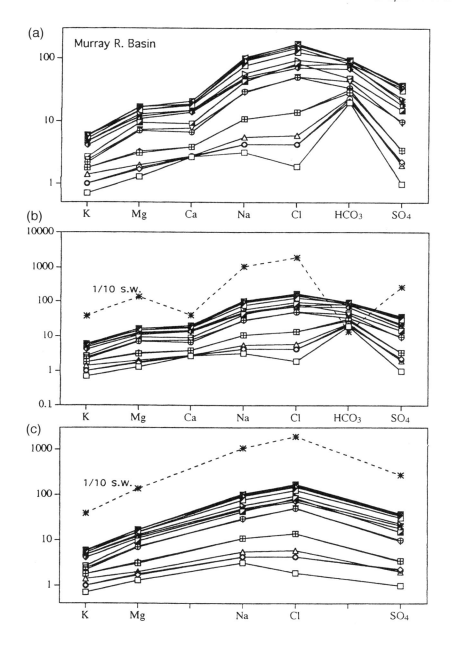

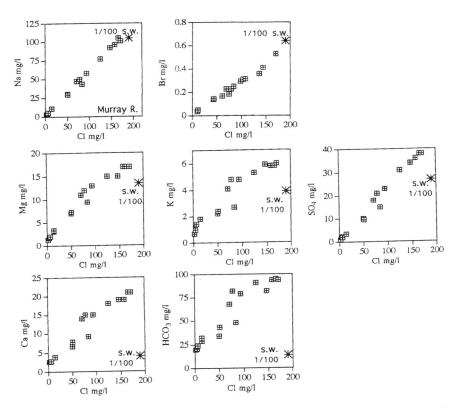

Fig. 4.4 Composition diagrams of water samples collected along the flow path of River Murray: Chlorine, Br, and Na are linearly correlated, maintaining the seawater (s.w.) abundance ratios. Magnesium, K, and SO_4 are slightly enriched over the seawater abundance ratio, reflecting contributions by CO_2-induced water-rock interaction. Significant enrichment of the HCO_3 and Ca concentrations, as compared to the seawater abundance ratio, indicates the main CO_2-induced water interaction is with calcium carbonates. (Data from Herczeg et al., 1993.)

Fig. 4.3 Composition fingerprint diagrams of water samples collected along the flow path of the River Murray: (a) the water is gradually more saline downstream, changing from $HCO_3 \gg Cl$ to $Cl \gg HCO_3$; (b) in comparison to seawater (marked as s.w.), the river water is rich in HCO_3 and Ca; (c) leaving HCO_3 and Ca out, it is easily seen how the river water becomes similar to seawater as the salinity increases, demonstrating the increasing contribution of sea-derived salts.

2. The rain falling on the land evaporates during the dry seasons, and the salts accumulate on the land surface. To these are added sea spray salts carried by the wind and precipitated as dust. Extra-strong rain events cause effective runoff that redissolves the salts and washes them into the river system.
3. The highly soluble sea spray salts, i.e., the chlorides, bromides, and to a large extent also the sulfates, are fully redissolved and reach the river along with the accompanying marine Na, K, and Mg.
4. The story of the HCO_3 and accompanying ions, often Ca, is different. In the dry seasons they are precipitated as carbonates, and these are poorly redissolved, their bulk staying behind and forming caliche-type crusts. Thus in the more saline surface waters the role of the nonmarine ions, stemming from secondary water–rock interactions, is minor.

The isotopic composition of the river water preserves its meteoric signature. The δD and $\delta^{18}O$ values were measured in April 1989 along the river and are plotted in Fig. 4.5. Along most of the river length the isotopic values are significantly negative and plot on the global meteoric water line (GMWL), disclosing very little evaporation. At the lower part of

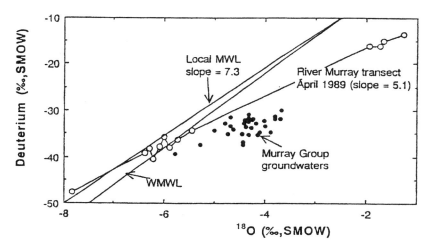

Fig. 4.5 Deuterium and oxygen-18 of water samples from a transact along the River Murray, collected in April 1989. The data (open circles) group mainly around very negative values, revealing minor evaporation, a conclusion supported by their plotting along the world mean water line (WMWL). This key observation disclosed that the salts and water are decoupled (see text). Some water is observed in the terrain to be contributed by springs issuing from the Murray Group (dots), but their quantitative influence on the river is minor. (From Herczeg et al., 1993.)

the river system somewhat less negative values were observed, plotting slightly off the world mean water line, indicating some evaporation. Some of this increase in the isotopic values is caused by isotopically heavier rain in the more inland regions. On the whole, evaporation is minor compared to the large increase in the concentration of the sea-derived ions.

In summary, the salts reach the ground all over the catchment area and are washed into the river by runoff. The latter flows only during substantial rain events, leaving little time for evaporation, and direct evaporation of the river water is negligible compared to its large water flux.

4.3 Sea Spray Salts Concentrated in a Closed Lake System Within an Arid Zone—Yalgorup National Park, Australia

Rivers and open lakes contain in general fresh water. In contrast, the salinity of closed lakes can reach high concentrations, as demonstrated by data from shallow lakes of the Yalgorup National Park, Western Australia (data from Rosen et al., 1996). The reported case study encompasses three groundwater-fed lakes: L. Clifton, L. Hyward, and L. Preston, part of a system of 11 shallow lakes (~1 m deep), spread along a stretch 8 km wide and nearly 30 km long, up to 10 km inland. The composition diagrams of Fig. 4.6 disclose the following patterns:

Linear correlations between the concentrations of Cl, Na, K, Ca, Mg, Sr, Br, and SO_4 reveal a common external source of sea spray. Remarkable positive linear correlation lines are seen between the concentration of Cl and the concentrations of Na, K, Ca, Mg, Sr, Br, and SO_4. The seawater value plots on these lines, indicating the ions stem from sea spray.

Formation of highly saline sea spray–tagged water. The salinity is observed to vary over the year, and within the different lakes, from fresh water up to seven times seawater concentrations. The lakes almost dry up at the end of the dry season and are refilled by runoff water in the rainy season. The sea spray-derived ions are introduced by the inflowing runoff water. The dimensions, morphology, and distance from the Indian Ocean differ from one lake to another, but the relative abundances of the dissolved ions are similar and seawater-like, supporting a salt origin from sea spray, and salinity determined by degree of local evaporation.

Conservative behavior of the Cl-correlated ions. A slight discontinuity is seen in the Ca–Cl and SO_4–Cl linear correlation lines (Fig. 4.6), possibly re-

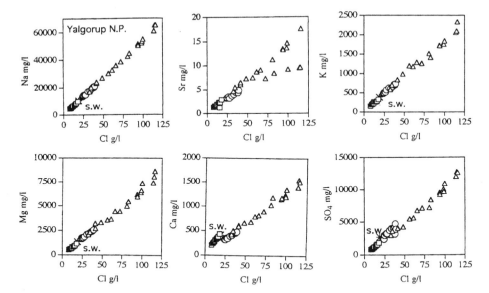

Fig. 4.6 Ionic concentrations measured during 1991 and 1992 in three lakes at the Yalgorup National Park, Western Australia. The following key observations emerge: (1) perfect positive correlation lines of the concentration of various ions as a function of Cl indicate a common external source; (2) the seawater value (s.w.) plots on these lines, reflecting the origin from sea spray; and (3) the salinity range is wide—from fresh water to 7 times the salinity of seawater. (Data from Rosen et al., 1996.)

flecting limited precipitation of gypsum. The linear correlation lines seen in Fig. 4.6 indicate that no quantitative mineral precipitation took place, nor was the composition modified by water–rock interactions.

4.4 Sea Spray Salts Concentrated in Unconfined Groundwater—Campaspe River Basin, Australia

Figure 4.7 portrays the composition of recent unconfined groundwater collected at shallow wells located in the Campaspe River Basin, northern Victoria, Australia, a few hundred kilometers inland (Table 4.3; data by Arad and Evans, 1987). The following patterns are observable:

The Cl is external. The Cl concentration is orders of magnitude below
   saturation with regard to halite, and hence no halite is present in the
   hosting rock system, and the Cl is external.

Shallow Cycling Groundwater

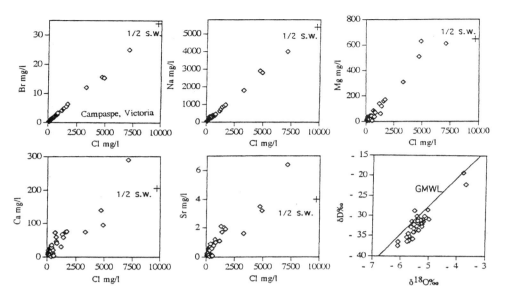

Fig. 4.7 Composition diagrams of unconfined groundwaters encountered in the Campaspe River drainage basin, Australia. The positive correlation lines and the match to seawater (s.w.) abundance ratios disclose how the salts originate from sea spray. The concentration of Ca, Mg, and Sr deviate slightly from the marine ratio, probably reflecting minor water–rock interactions. The δD and $\delta^{18}O$ values are negative and plot on the global meteoric water line, indicating origin from nonevaporated rains. (Data from Arad and Evens, 1987.)

Correlated concentrations of Cl, Na, Ca, Mg, Sr, and Br indicate sea spray tagging. Groundwater samples from adjacent wells reveal positive linear correlations between the concentrations of Br, Ca, Na, Mg, and Sr as a function of the Cl concentration, and the seawater value plots on these lines. A common source from sea-spray is thus demonstrated. The Mg, Ca, and Sr correlation lines are slightly disordered, and the seawater ratio does not plot exactly on the best-fit lines. These are indications that some secondary processes, e.g., water–rock interactions, slightly modified the initial sea spray signature.

Isotopic imprint of meteoric groundwaters. The isotopic composition of the Campaspe unconfined groundwaters is light, δD and $\delta^{18}O$ being negative, in the range of meteoric waters; and the data plot on the global meteoric water line, providing further evidence of an origin from ordinary rainwater (Fig. 4.7).

Table 4.3 Composition (mg/L) of Some Unconfined Groundwaters, Campaspe River Drainage Basin, Northern Victoria, Australia

Well no.		Ca	Mg	Na	Cl	SO$_4$	HCO$_3$	Br	Sr	δD (‰)	δ^{18}O (‰)
Elmore	8	11	17	71	100	16	110	0.4	0.29	−33.3	−5.53
	25	8	14	170	240	0.8	99	0.9	0.13	−32.9	−5.63
	26	58	37	380	770	4	15	2.7	1.02	−36.5	−6.08
	28	8	10	64	66	0.5	146	0.34	0.11	−28.8	−5.51
	30	20	17	50	91	2	119	0.46	0.28	−32.2	−5.37
	31	13	15	46	95	0.7	71	0.47	0.27	−30.7	−5.38
Goorong	37	140	510	2900	4700	1500	115	15.7	3.5	−33.7	−5.35
	38	94	630	2800	4900	1400	256	15.4	3.2	−34.2	−5.54
	39	30	140	640	1200	16	378	4.2	1.1	−37.4	−6.09
	40	13	15	120	200	14	77	0.98	0.16	−22.2	−3.7
	41	71	86	370	680	110	207	2.55	0.78	−31.7	−5.27
Diggorra	15	18	27	150	230	39	115	0.8	0.32	−19.3	−3.79
	61	6	5	74	83	1	96	0.32	0.14	−32	−5.27
	62	9	15	96	110	19	134	0.43	0.22	−32.7	−5.43
	64	13	20	79	110	16	134	0.42	0.28	−31.8	−5.36
	65	28	37	180	310	54	122	1.1	0.68	−31.9	−5.39
	69	23	30	180	270	54	122	1.06	0.49	−32.3	−5.52
Rochester	4	25	39	250	390	72	183	1.5	0.72	−31.1	−5.2
Rochester	W	2	10	270	330	3	195	1.2	0.04	−30.6	−5.11
	6	23	38	210	320	63	183	1.2	0.69	−32.8	−5.24
	7	28	47	290	390	130	231	1.6	0.72	−28.6	−5.03
	10	8	8	140	160	12	108	0.78	0.13	−30.2	−5.43
	15	5	15	180	260	2	90	0.89	0.14	−30.2	−5.15
	16	35	45	250	390	80	195	1.5	0.91	−31.4	−5.29
	17	21	36	190	290	54	171	1.1	0.61	−31.6	−5.19
	10029	17	28	160	220	40	171	0.76	0.5	−31.2	−5.26

Source: Data from Arad and Evans (1987).

4.5 Sea Spray-Tagged Fresh and Saline Groundwaters in the Unconfined Groundwater System at the Crystalline Shield of the Wheatbelt, Australia

Concentrations of dissolved Cl, Br, SO$_4$, Na, K, Ca, and Mg were measured in groundwater encountered in shallow wells, 2 to 46 m deep, located in six study areas in the flat Wheatbelt of Western Australia (Mazor and George, 1992).

The terrain, around 300 km inland, has a semiarid climate and is composed of crystalline rocks, Archaean granite and gneiss, weathered at the surface. The data, portrayed in Fig. 4.8 as functions of the Cl concentration, display the following patterns:

Large range of salinities. The Cl concentrations vary from very fresh water, of around 40 mg/L Cl, up to very saline, i.e. 63,000 mg/L (please note that the Cl concentration is given in grams per liter in Fig. 4.8). Similarly large concentration ranges are observed for the other dissolved ions.

The Cl is external to the water-hosting crystalline rocks. The local crystalline rocks contain no halite, so the observed Cl must be of an external source. The concentration of Cl is significantly below the saturation concentration with regard to halite, supporting the conclusion of an origin outside the hosting rock system.

Concentrations of Br, Na, and Mg, and in cases also SO_4, K, and Ca are positively correlated to Cl and reveal marine relative abundances. The Wheatbelt groundwaters reveal in all the six study sites a positive correlation between the concentrations of Br, Na, and Mg to that of Cl, over the wide range of observed salinities. In addition, the marine value (dashed line or solid triangle in Fig. 4.8) plots on the correlation lines. This clearly demonstrates tagging by airborne sea spray. In some of the cases, e.g., at the North Baandee site, also SO_4, Ca, and K are correlated to Cl and reveal a marine origin.

The conservative behavior is here attributed to two reasons: (1) the composition of the crystalline rocks and (2) the high ion concentration in these waters, which masks the small amount of ions that may have been involved in CO_2-induced water–rock interactions.

A large range of salinity observed between adjacent wells indicates groundwater moves in separated fracture systems. Large variations in the concentration of dissolved ions are often observed in the Wheatbelt between adjacent wells or at different depths within the same well. This observation, known from crystalline regions in arid terrains all over the world, indicates the unconfined groundwater moves along hydraulically separated fracture systems.

Rain and runoff water saturate the soil and the upper part of the aerated zone, and varying portions of the water are returned into the atmosphere by evaporation and plant transpiration. At the end of dry periods salts accumulate at and near the land surface, until a strong rain washes them effectively into the water table zone. The degree of water retardation and exposure to evapotranspiration varies from one land cell to another, according to local permeability and topography.

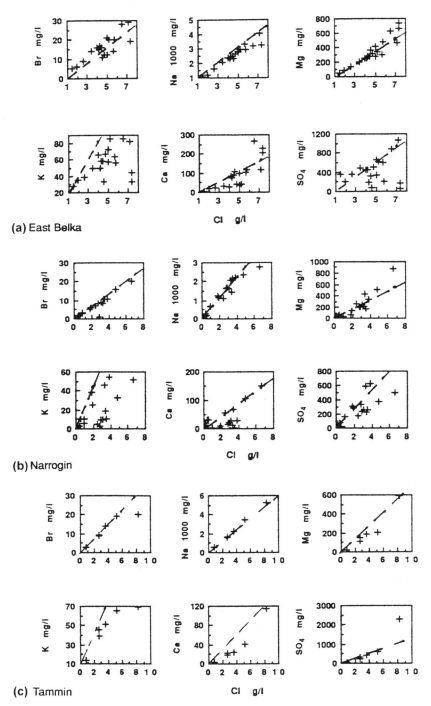

(a) East Belka

(b) Narrogin

(c) Tammin

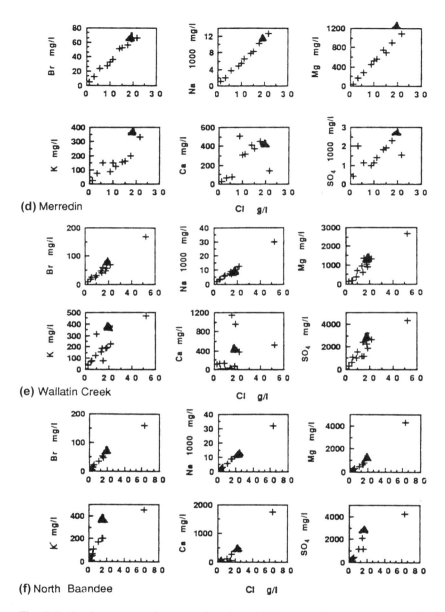

Fig. 4.8 Ionic concentrations as a function of Cl in groundwater encountered in six areas of the Wheatbelt, Australia. Linear correlations with Br, Na, and Mg are seen in all the areas, and the marine relative abundances (dashed lines) and the sea value (solid triangle) fall on these lines. Thus tagging by sea spray is obvious. Correlation with K, Ca, and SO_4 is seen only in some of the cases, e.g., at North Baandee, indicating these ions have been involved to different degrees in secondary water–rock interactions. (Data from Mazor and George, 1992.)

4.6 Sea Spray Versus Brine-Spray Tagging

Water tagging by sea spray is typical for the currently protruding continental reliefs, in contrast to brine-spray tagging, which is typical for formation waters indicating flat and low paleo-landforms.

The entire active water cycle is tagged by sea spray. In rain, snow, fresh surface water, and fresh groundwater the marine Cl and SO_4 are masked by HCO_3; the Br is too low to be measured; and Ca often dominates over Na. But in more saline surface water and groundwater the sea-derived salts are evident, e.g., by a marine Cl/Br ratio of around 290, Cl being the dominant anion and balanced by Na; and often the marine relative abundances of even Mg, K, Sr, Li, or SO_4 are preserved.

The tagging of recent water by sea-derived ions is observed throughout all the continents. In contrast, old entrapped formation waters are by and large tagged by brine-spray (Chapter 7). The Cl/Br ratios are of evaporative values, often reaching the range of 80 to 120, and Cl is partially balanced by Ca. Most important in this context is the isotopic composition of the old entrapped formation waters; they have negative δD values and in many cases also negative $\delta^{18}O$ values that are distinctly meteoric. Thus both recent cycling groundwaters and old formation waters were formed by rain and snowmelt, falling on the exposed continent. The landscape relief is at present distinctly protruding above the oceans, whereas the marginal marine sediments, in which the old formation waters are stored, were formed in an entirely different landscape of large-scale flatlands frequented by alternations of oceanic transgressions and regressions. These were large-scale evaporitic environments, a topic discussed in detail in Chapter 6.

4.7 Sea-Derived Ions Serve as Benchmarks Identifying Water–Rock Interactions

A look at the composition diagrams of the different groundwaters encountered at Campaspe, north Victoria (Fig. 4.7), or at the North Baandee site of the Wheatbelt (Fig. 4.8) reveals that the bulk of the observed dissolved salts is in these cases of an external sea-derived origin. In other cases, e.g., in the water sampled along the River Murray (Fig. 4.3) the contribution of water–rock interaction is seen to be HCO_3 and Ca, whereas Cl, K, Mg, Na, and SO_4 are quantitatively external.

Systematic scanning of available data of recent groundwater composition reveals that the role of water–rock interactions is limited to CO_2-induced reactions but not much more. Groundwater preserves its original composition and as such provides essential information on the environmental conditions of its formation.

4.8 Gravitational Flow in the Unconfined Groundwater System and Static Water Storage Beneath

4.8.1 Observations in Wells Revealing General Principles of Groundwater Hydrology

Aerated zone, saturated zone, and water table. The upper portion of a well is observed to be dry, and from a certain depth on the well contains water (Fig. 4.9). This sequence is observed in all wells, leading to the conclusion that below the land surface the rocks constitute an upper zone in which air is present in the voids within the rocks and a lower zone in which water fills all the voids. These zones are accordingly called the *aerated zone* and the *saturated zone*, and the top of the saturated zone is the *water table* (Fig. 4.9). A drill hole gets filled with water that flows from the saturated zone into the free space created by the drilling operation.

Rises of the water table indicate infiltration and recharge. Almost at any location the water table in nonpumped wells is seen to rise seasonally, following rainfall and snowmelt events. This is a direct indication that water infiltrates through the aerated zone and is added on top of the existing water table, causing the latter to raise.

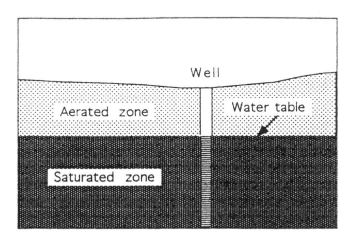

Fig. 4.9 Observations in a well. The upper part is dry and the lower part is filled with water, depicting that an aerated zone overlies a saturated zone. Water flows from the saturated zone into the free space of the borehole.

Drops of the water table indicate discharge. The water table in nonpumped wells is also observed to drop following dry seasons. This provides evidence that the recharged water flows away and gets discharged somewhere.

Base flow. Reaching the water table, the downward path of the infiltrating water is blocked, and the arriving water spreads above the water table (Fig. 4.10). In other words, local recharge infiltrates at every point of an unconfined system, and this local recharge forms a *base flow* that flows in free space at the base of the aerated zone, draining toward free space available on top of the ocean surface (Fig. 4.10).

Inclination of water tables provide another indication of discharge and flow in the direction of the sea. Measurements in many wells reveal that in nondisturbed areas the water table is everywhere nearly horizontal, with a minute inclination toward the nearest sea (or another large surface-water body). This is a direct indication that groundwater flows through the rocks, discharging into the ocean.

Sea-level altitude of the water table in near-shore wells indicates that the sea surface acts as the base of drainage. The water in nondisturbed coastal wells is fresh, and the water table is at mean sea level in boreholes drilled within the region of the shore waves. This provides direct evidence that the sea surface serves as the terminal base of drainage of continental groundwater.

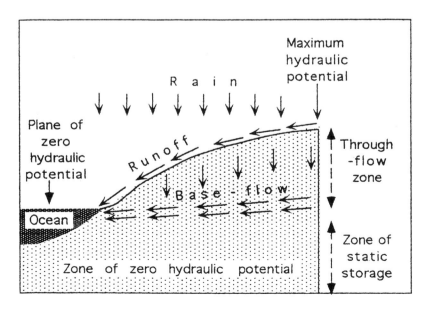

Fig. 4.10 Basic elements of groundwater systems.

The water table never drops below sea level, indicating deeper groundwater is stored in a static mode. A vessel filled and heaped with sand exposed to rain fills up, a process reflected in a rise of a water table (Fig. 4.11a). At a steady state all newly arriving water flows directly out *in a lateral base flow*, whereas the water inside the vessel is stored in a static mode (Fig. 4.11b). The brim of the vessel acts as the base of drainage. Once the rain ceases, the lateral base

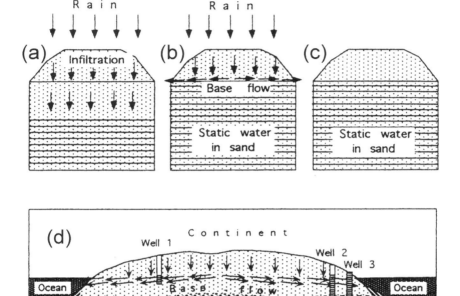

Fig. 4.11 The base flow and the underlying zone of static groundwater: (a) rising water table observed in a vessel filled and heaped with sand and exposed to rain, (b) at steady state all arriving water flows out at a base flow in the direction of free space available outside the brim of the vessel; (c) when the rain ceases the outflowing base flow ceases as well, but the water inside the vessel remains, demonstrating it is static; (d) the water table and base flow never drop below mean sea level, demonstrating all deeper groundwater is static, confined in rock-compartments. Wells 1, 2, 3 reveal different pressures, and they are hydraulically disconnected.

flow ceases as well, but the water inside the vessel, beneath the base of drainage, is stored permanently, demonstrating it is static (Fig. 4.11c).

The space within the vessel remains full with water, as there is no free space for flow in a direction with a downward vector, required for gravitational flow.

The nondisturbed water table is observed in wells to fluctuate, but never has it been observed at any place to drop below sea level, not even during successions of exceptionally dry years. This universal observation indicates that all the groundwater that is encountered in wells at depths greater than the base flow is static. Or, in a more generalized way, groundwater stored at a depth greater than sea level is stored in a static mode (Fig. 4.11d).

4.8.2 Flow and Static Storage of Water Determined by the Availability of Free Space for Gravitational Motion

The motion of groundwater can be explored in light of the gravitational pull vertically downward and the requirement of free space for flow.

Motion of Runoff Water. Rain condensed in clouds falls vertically down in freefall. Upon reaching the ground, the water is split between *runoff* and *infiltration,* the ratio between the two components being determined by the limited free space available for flow in interconnected voids within the soil and rocks.

The runoff flows along the steepest routes of available free space on the landscape relief. Meeting topographic depressions, runoff gets collected in them until they get filled up and the downflow course of the water is resumed. Reaching the sea surface, no free space for further downward movement is available, as all the space is occupied by seawater. Thus the discharged runoff spreads at the free space available on top of the sea surface. Runoff operates from the point of maximum hydraulic potential at the highest topographic point to the *plane of zero hydraulic potential at the sea surface* (Fig. 4.10).

Flow of groundwater—infiltration, base flow, and through-flow. Part of the infiltrating water is returned into the atmosphere by evapotranspiration, and the other part flows down in free voids in the aerated zone until the water table is reached. Within the saturated zone all voids in the rocks are filled with water and there is no free space for flow. Thus, the arriving infiltrating water is diverted to a lateral flow in the direction of the terminal base of drainage, generally the sea surface. The lateral seaward flow, the base flow, takes place in free space available in voids in the rocks of the aerated zone, just above the water table, causing the latter to rise (Fig. 4.10).

Base flow. The water flowing out from the vessel shown in Fig. 4.11b forms a layer of flowing water. This is the base flow. The thickness of the base flow is proportional to the flux of through-flowing water. The location of the base flow is determined by the availability of free space for flow; the base of the base flow is exactly at the lowest point above the base of drainage, so that the entire thickness of the base flow moves in the free space available above the base of drainage.

Vertical freeflow downward and lateral base flow are the dominating flow directions of groundwater. In natural systems built of highly permeable rocks, water can flow only vertically downward (infiltration) or laterally toward a base of drainage (base flow). The flow path of groundwater is occasionally diverted by impermeable rock barriers to follow an oblique track.

The base of drainage is the lowest available free space for water flow. This short sentence summarizes the essence of the gravitational flow of terrestrial water. Like the rolling ball (see Fig. 2.1), runoff and rivers flow at the free space above the landscape, following the steepest downward paths available. Reaching the sea, the water spreads in the free space above the sea surface. Similarly, recharged water infiltrates down in the free space available in the voids of the aerated zone; upon reaching the saturated zone, the route further down is blocked. The water then flows in a base flow in the free space above the water table, until the sea is reached, where the water spreads in the free space above the ocean surface.

4.8.3 Permeability—A Necessary but Insufficient Condition for Flow

Fig. 4.12a depicts infiltration experiments that serve to demonstrate the property of permeability. Tubes are filled with different rock types, water is let through via funnels placed at the upper part of each tube, and the time it takes a given volume of water to flow through each rock type is measured. Permeability is determined by various laboratory techniques and deduced from pumping tests. The results show the following general permeability pattern:

gravel > sand and fractured limestone > siltstone > clay and shale

Figure 4.12b depicts the infiltration experiment repeated with the tubes closed at the bottom. This time no through-flow takes place and all rock types appear impermeable. This result brings up a most important feature: *permeability is a potential property*, and through-flow is *conditional* on the availability of free space for motion to a base of drainage *outside the rock system.*

Compaction can make plastic rocks impermeable. Clay, shale, and marl reveal a very low hydraulic conductivity, measured on small samples collected on the land surface or at shallow depth. Subsidence and resulting compaction

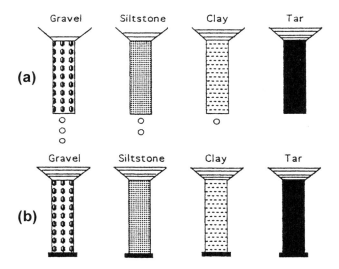

Fig. 4.12 Permeability is a condition for through-flow, but through-flow is activated only if free space for flow is available: (a) infiltration through tubes open on both ends; gravel manifests highest permeability and tar is impermeable; (b) when the lower end of the tubes is closed, no through-flow takes place, regardless of the potential permeability of the rocks. The condition of availability of free space for flow is met only in the rock zone of the continents above sea level altitude.

close in these plastic rocks' microfractures and any existing interconnected pores. The increased temperature increases the plasticity and intensifies the compaction-induced sealing. These processes make these rocks impermeable, explaining the storage of fluids within deep confined rock-compartments for extremely long durations, in certain cases going back to the early Paleozoic.

4.8.4 Storage of Static Groundwater in the Zone of Zero Hydraulic Potential—The Zone of No Flow

Hydrostatic and lithostatic pressures cannot operate simultaneously. Figure 4.13a depicts a hypothetical partially confined system that is hydraulically open across an entire basin. Through-flow induced by hydrostatic pressure will theoretically go on. In reality, such a system cannot exist, as compaction will compress the rocks and squeeze the water out from the open edges (Fig. 4.13b), causing collapse of the rock system at various sections and formation of isolated rock-compartments (Fig. 4.13c). The water remaining in the closed sections will be pressurized by compaction.

Pressure difference—a necessary but insufficient condition for flow. Water flows from high pressure to low pressure, and a pressure gradient can be

Shallow Cycling Groundwater

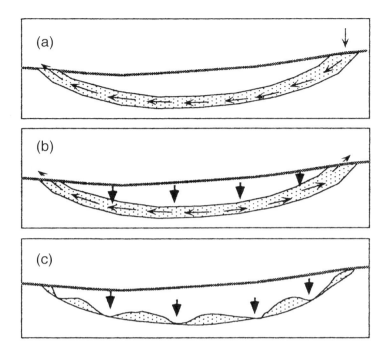

Fig. 4.13 Compaction impedes basin-wide through-flow: (a) a hypothetical semi-confined rock bed that is permeable extends through an entire sedimentary basin and lets groundwater flow through; (b) compaction would squeeze the groundwater out through the open ends, (c) resulting in collapse-induced cinfinement in isolated rock-compartments.

observed along every flow path. However, pressure difference is not a sufficient condition for flow—the other condition is *hydraulic connectivity* between the locations manifesting the different pressures. For example, the pressure in well 3 is higher than the pressure in well 2 in Fig. 4.11, but the two boreholes tap two separated systems, and hence no water flows from the zone tapped by well 3 to the zone tapped by well 2.

This may sound trivial, but groundwater flow directions are often marked on published cross-sections from the location of wells with high measured pressure to the location of wells with lower measured pressure, the hydraulic interconnections being taken for granted and not checked.

Abrupt changes of water heads in adjacent wells in confined systems indicate water storage in rock-compartments. Artesian wells in many regions over the world, at depths reaching up to 2000 m, are characterized by significant differences between the pressuried water heads observed in adjacent wells

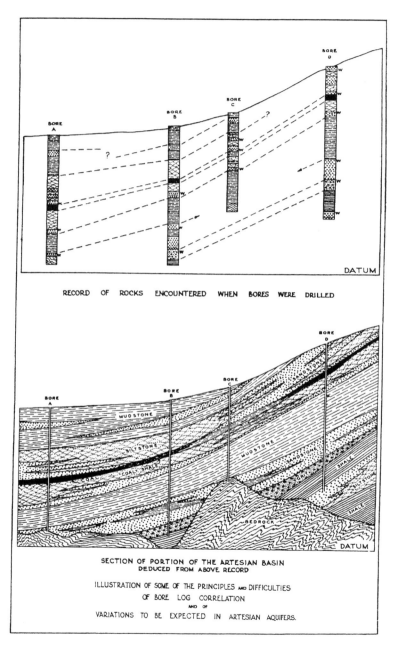

Fig. 4.14 Drilling data and deduced cross section through a portion of the Great Artesian Basin, Australia. The highly compartmentalized rock structure is evident.

Shallow Cycling Groundwater 91

(Mazor, 1995; Mazor et al., 1995). Similar abrupt head differences are observed in neighboring petroleum wells at depths of up to 7000 m (Ortoleva, 1994a,b). Pressure differences between confined systems can exist only if the systems are hydraulically isolated from each other by impermeable rocks (if interconnected, the pressure would be the same all over). Thus, pressure differences observed in the deeper parts of sedimentary basins indicate a structure of permeable rock-compartments that host fluids and are engulfed by impermeable rocks (Fig. 4.11d).

Lithological evidence for compartmentalization. Drilling data reveal a most common picture of lithological diversity in all three dimensions of a basin. Often rock sections compiled for boreholes a few kilometers to tens of kilometers apart are significantly different, implying facies changes, interfingering, unconformities, hiatuses, folds, and faults. Figure 4.14 is from a historical document entitled "Artesian Water Supplies in Queensland," issued in 1954 by the "Committee appointed by the Queensland Government to investigate certain aspects relating to the Great Artesian Basin with particular reference to the problem of diminishing supply." The cross-section depicts data from wells and their interpretation. A highly compartmentalized structure is concluded.

The emerging picture in all case studies is that permeable rock beds containing water actually have lentil-shaped structures, and they are confined by impermeable rocks.

4.9 Summary Exercises

Exercise 4.1: Tagging of surface water by sea spray is observable everywhere; why is the study case of the Murray River so impressive and convincing? From which direction can sea spray be brought inland in this case?

Exercise 4.2: Is there any fractionation happening between the ions carried along the far inland airborne transportation and the long water flow paths? How can we know?

Exercise 4.3: What would groundwater that encountered rock salt look like?

Exercise 4.4: Compare the data given in Table 4.2 for the Jingellic location and the Tailem location along the River Murray. Where are these points along the river system? What can be learned from the pronounced difference in composition?

Exercise 4.5: Let us have a look at Table 4.3. Can it be that the saltier groundwater of the Gorong 37 well is connate seawater?

Exercise 4.6: Which system is described by the composition diagrams of Fig. 4.6? Are these lakes filled by seawater? Are these totally closed lakes?

Exercise 4.7: What do the composition graphs of Merredin in Fig. 4.8 tell us? Why are the groundwaters as salty as seawater hundreds kilometers inland?

5
INTERSTITIAL WATERS IN ROCK STRATA BENEATH THE OCEANS

5.1 Extending Our Hydrological Curiosity to Beneath the Oceans

There is a long list of groundwater properties and a very large number of different groundwater types: shallow and deep; recent and old; flowing and static; fresh and saline; of different ionic ratios; of negative (light) δD and $\delta^{18}O$ values, depicting the meteoric origin, and of heavy evaporative signatures; cold and warm; and so on. All these are continental groundwaters, and in the previous chapter we observed that despite the high variability, there are some common threads: All the water sources that are part of the ongoing flow of the active water cycle are sea tagged and have a meteoric isotopic imprint. Along these lines, let us ask ourselves what kind of groundwaters do we expect to find beneath the oceans?

There are several hypothetical possibilities. The rocks beneath the open oceans may contain very little interstitial water because the pressure of the overlying few thousands of meters of seawater exert compaction and squeeze out any initial interstitial water. Or the opposite: beneath this great body of water the rocks may be very "wet." There may be active flow of groundwater in the rocks underlying the oceans, but actually what force might propel them? Might groundwater beneath the oceans be static? Has this groundwater the exact chemical composition as the overlying seawater? Or are there surprises?

The oceans cover two-thirds of the Earth's surface, so the amount of groundwater beneath them may be a very important slice of the global groundwater inventory. We will get acquainted with the large amount of data already available, make observations, and provide answers for the listed questions (and more).

5.2 The Deep Sea Drilling Project

The investigations of the Deep Sea Drilling Project (DSDP) accomplished hundreds of research drillings in all the oceans and smaller seas. The research project is conducted on board special ships, geared for undersea (beneath thousands of meters of seawater) retrieval of drilled cores, which are carefully collected and preserved. Individual drillings reached hundreds to over a thousand meters of rock! On-board laboratories conduct initial measurements, samples are sent to different laboratories, and representative samples are preserved in special repositories for future analysis. From some of the obtained cores the pore water has been extracted and analyzed, and the results are included in the many volumes of the DSDP Initial Reports (U.S. Government Printing Office).

The accuracy of the data is discussed in detail in the original reports, and compositional variations observed along depth profiles are occasionally explained by methodological artifacts, e.g., certain sediments strata contain gas hydrate, possibly disintegrating during core recovery, adding some water, and thus slightly lowering the Cl concentration and raising the isotopic values. However, the bulk of the data seems reliable and adequate for our purposes.

Deviations of the chemical and isotopic composition of the sub-marine interstitial waters from the composition of current seawater triggered many explanations related to the natural processes; these have been explained in the original DSDP reports and in a large number of papers. These include suggested variations of the composition of seawater during the geological history; diagenetic biological processes; water–rock interactions, e.g., dolomitization, de-dolomitization, ion exchange, and lithification; or upward diffusion of water in contact with halite or gypsum beds. These topics will occupy scientists for many years to come. The following sections give a glimpse of the general patterns that typify the suboceanic groundwaters, commonly termed interstitial waters, and reach a number of key conclusions.

5.3 Water Content in Suboceanic Sediments

The water content was determined in various cores; the following are a few examples:

Drill holes from the South Atlantic Ocean revealed a water content of 25 to 67% of the rock volume, in a number of cores, at depths up to 419 m (data from Sayles et al., 1970). The water content tends to be high at the uppermost meters, but an important control is provided by the lithology: The water content increases in the order clay → silt → sand. In

Hole 23 the water content was 49% at a depth of 2 m and it decreased to 39% at a depth of 120 m. In contrast, in Hole 29 the water content gradually increased from 34% at a depth of 8 m to 67% at a depth of 213 m.

Examples of the North Pacific Ocean (data from Manheim et al., 1970) reveal in Hole 32 fluctuations in the range of 53 to 59% in the depth range of 4 to 206 m. Whereas in Hole 40 the volume of the water content increases from 62 to 81% over the depth range of 4 to 125 m.

Other cores from the North Pacific (data from Manheim and Sayles, 1971) revealed the following patters: In Hole 54.0 the same water content of 42 ± 2% is observed from 89 m down to 264 m; in Hole 55.0 the content was 43 % at a depth of 4 m, and it dropped slightly to 39% at a depth of 334 m; and in Hole 56.2 the water content was 36% at 82 m, and it increased to 40% at 229 m.

Cores retrieved at the Gulf of Mexico revealed water contents of up to 85% in the uppermost few meters, and 58 to 29% at depths from 10 to 761 m, the decrease with depth being small and not consistent (data from Manheim and Sayles, 1969).

Drillings at the bottom of the Black Sea revealed the following pattern (data from Manheim and Schug, 1978); In Hole 379A the water content at a depth of 32 m was 34%, and from 60 m down to 605 m it fluctuated nonsystematically between 15 and 27%. Similarly, at Hole 380 the water content was 50% at a depth of 8 m, and from 92 m down to 975 m the content fluctuated in the range of 21 to 34%.

To sum up, interstitial water content is somewhat higher in the uppermost meters, but there are cases with the highest values at greater depths. *Beneath a depth of 10 to 20 m no depth effect is observed in the water content—the latter seems to be controlled by the lithology.*

5.4 The Widespread Marine Facies of Interstitial Water (Cl ~19 g/L, Cl/Br ~300, Diagenetic Changes Are Common)

A dozen drill holes from the South Atlantic Ocean (studied by Sayles et al., 1970) revealed constant Cl concentrations of 19 ± 1 g/kg in profiles up to 480 m deep. Rather constant concentrations were observed also of Na, K, and Mg. SO_4 occurs in many core sections in the seawater concentration, but in other sections it is significantly lower and manifests abrupt changes in certain depth profiles.

Examples of the North Pacific Ocean (data from Manheim et al., 1970) reveal in 11 cores, up to 369 m deep, the marine value of Cl = 19 ± 0.5 g/kg and, similarly, constant concentrations of Na, Ca, and Mg. SO_4 occurs in many sections in the seawater concentration, but in other sections it drops significantly.

Stratification is well marked by abrupt changes in ion concentrations. A similar pattern is portrayed by Manheim and Sayles (1971) from another group of cored holes in the North Pacific.

Now let us explore Table 5.1. It contains data obtained for interstitial water extracted from rock cores in Bores 533 and the nearby Hole 533A, Blake Outer Ridge, NW Atlantic Ocean (data from Jenden and Gieskes, 1984). The best way to get an initial view of the composition of water encountered in the rocks underlying the ocean floors is to compare the data, obtained at the different depths, with the respective seawater values. The uppermost line in Table 5.1 provides the composition of the overlying seawater to serve as a benchmark.

The following questions come up:

Are the below-ocean waters in this core similar to nonaltered seawater?
Which ions are of concentrations similar to the corresponding values of the overlying seawater?
Which ions deviate substantially from the seawater concentration?
Is the water composition the same at all depths?
Which patterns reveal the δD and $\delta^{18}O$ data?

Let us answer these questions and sum up observations based on Table 5.1:

1. *The Cl concentration indicates an origin from seawater.* The Cl concentration down to a depth of 92 m is precisely the seawater value of 19.3 ± 0.1 mg/L. Down to 396 m the Cl concentration is slightly different, but similarity to seawater is still revealed.
2. *Potassium and Sr also reveal close to seawater concentrations along the entire core.*
3. *The SO_4 concentration is close to seawater at the depth of 3 m and then decreases* until at a depth of 193 m it is below the limit of detection.
4. *The HCO_3 and NH_4 values increase significantly downward*, revealing a pattern that is opposite to that of SO_4; their concentration increases significantly downward.
5. *Magnesium and Ca decrease downward.*
6. *The $\delta^{18}O$ value supports an origin from seawater.* The oxygen-18 isotopic value is very close to the seawater value of 0 ± 0.5‰ down to a depth of 92 m. At a greater depth the fluctuations are slightly higher but still close to seawater.
7. *The δD value is lighter (more negative) than the present seawater.* The δD values range between −18.9 to 0‰. The more negative

Table 5.1 Chemical and Isotopic Data Reported for Interstitial Waters Extracted from Rock Cores in Holes 533 and 533A, Blake Outer Ridge, NW Atlantic Ocean

Depth (m)	Cl (g/L)	SO$_4$ (mg/L)	HCO$_3$ (mg/L)	NH$_4$ (mg/L)	Si (mg/L)	Mg (mg/L)	Ca (mg/L)	K (mg/L)	Sr (mg/L)	δD (‰)	δ^{18}O (‰)
Seawater	19.3	2710	145			1294	412	399	7.9	0.0	0.0
1	19.3	2294	518	9	12	1378	347	433	6.5	−4.9	−0.14
3	19.4	1805	725	14	16	1147	291	346	5.1	−3.3	0.09
10	19.5	950	1127	25	19	1137	154	445	3.8		
14	19.5		1501	34	11	1050	93	438	4.0	−1.3	−0.13
19	19.2	106	1609	41	13	1033	81	443	2.3	−2.1	0.02
22	19.5		1618	42	15	1013	89	459	4.1	−2.2	0.22
28	19.5	250	1628	52	19	991	80	475	2.3	−4.3	0.25
32	19.5		1559	61	10	938	55	465	2.3	0.3	−0.02
37	19.5	106	1674	63	17	936	86	423	4.1	−0.5	0.48
40	19.5		1526		16	955	76	420	2.3	−4.4	0.27
46	19.3	125	1557	76	11	911	63	499	2.3	−1.8	−0.60
48	19.5	106	1503	74	12	889	82	450	4.0	−3.3	−0.24
58	19.4		1465	92	9	945	66	531	4.0	−3.2	0.14
65	19.4	106	1562	94	13	945	76	423	4.1	−1.9	−0.72
76	19.3		1521	98	16	836	86	483	4.0	−2.7	−0.19
83	19.2	106	1404	112	14	785	70	527	4.0	−3.2	0.18

92	19.2	106	1410	120	16	722	74	515	4.0	−3.3	0.17
101	18.9	106	1366	124	12	676	76	460	3.7	−3.7	−0.21
112	18.8	106	1416	133	8	603	84	475	5.7	−3.5	0.35
119	18.9	278	1559	147	16	615	91	477	5.8	−6.6	−0.19
130	19.0	106	1543	163	15	542	64	578	4.0	−5.1	−0.29
137	18.9	115	1659	168	14	556	57	552	5.1		−0.19
148	18.9	96	1651	177	11	554	54	539	5.1	−4.9	−0.24
160	18.7		2002	198	17	552	75	346	6.1	−3.9	−0.66
193	18.7	<180	4186	299	29	668	159	567	5.4	−9.2	−0.45
216	18.6		4322	288	26	678	202	583	7.8	−16.5	−1.03
243	18.1	<180	2949	358	38	598	170	622	10.9	−18.9	−1.88
258	18.0	<180	2487	279	18	608	181	508	6.5	−6.5	0.23
272	18.5		3416	268	24	598	192	387	5.0	−12.8	−0.55
294	17.2	<180	2938	160	20	467	128	418	4.8	0.0	2.80
307	18.3		3276	346	20	537	190	504	4.9	−10.0	−0.54
332	17.7	<180	2586	302	21	513	137	461	6.6	−10.3	−0.26
349	17.9		2942	598	31	484	152	735	8.9	−10.7	0.06
365	18.3	<180	2994	533	32	498	159	778	8.9		
396	18.4	<180	2501	383	20	537	204	489	8.4	−9.5	0.06

Source: Jenden and Gieskes (1984).

values, seen in the deeper part of the core, are accompanied by a slight decrease of the Cl concentration. The explanation to this pattern is an open question.

8. *A stratified structure of the interstitial waters is indicated.* Changes of the concentration of the different ions and the isotopic values along the core are occasionally abrupt, disclosing stratification, marked in Fig. 5.1 by horizontal lines. The locations of the changes along the depth profile vary from one ion to the other, a feature seen in practically all the suboceanic cores.

A deeper profile, 558 to 1104 m, is given in Table 5.2 and Fig. 5.1, reporting data from Hole 534A, at the Blake-Bahama Basin (following data from Jenden and Gieskes, 1984). The following pattern is observable:

1. *Chlorine values are slightly higher than the current seawater concentration.* The Cl concentration is in the range of 18.8 to 22.4 g/L, i.e., close to the marine value or slightly higher.
2. *The δD and $\delta^{18}O$ values are meteoric and not marine.* δD varies in the range of -17.4 to $-6.6‰$, and the $\delta^{18}O$ values are in the range of -5.8 to $-1.5‰$. These values are of continental meteoric water.
3. *A layered structure is well seen in the depth profiles.* This is similar to the above-discussed case studies.

The combination of a Cl concentration slightly enriched relative to recent seawater and light meteoric isotope values is for the time being an enigma.

5.5 Continental Brine-Tagged Facies: Salinity Higher than Seawater Cl/Br 200 or Lower, Ca–Cl Present

5.5.1 The Hellenic Trench, Holes 127 and 127A

These assemblages of interstitial water include strata of brine-tagged continental water. Table 5.3 summarizes data obtained in drill holes located at the NE margin of the Hellenic Trench, at the Mediterranean Sea, beneath a water depth of 4654 m (data from Sayles et al., 1970; Presley et al., 1973). The data are presented as depth profiles in Fig. 5.2. The following pattern is seen:

1. *Stratification is displayed by the depth profiles.* Significant changes of the concentrations of the different ions along the depth profile depict stratification of the water-hosting sediments (marked by lines in Fig. 5.2). The different ions reveal somewhat different break marks in the depth profile, emphasizing the need for data of many parameters in order to get the complete picture and a good resolution of the stratification.

Interstitial Waters Beneath the Oceans

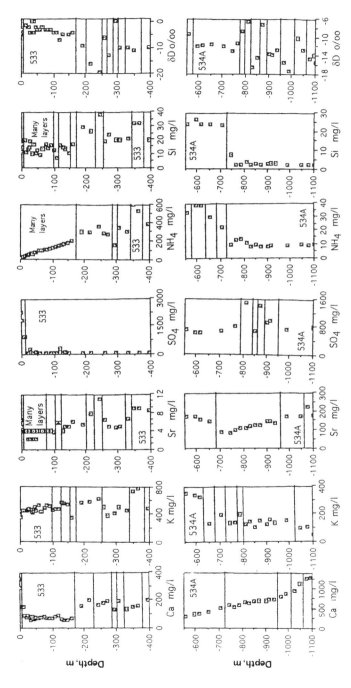

Fig. 5.1 Depth profiles of two drillings in the North Atlantic Ocean: (top) Holes 533 and 533A (combined, Table 5.1), (bottom) Hole 534A (Table 5.2). Each parameter discloses a number of abrupt changes in concentration (marked by horizontal lines), indicating presence of different water types entrapped in the cored sequence of rocks. Putting together the compositional breaks identified by the individual chemical and isotopic markers, at least 20 different water-containing strata are identifiable along the combined depth of 1100 m of cored sediments (see text).

Table 5.2 Chemical and Isotopic Data Reported for Interstitial Waters Extracted from Rock Cores in Hole 534, Blake Bahama Basin, NW Atlantic Ocean

Depth (m)	Cl (g/L)	SO$_4$ (mg/L)	HCO$_3$ (mg/L)	NH$_4$ (mg/L)	Si (mg/L)	Mg (mg/L)	Ca (mg/L)	K (mg/L)	Sr (mg/L)	δD (‰)	δ^{18}O (‰)
558	20.7	701	537	33.0	23.5	419	304	356	173	−11.7	−2.20
596	20.1	605	495	38.8	26.3	415	350	340	168	−11.5	−1.82
617	20.3	605	544	38.7	23.7	394	378	325	156	−12.0	−2.32
653	20.4		448	29.7	23.7	401	412	129	145	−12.9	−2.41
708	20.2	643	240	22.2	23.4	416	508	192	87	−11.4	−2.39
744	19.4		198	9.7	7.6	411	566	137	84	−8.7	−1.98
766	18.8		128	12.7	2.0	380	536	141	99	−6.6	−1.52
786	20.2	816	159	14.2	2.2	426	582	199	110	−16.4	−2.67
813	20.5		140	11.2	3.6	467	644	121	111	−14.5	−2.83
832	19.6	1488	140	8.6	2.2	402	630	145	117	−7.1	−1.85
853	29.9	653	134	9.7	2.5	416	670	109	125	−13.6	−3.87
875	20.6	1402	106	8.6	2.2	461	682	152	127	−13.8	−2.83
902	19.7	902	76	8.6	2.5	424	698	125	142	−12.8	−2.86
915	21.3	979	57	8.6	3.1	443	732	160	142	−15.5	−3.01
935	21.5		47	9.0	2.9	438	734	137	135	−17.4	−2.84
966	21.7					449	790			−10.3	−2.10
990	21.6	730	20	8.8	2.0	442	872	152	173	−12.7	
1028	21.6					448	952			−14.6	−2.80
1047	21.9			9.4	2.1	442	1126	94	170	−14.1	−5.78
1074	22.4			9.0	2.2	427	1262	109	226	−15.4	−2.53
1092	22.4					436	1276			−15.1	−2.97
1104	21.9	787		9.4	2.2	425	1288	51	177	−17.0	−3.43

Source: Jenden and Gieskes (1984).

Table 5.3 Chemical Data for Holes 127 and 127A, Located at the Northeast Margin of the Hellenic Trench, Mediterranean Sea, at a Water Depth of 4654 m

Depth (m)	Cl (g/L)	Br (mg/L)	Cl/Br	SO$_4$ (mg/L)	HCO$_3$ (mg/L)	Na (g/L)	Mg (mg/L)	Ca (mg/L)	K (mg/L)	Li (mg/L)
22	21.6	75	288	2640	320	12.0	1300	330	370	0.18
42	22.3	78	286	870	370	12.3	1100	230	370	0.17
50	24.0	85	282	<50	460	13.1	1000	200	360	0.15
77	30.0			<50	120	16.6	1200	350	420	
91	29.2			<50	120	16.0	1200	350	400	
108	36.1	122	295	<50	210	19.6	1500	550	450	0.31
172	64.4			180	60	36.0	1300	1000	550	
233	76.8	358	214	2400	40	42.3	2300	3300	710	0.7
284	82.1	405	202	3100	70	44.6	2300	3600	800	2.3
308	78.2			2800	50	42.3	2300	3200	730	
336	82.9	400	207	2800	40	46.3	2500	2900	800	1.7
427	89.4			2500	70	51.8	2000	2200	790	

Source: Data from Sayles et al. (1973) and Presley et al. (1973).

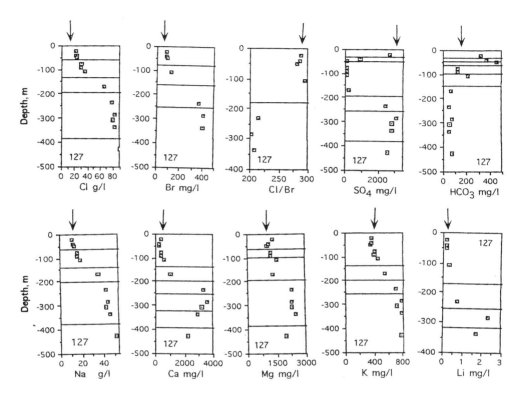

Fig. 5.2 Depth profiles of Holes 127 and 127A (Table 5.3). Arrows denote the respective concentration in the overlying seawater. Between a depth of around 200 m to the end of the core, at 427 m, reside waters that are interpreted as groundwater fed by brine-spray (see text).

2. *A section of interstitial water resembles slightly modified seawater.* The upper 90 m of the rock section hosts interstitial water with a composition that resembles that of little modified seawater.
3. *A section of the medium-saline interstitial water strata has an evaporitic imprint.* At the depth interval of 100 to 427 m the Cl concentration increases up to four times the concentration in seawater, Br concentration is high as well, the weight ratio of Cl/Br reaching the value of 202, resembling residual evaporitic brines. The presence of Ca balanced by Cl is an independent indicator of an imprint by a brine that evaporated beyond the point of halite precipitation.

4. *The medium-saline waters are dilute compared to residual evaporitic brines.* The concentration of up to 89 g/L Cl is distinctly lower than the concentration of ~200 g/L Cl observed in evaporitic brines that precipitated halite.

This pattern fits well the scenario of continental meteoric groundwater tagged by brine-spray. In a way these interstitial waters are similar to the saline water in the Yalgroup National Park, Western Australia (see section 4.3) or the groundwater of the Campaspe River Basin, northern Victoria, Australia (see section 4.4). The difference is only that the earlier examples of recent waters are sea tagged, whereas the discussed interstitial water cases are brine tagged.

5.5.2 The Baleric Basin, Hole 372

These samples provide another example of sediment layers containing interstitial waters with the imprint of brine-tagged groundwater. Table 5.4 sums up data obtained from a core at the Baleric Basin, and Fig. 5.3 provides the depth profile graphs. What patterns are seen? Cl is high in the described core section of 120 to 579 m, reaching 3.5 times the concentration in seawater. Part of the Cl is balanced by Ca, providing an evaporitic brine imprint. Magnesium is all the way through close to the seawater concentration, whereas SO_4 drops downward. Abrupt concentration changes are observed, each parameter marking its own breaks (marked by horizontal lines in Fig. 5.3). The $\delta^{18}O$ values are slightly heavier than seawater, indicating some evaporation of the water phase.

This sequence of interstitial waters is another example of a facies of brine-tagged groundwater, disclosing a paleo-continental origin, to be further discussed in section 5.8.

5.5.3 The Ionian Sea, Hole 374

These samples show strata containing entrapped continental evaporation brines. In this core (Table 5.5, Fig. 5.4) the Cl concentration reaches the very high value of 211 g/L; SO_4 reaches as high as 9200 mg/L; and the concentration of Mg is more than three times higher than that of Na.

This outstanding composition of suboceanic interstitial water resembles a special type of an evaporitic residual brine, found in sabkhas, occurring in flat lowlands that are exposed to frequent invasions and regressions of the adjacent sea.

Here again a clear stratification of the interstitial waters is depicted by abrupt changes in the ion concentrations. Each stratum of sediment

Table 5.4 Composition of Interstitial Waters, Hole 372, Mediterranean Sea, the Belaric Basin

Depth (m)	Cl (g/L)	SO$_4$ (mg/L)	HCO$_3$ (mg/L)	Na (g/L)	Ca (mg/L)	Mg (mg/L)	K (mg/L)	Sr (mg/L)	NH$_4$ (mg/L)	Si (mg/L)	δ^{18}O (‰)
120	30.7	3216	145		1340	1890			34	5.8	
139	32.5	3600	98	17.1	1810	1810	290	53	30	6.1	
148	32.8	3859	83	17.1	2000	1840	328	61	31	5.4	+1.6
203	43.7	3274	69	22.0	3190	2080	348	60	58	9.9	
251	43.7	1728	46	22.5	2690	1700	332	97	60	4.8	
299	45.1	893	35	23.2	2960	1400	285	158	64	4.8	+2.6
348	49.7	720	46	26.2	2960	1446	290	211	66	6.4	
395	60.0	653	45	31.7	3750	1530	305	229	71	13.4	
471	58.6	480	52	30.4	3900	1630	262	238	74	11.5	+2.5
579	65.3	566	54	34.0	4440	1650	290	211	51	13.8	

Source: Data from McDuff et al. (1978).

Interstitial Waters Beneath the Oceans

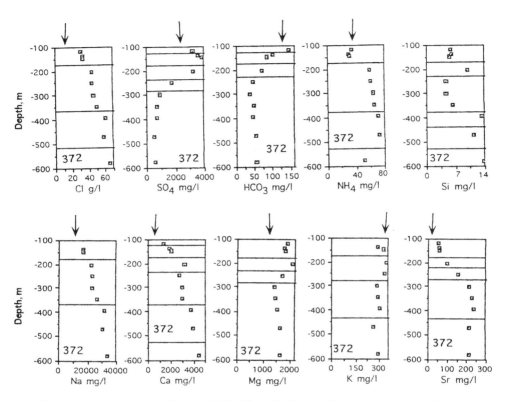

Fig. 5.3 Depth profiles of Hole 372 (Table 5.4). Arrows denote the concentration in the overlying seawater. Altogether around 10 compositional breaks are marked by the various parameters. The high Cl concentration is balanced mainly by Na.

contains a specific type, or facies, of interstitial water, and these have been well preserved since their confinement by the subsequently accumulated sediments.

5.5.4 A Case of Interstitial Water that Reflects Dissolution of Halite, Disclosing Formation During a Sea Regression Phase

Cores retrieved at the Gulf of Mexico revealed in a number of holes, up to 530 m deep, distinct deviations from seawater concentrations, a most

Table 5.5 Composition of Interstitial Waters, Hole 374, Mediterranean Sea, the Ionian Sea

Depth (m)	Cl (g/L)	SO_4 (mg/L)	HCO_3 (mg/L)	Na (g/L)	Ca (g/L)	Mg (g/L)	K (mg/L)	Sr (mg/L)	NH_4 (mg/L)	Si (mg/L)	$\delta^{18}O$ (‰)
160	35	557	226	17.8	0.6	2.2	348	30	38	6.7	+1.3
255	80	2420	54	22.7	5.1	12.8	547	123		3.1	+1.1
302	137	1280	25	23.5	10.1	28.2	1210	200	21	2.7	+0.9
322	129	900	21	19.8	13.7	25.0	1720	185	13	2.2	
338	129	1400	21	17.7	14.2	25.7	1920	185	12	2.4	
342	156	1270	95		18.6	32.5	1888	200	41	3.2	
345	154	1560		14.5	18.9	33.5	1720	190	31	3.2	
346	158	1770	105	16.7	19.2	33.2	1720	200	29	3.8	
350	155	2570	100	16.4	18.8	33.0	2230	190	31	3.2	
356	148	3280	124	15.8	17.2	32.0	1880	180	31	2.7	
367	166		180		18.2	38.9	2660	80	42	1.7	
378	168	2530			12.0	47.1					
381	164				11.4	52.2					
383	168	3990			7.0	53.7					
385	169	4280			3.5	55.4					
389	170	4530	300	13.1	0.4	54.6	1680	14	32	4.2	
405	178	9200	300	14.7	0.4	54.6	1920	12	24	6.1	

Source: Data from McDuff et al. (1978).

Interstitial Waters Beneath the Oceans

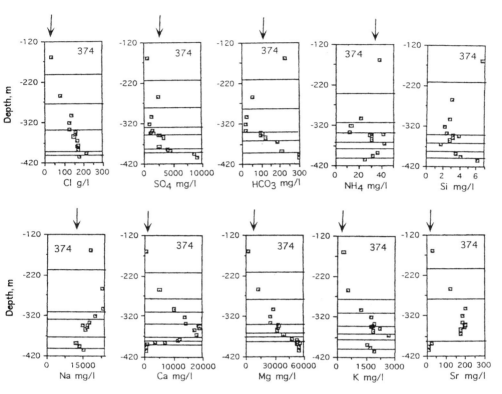

Fig. 5.4 Depth profiles of Hole 374 (Table 5.5). Arrows denote the respective concentration in the overlying seawater. Altogether around 10 compositional breaks are marked by the various parameters. The high Cl concentrations are remarkable, balanced here by Mg.

conspicuous example being observed in Hole 92. The following pattern was observed (based on data from Manheim and Sayles, 1973; Presley et al., 1973):

1. At a depth of 229 m the Cl concentration is as high as 104 g/kg, balanced by Na.
2. The other ions, including Br, reveal regular seawater concentrations.

In this case dissolution of some halite by seawater seems plausible. This conclusion is of key importance, as it discloses that there preceded a stage of sea desiccation, during which halite was precipitated.

5.6 Information Retrievable from Below-Ocean Interstitial Waters

The list of observations presented above leads to the following informative conclusions:

5.6.1 Ocean Water Does Not Flow Downward

It does not replace water stored in underlying sediments. This conclusion is based on the observation that interstitial water has been found to deviate from the composition of seawater even at a depth of a few meters, as may be seen in the example reported in Table 5.1. This is in good accord with the conclusion drawn from first principles considerations that the sea surface is a plain of zero hydraulic potential, beneath which, no gravitational water flow occurs. Beneath the ocean floor extends a *zone of no flow*.

5.6.2 Squeezing Out of Interstitial Water Due to Compaction is Limited and Only via Preferred Flow Paths

Compaction by overlying rocks has been suggested by geologists as a mechanism that squeezes out water from buried rocks due to the lithostatic pressure exerted by them. Large-scale mobilization of groundwater stored in rocks has been concluded. This hypothesis is questionable in light of several observations, including a large number of deep wells that sustain high yields of water and of oil, demonstrating that the fluids have often been well preserved, even at great depths.

Let us examine this issue in the suboceanic sediments. The water content discussed in section 5.3, expressed in percent volume, is seen to be 15 to 60% at studied cores, up to a depth of several hundreds of meters. Furthermore, in several examples the water content did not drop with depth, but increased downward. These observations reveal that compaction-induced water squeezing is minor or even negligible.

The pronounced stratification of sediment beds, disclosed in many cores by abrupt changes of ion concentrations and relative abundances, reveals independently that no water moved through these sediment beds. Hence, water out-squeezing was minor or negligible, and if some interstitial water was pressed upward, the motion was limited to preferred flow paths, such as open fault plains.

5.6.3 Sea Like Concentration of the Conservative Cl Ion Discloses that the Deep Oceans Prevailed Continuously for Long Geological Periods

In the majority of the examined cores the Cl concentration in the interstitial water is constant and close to the present seawater concentration of about 19 g/L. This key observation leads to two key conclusions: (1) Chlorine is a conservative ion, not involved in water–rock interactions and (2) the continuous Cl profiles indicate continuous prevalence of the sea.

5.6.4 Suggested Upward Diffusion of Halite-Saturated Solutions Are Ruled Out

In their initial report Sayles and Manheim (1973) interpreted the elevated Cl and Na concentrations of Holes 127, 127A, and a few others, as caused by diffusion of water that dissolved halite, suggested to be buried at a greater depth. However, applying the respective Br data (Presley et al., 1973) an origin from halite dissolution is ruled out by the following observations (Fig. 5.2):

1. The Cl/Br ratio is around 210, drastically different from the ratio of >3000 typifying halite.
2. There is a downward increase also in the concentrations of SO_4, Ca, Mg, K, and Li. The halite-saturated water diffusion hypothesis does not explain these observations, and hence a different explanation is required.

Upward diffusion of water that dissolved buried evaporites has been discussed also for the data of Hole 374 (McDuff et al., 1978), but a glimpse at the depth profiles (Fig. 5.4) reveals the following controverting observations:

1. Sodium reaches a maximum value at a depth of around 300 m and drops down to the depth of 400 m, so in which direction is diffusion taking place? And where is the diffusing source located?
2. Calcium reveals as well a pronounced maximum, but at a distinctly different depth than the Na, namely at 360 m.
3. Magnesium reveals a depth profile that is entirely different from the previously discussed ions; it increases stepwise until it is the dominant cation at a depth of 400 m.

Thus, diffusion of halite-saturated water is in this case ruled out as well.

5.6.5 Large-Scale Entrapment of Residual Evaporitic Brines is Ruled Out by Field Relation Considerations

The very saline, Mg–SO$_4$ and Ca–Cl brine, observed at the 220- to 350-m interval of core 374 (Table 5.3 and Fig. 5.4) is an outstanding and rare case of a directly entrapped residual evaporation brine. But in general the saline interstitial waters are not of such an origin, as is born out by the following observations and considerations:

1. The floor of an evaporating basin, or sabkha, must be impermeable, as the evaporating water has to be kept long enough to provide the time needed for the evaporitic cycle. The floors of recent sabkhas and lagoons are indeed covered with impermeable mud, to which are added the precipitating carbonates, gypsum, and halite. Thus, residual evaporitic brines had practically no chance to infiltrate downward and get entrapped in underlying rocks.
2. The interstitial water of muds underlying sabkhas would rather reflect dissolution of halite and other salts, but this is uncommon.
3. The concentrations of Cl, Na, and the other ions are distinctly lower than their concentrations in evaporitic brines that have a low Cl/Br ratio and a calcium chloride component.

5.6.6 Later Dilution of Entrapped Brines by Seawater or Meteoric Water is Improbable

Let us examine the hypothesis that the saline interstitial waters originated by entrapment of evaporitic brines, or water that dissolved salt deposits, and later on fresh water, or seawater, moved down and diluted the original brines. This is improbable in light of the following observations:

1. The depth profiles of interstitial water revealed, in all the studied cores, a distinct layering of water of different types, abrupt changes being common. Later penetration of different water types would wipe out these differences.
2. Most of the interstitial waters differ in one parameter (or several), from the composition of seawater, demonstrating that no seawater moved down into the underlying sediments.
3. In several cases pore water was extracted from the first meters of core beneath the sea bottom, and diagenetic changes of the original seawater are observed. This reveals that neither seawater nor fresh water penetrated down, and they did not replace even the most shallow interstitial water of the underlying sediments.

The listed observations reveal that the interstitial waters are connate, a topic dicussed in the next section.

5.6.7 Preservation of Halite Beds Beneath the Sea Indicates this is a Zone of No Flow—A Zone of Zero Hydraulic Potential

Preserved Messinian halite beds, found all over the Mediterranean basin, testify that neither seawater nor meteoric water flowed through the rock systems beneath the ocean floor. Thus the domain beneath the oceans is indeed a zone of zero hydraulic potential, at which all groundwaters are at rest.

5.7 Interstitial Water is Connate Water, Entrapped in Its Host Rocks Since the Initial Stage of Sedimentation

The structure of well-preserved strata, containing interstitial water of different salinities, different ionic abundances, and varying isotopic compositions, rules out cross-formation flow, as such a process would wipe out the observed abrupt composition variations. Hence, every interstitial water was stored in its host rock when the latter was sedimented on the ocean bottom and was confined and isolated by subsequently accumulated sediments. Clay beds, included in sediment sequences, provide efficient hydraulic isolation, preventing vertical and lateral exchange of fluids between adjacent sediment strata and neighboring rock-compartments.

Thus, the suboceanic interstitial waters are connate groundwaters in the true sense, as suggested by Lane (1908). The term connate was suggested for water entrapped in a rock body since the latter was formed. For a full description Lane also used the synonym "singenetic water." This concept, suggested in 1908, was much forgotten or misused as a general term for old formation water. The results of the Deep Sea Drilling Project provide new evidence for the vast occurrence of connate groundwater.

5.8 Interstitial Waters Tagged by Brine-Spray Disclose that the History of the Mediterranean Sea Basin Included a Continental Stage

Intensive evaporation of seawater takes place in sabkhas and lowlands that are frequented by sea invasions and regressions. Hence, saline interstitial waters that have the signature of residual evaporitic brines must have been formed on flat and low continental landscapes bordering the oceans.

Residual evaporitic brines, enriched in Br and Ca–Cl, attain their composition after precipitating halite (Chapter 6). At this stage the Cl concentration is on the order of 200 g/L. The interstitial waters that are brine-tagged have by and large Cl concentrations that are significantly lower than the named concentration of saturation with regard to halite. This leads to the suggestion that one deals with meteoric groundwater that was tagged by brine-spray, recharged at stages of an exposed dry Mediterranean basin.

Groundwater formed in semiarid and arid regions at present is observed to reach a very high salinity due to intense evapotranspiration. During dry periods infiltrated water is dried up by evapotranspiration, leaving behind the salts (Mazor and George, 1992), and these are redissolved and washed down into the saturated groundwater zone by extra-strong rain events.

In a similar way it is inferred that the medium-saline brine-tagged interstitial waters encountered beneath the Mediterranean floor were formed as groundwaters on parts of the basin that were exposed, and the brine-spray originated from adjacent large evaporating seawater bodies.

A large variety of different brine-spray-tagged interstitial waters is encountered in the sediment section from 100 to 427 m depth (Table 5.3, Fig. 5.2), i.e., within a rock thickness of over 300 m. This seems to indicate that the exposure of the terrain as a shallow continent went on for some time, with repeated stages of drying up and limited inflow of new seawater, giving rise to a succession of evaporitic facies and repeated formation of saline groundwaters.

5.9 Geological Evidence Proves that the Mediterranean Sea Underwent a Phase of Drying Up

The conclusion that the Mediterranean Sea dried up for some time has been reached in the present chapter based on evidence recorded in a part of the interstitial water column, studied beneath the sea bottom. These conclusions can be tested in light of an impressive set of geological evidences obtained in the last four decades of intensive research, the most dramatic and convincing report being by Hsü (1972a,b). The following findings led the researchers to the conclusion that the Mediterranean basin was turned into a dry land:

1. Oil prospection drillings and drillings in the beds of large rivers, e.g., the Nile and the Rhone, disclosed that all around the Mediterranean Sea the bottoms of many rivers are filled with gravel and fill-sediments, hundreds to a thousand meters deep. This reveals that there were very deep canyons, disclosing that the common base of

Interstitial Waters Beneath the Oceans

drainage was substantially deeper, i.e., the water level of the Mediterranean Sea was much lower, or it even dried up, some time in the geological past.
2. Geophysical measurements revealed that beneath the Mediterranean floor, at a depth of about 100 m, an acoustic reflector is observed, named the M reflector. Drilling revealed this is a hard bed, composed of evaporitic minerals, including anhydrite. Such a rock could be formed only in coastal lagoons and large deserts, i.e., in a continental lowland bordering a sea.
3. Stromatolites, carbonate laminae formed by blue-green algae, were found in cores of the M layer, indicating a phase of very shallow water.
4. Geophysical surveys disclosed the presence of a very large number of halite diapirs and a salt bed up to 2000 m thick! Such salt deposits could be formed only in drylands frequented by sea invasions and retreats.
5. Several alternations of marine sediments and evaporites were encountered, indicating repeated drying up and flooding of the sea basin during an interval of about 1 million years.
6. The evaporitic sediments are covered by marine sediments, indicating that about 5.5 million years ago the Mediterranean basin was again a deep sea.

These and additional observations led to the now well-known theory that during the Messinian stage, some 6 million years ago, the Mediterranean was disconnected from the Atlantic Ocean and as a result dried up. The plate of Africa was pushed against the European plate, and the Strait of Gibraltar was closed. As already mentioned, the strait was reopened and closed several times. The Mediterranean was at its dry stage an evaporitic flatland with large evaporitic lakes, the Baleric Basin being the largest. Groundwaters formed at this paleo-environment were brine tagged and saline and were incorporated in the evolved sediments.

Thus the geological information confirms the conclusions deduced independently from the salinity and composition of the suboceanic interstitial water types. The sealike interstitial water facies indicate prevalence of the open sea environment, whereas the evaporitic brinelike facies indicate regression of the ocean and a continental arid environment.

5.10 Summary Exercises

Exercise 5.1: What do the observations reveal; is seawater flowing in the rocks beneath the ocean floors?

Exercise 5.2: Does a first principles analysis predict that seawater cannot flow into the underlying rocks?

Exercise 5.3: What is meant by the term *connate water*?

Exercise 5.4: Table 5.1 represents the data obtained from the majority of the DSDP drillings; What do the Cl concentrations along the depth profile tell us? What do the SO_4 concentrations tell us?

Exercise 5.5: The depth profiles reported in Tables 5.3 to 5.5 reflect the interplay of continental and marine environments; on what is this conclusion based?

Exercise 5.6: The interstitial water in Hole 127 has a brine-spray imprint: Which compositional markers reveal this character? At which depth interval is it seen?

Exercise 5.7: By what is the profile of Hole 374 unique?

6
SALT, GYPSUM, AND CLAY STRATA WITHIN SEDIMENTARY BASINS DISCLOSE LARGE-SCALE EVAPORITIC PALEO-LANDSCAPES

Rocks and formation water facies are two independent records of geological paleo-environments. The rocks, limestone, chalk, and certain clays reflect sedimentation in the open sea; they record sea transgression phases. In contrast, halite, gypsum, and certain clays (muds) are formed in continental evaporitic environments; they record sea regression phases and thus complement the record of brine-tagged meteoric formation waters, addressed in the following chapter.

6.1 Minerals Formed Along the Continuous Evaporation Path of Seawater and Notes on the Composition of the Residual Brines

As seawater evaporates, it is gradually saturated with regard to different salts that are accordingly precipitated. The general order of precipitation is

$$CaCO_3 \to CaSO_4 \to NaCl \to \text{ complexes of K and Mg}$$

Thus, as seawater evaporates the following succession of minerals is precipitated (McCaffrey et al., 1987):

Calcium carbonate precipitates when seawater is evaporated by a factor of 1.8.
Gypsum begins to precipitate at a brine concentration of 3.8 times seawater.
Halite precipitates in the range of 10.6 to 70 times seawater concentration.
Magnesium sulfate begins to precipitate at a concentration factor of around 70.

Potassium minerals precipitate at a concentration factor of around 90.

This order of minerals precipitating from evaporating seawater has been observed in controlled experiments by many investigators, and it is given here only as a general guideline.

Under natural conditions seawater in the open ocean never reached the precipitation points of gypsum or halite, the open ocean water was always close to its present salinity and composition.

Closed inland lakes become saline due to evaporation, and their Cl, Na, Br, and other ions are in most cases from sea spray, as discussed in the previous chapters.

Very high Cl/Br weight ratios are found in the precipitating halite. The solubility of Br salts is significantly greater than that of Cl salts. As a result, very little Br is incorporated in the precipitated halite. The Cl/Br ratios of halite deposits are found to be >3000, i.e., very different from the seawater ratio of 293.

Bromine enrichment relative to Cl in the residual brine is expressed by the lower Cl/Br weight ratios, which are commonly in the range of 80 to 200 (compared to 293 in seawater). This is an outcome of the solubility of bromides that is greater than that of chlorides. When NaCl precipitates, Br stays behind and is enriched in the residual brine as well as in later precipitating salts.

Also observed is $CaCl_2$ in residual evaporitic brines and late precipitates. Enrichment of Ca relative to Na in the residual brine and among late precipitates is explained by dolomitization of limestone components (e.g., from detital dust entering the brine). In this way Mg that was enriched in the early precipitation stages, being balanced by Cl, is taken out of the residual brine, and Ca is enriched. The common occurrence of dolomitic rocks in evaporitic rock facies supports this explanation.

6.2 Formation of Halite and Gypsum Deposits Necessitated Evaporation of Tremendous Amounts of Seawater During Extended Time Intervals

Halite precipitation is a straightforward calculation: One liter of seawater contains about 30 g halite (NaCl). Thus, evaporation of a seawater column of a cross section of 1 cm^2 and 10 m high (i.e., a volume of 1 L) will result in the precipitation of 3 g halite. Or, dividing by the density of ~2.5, we find that such a column of seawater will precipitate about 1 cm of halite. In other words, *an evaporating body of seawater will precipitate around 1 cm halite with evaporation of every 10 m of its water.*

Therefore, for the formation of a 1-m halite bed, 1000 m of seawater had to be evaporated. And a 100-m halite section (observable in many locations) discloses that at least $100 \times 1000 = 100,000$ m *of seawater* evaporated!

A long time span is required for halite accumulation. Seawater evaporation is in arid climates on the order of 1 m per year. Thus, to evaporate the mentioned 100,000 m of seawater (to form 100 m of rock salt), a time span of at least 100,000 years was required. In other words, the conditions of halite precipitation had to prevail for extended time periods.

These are minimal figures! Salt accumulating on the ground is easily washed away. Thus, the preserved halite is a small portion of the amounts formed. The real figure of evaporated seawater is, therefore, even larger, and the time intervals of halite accumulation were longer.

The figures are even higher for gypsum or anhydrite. The concentration of $CaSO_4$ in seawater is about 1/20 of the concentration of NaCl. Thus, the above figures given for halite formation would be 20 times higher for gypsum and anhydrite formation.

6.3 Evaporitic Paleo-Facies: Information Recorded by Associated Formation Waters

A variety of evaporitic facies is described and discussed in the literature (e.g., Hardie, 1984, with comprehensive references). The large number of studies dealing with evaporitic deposits and recent evaporitic systems are based mainly on the associated rocks and composition of the residual brines.

A strong tool to be added to this research is the chemical and isotopic composition of the associated groundwaters in recent systems and the characteristics of associated formation waters in ancient systems. The main evaporitic facies are as follows:

Epicontinental sabkhas and lagoons. i.e., flat lowland coastal areas that are frequently invaded by the sea for short periods. The landward invading sea flushes residual brines and supplies new seawater to the shallow evaporation sabkhas and lagoons. The rocks are mud, clay, sand, gypsum, halite, and other evaporites. The rocks are thus both marginal marine and continental—a distinction that is often hard to make as clay-rich rocks may be sedimented in the shallow sea or may originate from mud accumulated on land. Similarly, sand accumulates on land and is washed into the sea and vice versa.

At time intervals of sea regression the low landforms act as through-flowing groundwater systems; the water later found confined in the rocks is brine tagged; and, most important, it has meteoric δD values that are negative, well in the range of meteoric groundwaters. The $\delta^{18}O$

values are meteoric as well in all shallow systems, but in formation waters of deep sediments their values plot in some of the cases to the right of the main global water line (MGWL) and reflect temperature-induced ^{18}O exchange with the host rocks.

Temporarily closed seas. The best known example is the Mediterranean Sea that was separated from the open ocean system during the Messinian, as is demonstrated by thick halite and gypsum strata. Meteoric isotope composition of associated formation waters indicates the host rock was exposed subaerially during a sea transgression phase. It turns out that this is the common case whenever isotopic measurements were conducted. It is recommended that the isotopic composition of studied formation waters is always analyzed.

Inland lakes. Evaporitic conditions are provided by closed inland lakes, common, for example, in rift valleys. In these cases, too, the isotopic composition of associated formation waters reflects a meteoric origin.

6.4 The Permian "Saline Giant" of the Salado Formation—An Ancient Evaporitic Megasystem

The following data are from Lowenstein (1988): The Salado Formation is a giant halite-dominated evaporite deposit with potash-bearing intervals. The complex has been traced over a huge area of 150,000 km^2, reaching a total thickness of as much as 700 m. The Salado Formation is a part of the Permian Ochoan Series of west Texas and southern New Mexico, an evaporite sequence altogether as much as 1300 m thick. The complex is mainly of flat beds composed of halite, muddy halite, anhydrite, polyhalite, dolostone, and mudstone.

The depositional sequences that developed in the center of the large enclosed Salado basin reveal repetitive stratigraphic cycles that can be recognized by changes in mineralogy and sedimentary textures and structures.

The Salado basin is envisaged as a shallow evaporating brine body, of a size of 250 × 250 km, at which halite accumulated, surrounded by dry land from which meteoric runoff and groundwater drained into the marginal marine evaporitic lagoon.

There is a possibility that mixed seawater and nonmarine inflow produced alternations of marine lagoons and continental salt-pans. Saline mudflat deposits may apply to other ancient "saline giants" formed in isolated satellite basins, especially those that contain evidence of marine inflow and continental-derived detritus.

The above discussed information, extracted from Lowenstein (1988), has direct implications to the understanding of the composition of groundwaters encountered in deep basins, a topic discussed in the next chapter. The main points of interest to be learned from the Salado complex are

1. Large-scale complexes of rock beds of halite, gypsum, and other evaporites indicate that during certain geological epochs the evaporitic environmental conditions prevailed in much greater scales than at present.
2. The soluble salts were accumulated in rock-compartments, engulfed with clay-rich sediments that preserved them for geologically very long periods.
3. These observations confirm that no groundwater flow occurred through the evaporitic deposits, as is expected from their being buried below sea level, i.e., within the zone of no flow—the zone of zero hydraulic potential.

Lowenstein (1988) summed up his detailed lithological study of the vast (150,000 km^2) Permian Salado evaporites, concluding

> They record a temporal evolution of environments from a shallow saline lake to an ephemeral salt-pan–saline mud flat complex and are interpreted as continental-dominated sequences sourced by meteoric inflow from surrounding land areas that mixed with variable amounts of sea water, either residual or introduced into the Salado basin by seepage...The vertical stacking of...cycles is best explained by periodic invasions of sea water into the Salado basin coincident with eustatic sea-level rises.

6.5 Evaporite Deposits Are Common in Sedimentary Basins

Zechstein, late Permian. Thick deposits of halite, potash salt, anhydrite, and gypsum, along with other marginal marine sediments of Permian age, are encountered in Greenland, the southern part of the North Sea, and north Europe (Taylor, 1985). The evaporitic sequence is composed of several stages that together reach 2000 m at several locations. The southern salt basin extends from eastern England through the Netherlands and Germany to Poland and western Russia, and has been well studied in the context of petroleum exploration.

Large accumulation of rock salt and gypsum beneath the Mediterranean Sea. As mentioned, over 2 km of evaporites are found beneath the floor of the Mediterranean, precipitated 5 to 6 million years ago, during the Messinian event of the drying up of this sea.

The Dead Sea Rift Valley is filled by thousands of meters of Miocene and younger sediments. Mount Sedom, at the Southwest corner of the Dead Sea Basin, is a halite diapir, and so is the Lisan Peninsula of the Dead Sea. At the Zemach drill hole, south of the lake Tiberias in Israel, thick halite beds were encountered at a depth of around 4 km; and a large number of halite diapirs have been disclosed by geophysical surveys. Gypsum is quarried in the Jordan Valley as well.

6.6 Silurian Salt Deposits Were Not Dissolved by the Nearby Formation Water

The composition of highly saline groundwater found in contact with salt deposits is often regarded as a result of halite dissolution. For this reason there is special interest in the analyses of five samples of highly saline groundwaters collected from boreholes that reached different Silurian salt-bearing strata (data from Dollar et al., 1991). The data (Table 6.1) reveal the following pattern:

1. A very high Cl concentration of up to 232 g/L. In three cases, Na is the dominant cation, but in the other two cases Ca is dominant. These last two cases are clearly groundwaters with an affinity to residual evaporitic brines and are in no way the product of halite dissolution.
2. The Cl/Br ratio in the Na-rich samples is 350 to nearly 600, a ratio that is significantly lower than the ratio in halie (>3000), ruling out halite dissolution. The Cl/Br ratio in the two Ca-rich samples is 61, i.e., well in the range typifying evaporitic brines that passed the halite deposition stage.

Table 6.1 Composition of Formation Waters from Silurian Salt-Bearing Strata, Ontario

Sample	Na (g/L)	Ca (g/L)	Mg (g/L)	K (g/L)	Cl (g/L)	Br (g/L)	Cl/Br	SO_4 (g/L)	δD (‰)	$\delta^{18}O$ (‰)
SF-1	100	8.2	2.8	2.6	207	0.59	352	0.75	−55	−5.5
SF-2	94	10.3	3.1	2.8	194	0.39	497	0.51		
SF-3	94	9.6	3.3	2.6	193	0.32	593	0.60	−52	−4.7
SA2-1	33	48.4	16.6	5.0	232	3.22	61	0.11	−52	2.9
SA2-2	37	46.8	16.2	6.4	232	3.21	61	0.11	−48	3.2

Source: Data from Dollar et al. (1991).

3. The δD values, of −55 to −48 ‰, clearly indicate an origin as non-evaporated meteoric water tagged by brine-spray salts. The $\delta^{18}O$ values reveal two types of groundwater; little-modified meteoric water and water that reveals a significant ^{18}O shift.

The discussed groundwaters are not the result of halite dissolution, but are fossil brine-spray-tagged groundwaters that are stored in hydraulically isolated rock-compartments in the vicinity of the salt deposits, but not in contact with them. They have a Silurian confinement age.

The above example has been brought up as evidence that buried salt deposits are hydraulically isolated and well preserved for extended geological periods.

6.7 Recent Lowering of the Dead Sea Lowered the Coastal Groundwater Base Flow and Initiated Rapid Dissolution of a Buried 10,000-Year-Old Halite Bed

Large sections of the narrow western coast of the Dead Sea have undergone dramatic and large-scale formation of collapse-sinkholes (Arkin and Gilat, 1999; Gavrieli et al., 1997). Observations and conclusions provide an overall picture of the dynamics of the formation of collapse-sinkholes including the key role of a buried halite bed and patterns of groundwater flow and stagnation:

Lowering of the Dead Sea level resulted in the lowering of the regional base of drainage. *Observation*: The Dead Sea receded from a level of −396 masl in 1970 to a level of −413 masl in 2000. *Conclusion*: The regional base of drainage was lowered by 17 m in a period of 30 years. This must have had a profound impact on the dynamics of both the runoff and groundwater systems.

Lowering of the regional groundwater table followed the drop of the level of the Dead Sea. *Observation*: Research wells 300 m west of the Dead Sea shore, at the location of the Samar Springs, revealed that between 1971 and 1999 the local groundwater table dropped gradually by about 6 m, in parallel to a drop of 16 m of the Dead Sea in the same time interval. A similar drop was observed in other coastal wells. *Conclusion*: The groundwater table and base flow are rapidly adjusting to the lowering of the lake level. In other words, the recession of the lake deepened the acrated zone along the shore, and the groundwater base flow was lowered accordingly, as expected for gravity flow.

Sudden onset of collapse features. *Observations*: Since 1990 collapse-sinkholes have formed at an increasing rate; the lake level was at that time −407 masl. *Conclusion*: At that time there commenced subterranean dissolution of a highly soluble rock, causing collapses of overlying soft sediments.

Sudden dissolution of a buried halite bed triggered the formation of the collapse-sinkholes. *Observation*: A halite bed was encountered in a drill hole at Nahal Zeelim, 400 m west of the shore. The top of the halite is at −418 m, and it is 6.5 m thick (Yechieli et al., 1993). The halite bed was encountered in other wells as well. *Conclusions*: (1) The halite bed seems to extend along stretches of the western Dead Sea coast, being slightly inclined eastward; (2) the halite discloses a stage when the lake level was relatively low and the Dead Sea extended close to the rift escarpment; (3) at that stage the Dead Sea precipitated halite; (4) thereafter the lake level rose, and the overlying clay, sand, and pebble beds accumulated from washed-in erosion products; and (5) groundwater through-flow suddenly reached the halite bed and started to dissolve it, flushing the salt to the lake, creating dissolution cavities that triggered collapse of the overlying soft sediments.

The halite bed was well preserved for 10,000 years. *Observation*: The age of the halite bed at the Nahal Zeelim drill hole was determined by dating organic material found above it and beneath, the deduced age being around 10,000 years (Yechieli et al., 1993). *Conclusion*: For 10,000 years the halite was well preserved, disclosing that no fresh groundwater reached it. Or, in other words, the groundwater base flow was throughout this long period restricted to the zone above the top of the halite bed.

The preservation of the halite bed discloses that for millennia it was within the zone of static groundwater. The conclusion that the halite was preserved for 10,000 years discloses that only static groundwater was in contact with it, saturated with respect to halite. Thus, beneath the shallow groundwater base flow there extended a zone of static groundwaters.

6.8 The Many Preserved Salt Beds Manifest the Preservation of Connate Groundwaters

Salt beds have been stored in the static groundwater zone since their burial and confinement. The young 10,000-year-old salt bed of the Dead Sea coast, as well as halite beds preserved elsewhere, including those of Precambrian age, all were stored in the static groundwater zone. No flowing groundwater came in contact with them throughout those long time periods. The halite beds were preserved in rock-compartments engulfed by impermeable clay-rich rocks that confined them.

The same holds true for groundwater stored in rocks that are buried beneath the respective terminal base of drainage. The rather common preservation of halite beds within sedimentary basins demonstrates that all the accompanying rock units as well are stored in the zone of static groundwater. Thus, the preserved salt beds disclose preservation of connate groundwaters.

6.9 Limestone–Clay Alterations Reflect Alternating Sea Transgressions and Regressions

Every field geologist is familiar with exposures of alternations of limestone and clay-rich rock strata. There are regions with tens to hundreds of alternations, the clay beds ranging in thickness from a few centimeters to several meters.

A traditional way to explain these alternations was that they reflect changes in the materials supplied to the respective paleo-sea from the nearby continent. But what kind of changes could that have been? And why were they so abrupt?

A more convincing mechanism is that the limestone beds were formed during sea transgression phases, and the clay beds reflect formation at evaporitic flatlands during very shallow to full regressive phases. Part of the formed clay beds were possibly eroded and not preserved, but the regressive event is recorded by the structure of separated limestone beds.

This fits well with the record displayed by formation waters, that are meteoric but tagged by brine-spray, and thus display phases of land exposure.

6.10 Summary Exercises

Exercise 6.1: How many meters of seawater had to be evaporated to form 10 m of halite? How many years did it take to evaporate this amount? This is a minimal estimate—why?

Exercise 6.2: What is the most probable source of Cl, Na, and Br in saline closed inland lakes?

Exercise 6.3: Which observations lead to the conclusion that during past geological periods there prevailed large-scale evaporitic flatlands?

Exercise 6.4: Is a "saline giant" known also from Europe?

Exercise 6.5: The Deep Sea Drilling Project and geophysical studies revealed salt deposits within the rock sequence beneath the Mediterranean Sea. Are they thick? How were they formed?

Exercise 6.6: Does the composition of formation waters encountered close to the contact with Silurian salt deposits reveal halite dissolution? Discuss.

Exercise 6.7: A substantial recent lowering of the Dead Sea level caused sudden dissolution of a buried salt bed. What is the story?

PART III
DEEP GROUNDWATER SYSTEMS—FOSSIL FORMATION WATERS

*The no-flow zone of
zero hydraulic potential*
hosts
fossil meteoric brine-tagged formation waters

7
THE GEOSYSTEM OF THE FOSSIL BRINE-TAGGED METEORIC FORMATION WATERS

Having looked at the unconfined through-flowing groundwater, let us now go further and examine the deeper confined groundwater geosystem, hosting the formation waters. The latter have been encountered at a depth range of up to 9 km—as deep as research drillings have been done. The volume of formation water exceeds the volume of water residing in the upper unconfined system; it is huge.

Formation waters are found within the large sedimentary basins and rift valleys and even within crystalline shields; they are everywhere. Are they static? Are they connate? Are they tagged by sea spray? Or are they brine-tagged? What can be learned from them about the paleo-landscape that prevailed during sediment collection in the continent-wide or local small basins? How can they be managed?

7.1 Formation Waters Within Sedimentary Basins

The term *formation water* relates to groundwater encountered in boreholes drilled to a depth that is greater than sea level, i.e., formation waters are within the zone of no flow, or the zone of zero hydraulic potential. Let us have a look at a number of case studies in sedimentary basins. The amazing overall picture is that on the one hand there is a vast variability of salinity, ionic relative abundances, and isotopic composition, and yet certain patterns turn out to be very clear and seen in all studied cases.

7.1.1 The Alberta Basin: Formation Waters with a Brine-Spray Composition and Meteoric Isotopic Values

Data of the sedimentary Alberta Basin, western Canada, have been published by Connoly et al. (1990a,b) and serve as a base for the following set of observations and conclusions:

The regular temperature increase with depth reveals the samples were properly collected. The temperature depth profile (Fig. 7.1) indicates the researches had access to well-defined points in the wells, ruling out dilution by shallow water during sampling, or short-circuiting in the well between different water bodies (which would result in an erratic temperature–depth graph).

The composition differs from recent groundwater, indicating that the formation waters are fossil. The high salinity and ionic relative abundances indicate these formation waters were formed under environmental conditions different from the present conditions in the respective locations, and thus they are fossil.

The Cl concentration is significantly below the value of saturation in regard to halite indicating no halite has been encountered. Hence, the Cl is allochthonous, i.e., of an external origin and not the result of any water–rock interactions.

Linear correlations between the concentrations of Cl, Br, Na, Ca, Mg, K, Sr, and Li indicate these ions are of external origin. The composition diagrams, seen in Fig. 7.1, reveal a positive linear correlation between the concentrations of Cl, Br, Ca, Mg, K, Sr, and Li, indicating they are all external, brought in together with the Cl.

The same ionic abundances occur in waters encountered in different host rocks, another indication of external origin. The Alberta Basin formation waters are encountered in a variety of rock types, including sandstone, limestone, dolomite, and varieties of clay. Yet all the sampled waters reveal the same ionic relative abundances, indicating that the intercorrelated ions did not stem from water–rock interactions. Hence, the respective ions, encountered in the Devonian to Cretaceous formation waters in the Alberta Basin, had an external origin.

Low Cl/Br ratios disclose the external source of salts had an evaporitic composition. The water samples from the research drillings in the Alberta Basin reveal Cl/Br ratios in the range of 70 to 260 (Fig. 7.1). This observation indicates that the external source of ions was from airborne spray of evaporitic brines (in contrast to sea spray, which is the common source of external ions in the shallow through-flowing groundwaters).

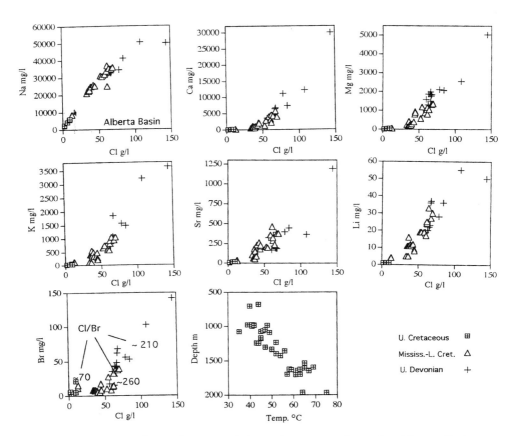

Fig. 7.1 Composition diagrams of formation waters from the Alberta Basin, Canada. The samples, collected from rocks of upper Devonian to upper Cretaceous at a depth interval of 660 to 1970 m, reveal linear correlations between the concentrations of Cl, Na, Ca, Mg, K, Sr, and Li, indicating an allochthonous origin of these ions. The Cl/Br weight ratio is in the range typical to evaporitic brines, and the importance of Ca besides Na is another marker of tagging by evaporitic brines. (Data from Connoly et al., 1990a,b.)

Significant concentration of $CaCl_2$, and $MgCl_2$—another evaporitic imprint. The relatively high Ca and Mg concentrations (Fig. 7.1), besides the common Na component, are an additional marker that the dissolved salts originated from evaporitic brines.

As halite precipitates from evaporating seawater, Na is taken out of the brine, whereas Mg stays behind, as $MgCl_2$ is more soluble. Thus, residual evaporitic brines are relatively enriched in Mg. However, the common observation is that evaporitic brines are enriched mainly in $CaCl_2$. The plausible explanation is that such brines came in contact with limestone, and a dolomitization process consumed part of the Mg in the brine and enriched it instead with Ca. A common source of limestone for this dolomitization process is detrital limestone particles carried into the evaporation lagoons by runoff and as windborne dust.

A large composition variety characterizes the formation waters, indicating they are entrapped within distinct rock-compartments. The compositional fingerprint diagram of Fig. 7.2 depicts the pronounced variability of the formation waters; they are scattered over a range of two orders of magnitude of salinity and differ from each other in their ionic abundances. Hence these formation waters are enclosed within hydraulically separated rock-compartments.

The δD values indicate unequivocally that these are meteoric waters formed at the continent. The isotopic composition of the discussed formation waters is depicted in Fig. 7.3 (data from Connoly et al., 1990b). The δD

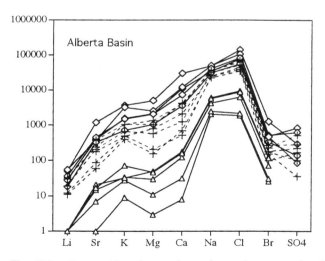

Fig. 7.2 Composition fingerprints of petroleum-associated formation waters encountered in upper Devonian ($\Diamond$), Messinian/lower Cretaceous (+), and upper Cretaceous rocks (Δ), in the Alberta Basin, Canada. The compositions differ drastically from the recent groundwaters, indicating the formation waters are fossil, and the distinct diversity manifests confinement in hydraulically separated rock-compartments. In spite of the variability a general pattern is observable, namely, Cl > Na > Ca > Mg, a pattern seen in formation waters worldwide. (Data from Connoly et al., 1990a,b.)

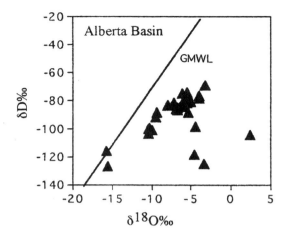

Fig. 7.3 Isotopic composition of the formation waters from the Alberta Basin, Canada. The δD values are in the range of meteoric groundwaters. Some of the $\delta^{18}O$ values are also negative, indicating an origin as meteoric groundwaters, and other values are distinctly more positive, indicating temperature-induced isotopic exchange with the host rocks. This indicates, in turn, subsidence to a depth in the range of the petroleum temperature window. (Data from Connoly et al., 1990b.)

values are in the range of −130 to −65‰, i.e., well in the range of meteoric groundwaters.

The $\delta^{18}O$ values are slightly heavier, reflecting temperature-induced isotopic exchange. Some of the $\delta^{18}O$ values are also negative, e.g., −15 to −6‰ (Fig. 7.3), indicating an origin as meteoric groundwaters. Other values are distinctly more positive (plotting to the right of the GMWL), indicating temperature-induced isotopic exchange with the host rocks. This indicates, in turn, subsidence to a depth equal to the petroleum formation temperature window.

Key conclusion; these are meteoric paleo-groundwaters tagged by brine-spray. The listed observations and deduced conclusions lead to a clear and simple key conclusion: The Alberta Basin formation waters were formed on an ancient continent that was exposed to large-scale evaporitic systems.

Variability of water at different depths, indicating entrapment in separated rock-compartments. Figure 7.4 represents the Alberta Basin data as a function of the depth from which the samples were collected. The following patterns are seen:

1. The concentration of each ion varies significantly with no systematic depth-related trend.

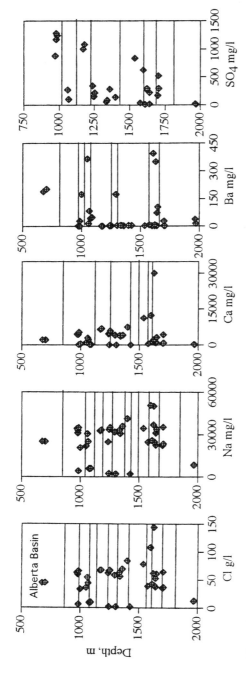

Fig. 7.4 Compiled depth profiles for various ions dissolved in petroleum-associated formation waters in a study area in the Alberta Basin. A pronounced variability with depth is seen, and abrupt changes in ion concentrations are marked by horizontal lines. Different ions serve as markers identifying different layers, hence the importance of detailed analytical data. This structure reveals that the formation waters are stored in distinct, hydraulically isolated rock-compartments. (Data from Connoly et al., 1990a.)

2. Concentration changes are often abrupt, suggesting different formation waters are stored in distinct rock-compartments.

The listed observations, demonstrated by the example of the Alberta Basin, reveal that in formation waters of the described type the allochthonous dissolved ions were brought in by the recharged water and stemmed from airborne brine-spray. This point is further discussed latter on in light of the connate nature of such formation waters and their common association with petroleum, shale, sandstone, gypsum, and halite.

Boundary conditions: the hosting rocks are marginal marine as well as of continental facies, but the contained formation waters are all meteoric, i.e., continental. As in many sedimentary basins, the rocks are varieties and combinations of sandstone, limestone, dolomite, and clay. These lithologies are either the product of shallow near-shore seas or deposits of evaporitic lowlands. The origin may be best defined by remains of fossil fauna and flora. The water is as a rule meteoric; it originated from continental precipitation. These observations provide very clear boundary conditions.

How were so many different water types introduced to their host rock-compartments? The sediments were accumulated in a flat lowland, frequented by sea invasions and retreats. During a regression the last-formed marine rocks were subaerially exposed and functioned as a groundwater through-flow system, tagged by brine-spray blown in from vast adjacent evaporitic lagoons. The following sea transgression produced new sediments that covered the former ones and confined them along with the groundwater they contained.

A note: The reader may feel that too much has been seen in the data of the discussed case study, and too much has been concluded. Let us have a look at additional examples in the present chapter and also later on in Chapter 10. It is amazing how the same patterns are observable time and again. There is a general facies of formation waters!

7.1.2 Niagara Falls Region, New York: Saline Formation Waters in Silurian Rock Units

A second case study, discussed here in some detail, addresses saline groundwaters that occur in bedded dolomites of the Silurian Lockport Group near Niagara Falls, New York. The bedrock strata are gently inclined southward, being truncated near the surface and covered by glacial sediments, except along a prominent east–west trending escarpment and in the north–south trending Niagara River Gorge (Tepper et al., 1990). Detailed hydrochemical data, reported by Noll (1989), are interpreted in light of the above discussed guidelines:

These are fossil waters, a conclusion based on a composition that differs from recent water. Waters of different compositions were encountered in a research well drilled to a depth of up to 77 m. Waters varied from slightly saline and Ca–SO$_4$ dominated to saline Na > Ca > Mg and Cl >> SO$_4$ > HCO$_3$ waters (Fig. 7.5). These compositions differ significantly from the local recent fresh Ca-HCO$_3$ groundwaters, indicating the former are fossil groundwaters.

An allochthonous origin of Cl and other ions. The concentration of Cl, Na, Ca, Mg, K, Sr, and Br are linearly positively correlated (Fig. 7.6), indicating a common allochthonous origin.

The Cl/Br ratios are in the evaporitic range. The Cl/Br weight ratios are in the range of 50 to 104, ruling out dissolution of evaporites and indicating an origin from residual evaporitic brines.

Light isotopic composition discloses meteoric water. The δD values are in the range of −100 to −66‰, and the δ^{18}O values are in the range of −14 to −7‰, indicating an origin from nonevaporated continental meteoric waters.

These are meteoric groundwaters tagged by brine-spray, as is revealed by the observed combination of isotopic evidence of a meteoric origin of the water phase and a dissolved ion composition that indicates tagging by an evaporitic brine.

The paleo-landscape was a marginal evaporitic lowland, frequented by sea invasions and retreats. The meteoric pattern of the discussed formation waters discloses that they were formed on the continent, and the evaporitic composition reveals that nearby were large-scale seawater evaporation systems, e.g., lagoons and sabkhas. The rocks hosting the formation waters are marginal marine, rich in clays, deposited mainly during marine phases. The meteoric groundwater was introduced during phases of sea retreat and continental subaerial exposure. During marine phases marginal marine sediments were sedimented, containing interstitial waters of marine origin (Chapter 5). During phases of land exposure, a continental water cycle operated that washed away the marine interstitial water and replaced it by meteoric groundwater, tagged by spray from surrounding evaporitic systems. Sediments of the following marine phase confined the previous sediments, along with their contained paleo-groundwater. The evaporitic setups were variegated, as is disclosed by the variability of chemical composition revealed by the formation waters.

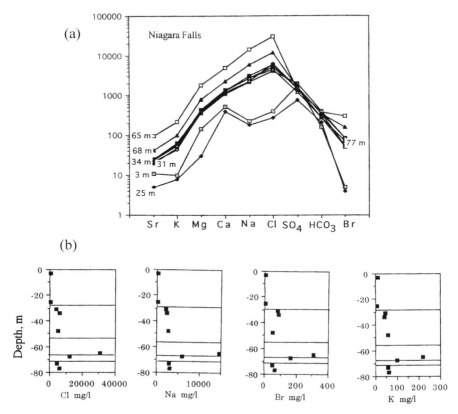

Fig. 7.5 Composition fingerprint diagrams (mg/L) of saline groundwaters encountered at the marked depths in a borehole in Silurian rocks, Niagara Falls. The two shallow water samples seem to belong to the local unconfined groundwater system, whereas the 34- to 65-m samples are saline and differ significantly from the present local groundwaters, indicating these are fossil waters. The distinct compositional variations indicate the fossil waters are stored within rock-compartments that are hydraulically sealed from each other. In this case the compartments are only tens of meters thick, and interestingly they are preserved, even at their present shallow depth. (Data from Noll, 1989.)

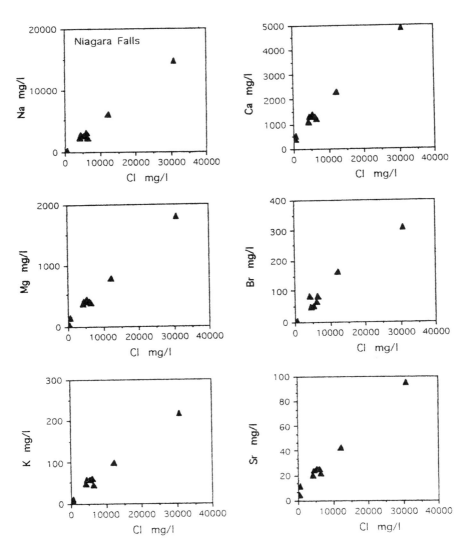

Fig. 7.6 Composition diagrams of shallow wells in Silurian rocks, Niagara Falls area. The two points near the zero values are of the unconfined groundwater system, and the rest are connate formation waters (well seen in the previous figure). Linear positive correlations are seen between the concentrations of Cl, Br, Na, Ca, Mg, K, and Sr, indicating a common allochthonous origin, which must have been brine-spray, as indicated by the Cl/Br ratios of 50 to 104 and the relative importance of Ca. (Data from Noll, 1989.)

The spatial variability indicates entrapment in isolated rock-compartments. The composition of the waters, encountered in wells that are only hundreds of meters apart, varied distinctly (Fig. 7.5), indicating that they are stored in distinct bedded rock-compartments that are hydraulically well separated.

These are connate waters of Silurian age. The samples plotted in Fig. 7.5 reveal that various types of groundwater reside at the research area, stacked one on top of the other. The only way these different waters could be introduced into their host rocks was when the latter were exposed on the land surface.

Thus the age of the various saline groundwaters in the Lockport Group dolomites must be the same as the age of the host rocks, i.e., Silurian.

The connate waters are in this case preserved at a shallow depth of 34 to 65 m, as marked in Fig. 7.5. This leads to the conclusion that the through-flowing groundwater system is at this site very shallow—at a depth of up to around 28 m.

7.1.3 Mersey Basin, England: A Depth Profile in a Research Borehole in Permian–Triassic Rocks

Tellam (1995) reported a detailed study of groundwater carefully sampled at a depth profile within a drill hole that penetrated Permo-Triassic sandstone, at the Mersey Basin, northwest England. The following pattern emerges:

Salinity and composition vary with depth within a single borehole, indicating entrapment in a sequence of rock-compartments. Figure 7.7 reveals different salinities (the vertical scale is logarithmic) and difference in ionic ratios, especially of HCO_3. The same data are plotted in Fig. 7.8 as depth profiles of the discussed well. Distinct water-hosting rock layers are noticeable, revealed by the various ions. Less saline and more saline water layers alternate, indicating confinement in separate rock-compartments (marked by horizontal lines). At least six separated compartments are noticeable in the short depth range of −80 to −190 ml.

Pronounced linear correlations between the concentrations of Cl, Na, K, Ca, Mg and SO, indicate they are of an external origin. The data of the research well are plotted in Fig. 7.9 in composition diagrams that reveal well-defined linear correlations of Cl and several ions, indicating they are of an external origin. However, HCO_3 is not correlated to the other ions (not plotted in Fig. 7.9), indicating an independent origin.

Permian confinement age of the waters. As mentioned, along the depth section from −80 to −190 masl at least six different water beds can be seen in

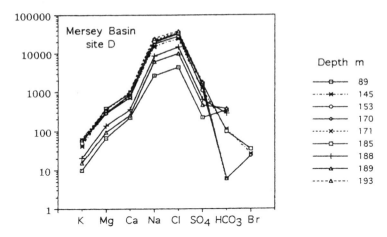

Fig. 7.7 Composition fingerprints of pore water and packer samples obtained from the Padegate (site D) observation borehole in the Mersey Basin, England. A pronounced salinity range is seen (the vertical scale is logarithmic) and slight compositional variability, mainly manifested by the HCO_3 concentration. (Data from Tellam, 1995.)

Fig. 7.8. The only way these different waters could have entered their host rocks was as paleo-groundwater formed during land exposure phases that alternated with sea invasion and formation of new sediments that covered the previous ones and confined them. Thus these are connate waters of a Permian Age.

7.1.4 Upper Silesian Coal Basin, Poland: Saline Formation Waters

A complex of saline groundwaters of different compositions was encountered in Carboniferous to Miocene formations in a study area in the Upper Silesian Coal Basin. Let us have a look at the data, retrieved from a paper by Pluta and Zuber (1995). The studied fossil formation waters include a variety of facies.

Brine-spray-tagged meteoric formation waters within Miocene rocks. The studied groundwaters are saline, differing from the presently formed local groundwater and, hence, are fossil. Their Cl concentration is rather high, in the range of 18 to 26 g/L. The Cl/Br weight ratio is around 160, thus these are not entrapped seawater but brine-tagged waters. The δD values are in the range of -7 to $+2‰$, and the $\delta^{18}O$ values are in the

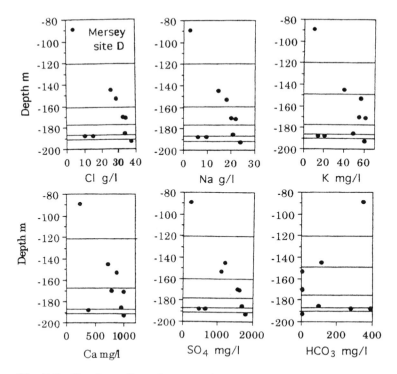

Fig. 7.8 Depth profiles of a research borehole, site D at Padegate, Mersey Basin, England. A pronounced bedding of different water types is well seen, indicating confinement of connate groundwaters. (Data from Tellam, 1995.)

range of −0.4 to +0.3‰. Thus these are meteoric waters that were partly evaporated prior to infiltration.

Saline meteoric formation waters, some tagged by sea spray and others tagged by brine-spray. A variety of compositions is reported for saline groundwaters of the older formations. For example, one group contains 5 to 138 g/L Cl, with a distinct Na–Ca–Cl composition, a Cl/Br ratio of 200 to 308, δD in the range of −28 to −12‰, and $\delta^{18}O$ in the range of −2.9 to +0.9‰. These waters resemble meteoric groundwater fed in some of the cases by brine-spray and in others by sea spray, and they were partly evaporated prior to infiltration.

Nonevaporated saline formation water tagged by sea spray. An entirely different example of formation water, encountered in a well in the same study area, contains 9 g/L Cl; the Cl/Br ratio is 333; and the isotopic

The Fossil Brine-Tagged Formation Waters

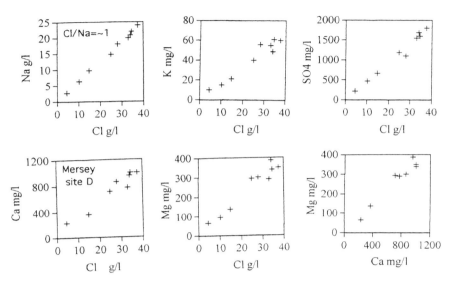

Fig. 7.9 Composition diagrams of the water samples collected at the Padegate research well. The plotted ions are positively correlated to the Cl concentration, disclosing these are external (allochthonous) ions. (Data from Tellam, 1995.)

composition is distinctly light, $\delta D = -74‰$ and $\delta^{18}O = -10.1‰$ (Pluta and Zuber, 1995). This seems to be sea spray–fed groundwater that was formed in an arid climate. The salts were concentrated by intense evapotranspiration and washed down by strong rain events, thus maintaining the light isotopic composition of rain.

Entrapment in hydraulically isolated rock-compartments. The multitude of different saline groundwater types encountered in the Upper Silesian Coal Basin (Fig. 7.10) manifests retention in a large number of vertically and laterally separated rock-compartments.

These are connate waters. The stable isotopes reveal that all these waters are meteoric, i.e., they originated as unconfined systems that subsided and got confined. Hence these saline groundwaters are connate and by and large of the age of their host rocks.

Fossil groundwaters of the described type are reported from many other north European countries, the Basin of Paris being another example. This occurrence of brine-spray-tagged fossil groundwaters is of relevance to the large-scale occurrence of the Zechstein complex of evaporitic deposits from the Permian to Jurassic and other evaporitic deposits occurring in Neogen rocks. In this light it is suggested that the

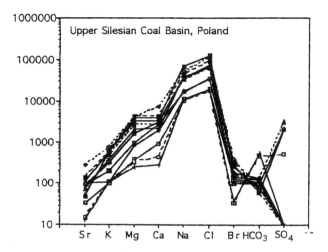

Fig. 7.10 Composition fingerprint diagrams (mg/L) of saline groundwaters of the Upper Silesian Coal Basin. The compositions differ distinctly from the respective recent groundwaters, indicating these are fossil groundwaters. The significant variability indicates storage in hydraulically separated rock-compartments. These are connate waters. (Data from Pluta And Zuber, 1995.)

discussed fossil groundwaters of the Upper Silesian Coal Basin are by and large of the age of their host rocks.

Coal is the inland twin-facies of petroleum. The saline fossil formation waters, encountered in sedimentary basins all over the world disclose formation in vast flatlands that were frequented by sea invasions and retreats, thus forming extended evaporitic paleo-landscapes. The coal of the Silesian Coal Basin seems to have been formed from forests that grew in a more inland periphery, with a less saline environment, but occasional sea transgressions reached these zones and covered them effectively by confining sediments. Upon subsidence to an elevated temperature zone, coal was formed.

7.2 Formation Waters Within Rift Valleys

Warm and cold water, issuing in springs and encountered in wells in the Dead Sea rift valley segments (Fig. 7.11), puzzled researchers over the last decades because they are significantly enriched in Br. These groundwaters have a Cl/Br weight ratio of around 100 (i.e., Br relatively three times more than in seawater); the Cl concentrations range from 1 g/L to as high as 17 g/L (in the

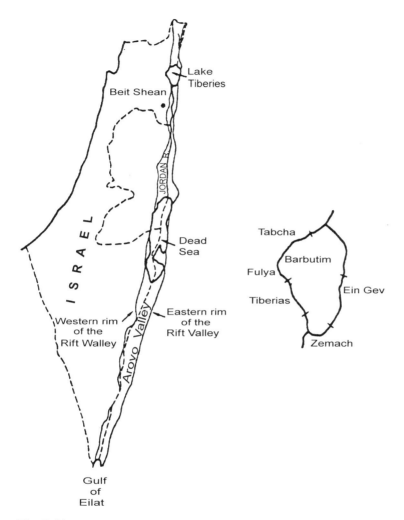

Fig. 7.11 Map of the Jordan–Dead Sea–Arava rift valley and geographical/geochemical subareas around Lake Tiberias. (Following Mazor and Mero, 1969a.)

Tiberias 61°C spring); the relative abundances of the different ions vary from one location to another (Mazor and Mero, 1969a,b; Mazor et al., 1969; Mazor and Molcho, 1972); and the stable isotopes of the water are distinctly lighter than seawater (Gat et al., 1969), disclosing their meteoric origin.

In parallel, Neogene gypsum and halite deposits (the salt plug of Mount Sedom being a well-known example) are exposed and exploited in the Dead

Sea rift valley, and many more evaporitic deposits were encountered in deep boreholes and in geophysical surveys. It seems, thus, inevitable to conclude that the Br-enriched rift valley groundwaters were formed within a large-scale evaporitic environment that prevailed all along the Dead Sea Rift Valley during the time of formation of the gypsum and halite rocks.

7.2.1 Lake Tiberias Basin, Dead Sea Rift Valley

A large number of research boreholes and operational wells have been drilled on the shores of Lake Tiberias. The composition of the encountered groundwaters revealed geographical groupings into small subregions (Fig. 7.11). The rather large body of available data (Mazor and Mero, 1969a,b) was retrieved four decades ago when the anthropogenic intervention was still limited.

Fossil formation waters. The formation waters differ from the recent groundwater and hence were formed during different environmental conditions, and these are fossil waters.

Elevated temperature of the Tiberias Hot Spring indicates deep storage and artesian pressure. The Tiberias Hot Spring (actually a very shallow well) has a temperature of 61°C. This is about 40°C above the local average ambient temperature. The region has no volcanic activity, and hence the elevated temperature indicates deep storage. The local heat gradient is around 25°C/km, thus the Tiberias hot water is stored at a depth of 40/25 = 1.6 km. The water ascends under artesian pressure, caused by the lithostatic pressure and possibly added tectonic pressure.

Elevated ^4He concentration indicates that the Tiberias Hot Spring water is old, possibly Neogene in age. Air-saturated water contains around 4×10^{-8} ccSTP He/cc water, whereas the Tiberias Hot Spring ^{4}He concentration of 2.6×10^{-5} ccSTP He/cc water was observed, indicating an age from the Neogene (Mazor, 1972).

Vertical and lateral variability discloses compartmentalization. The fossil waters reveal a conspicuous diversity in composition, disclosing confinement in hydraulically isolated rock-compartments. The circum-lake areas of mineralized waters were divided into geographical/geochemical subareas (Fig. 7.11). In some of these areas several water groups are present, as is revealed by the composition variability (Figs. 7.12 to 7.17).

Light δD and $\delta^{18}O$ values indicate these are meteoric waters. The water isotopes reveal light δD and $\delta^{18}O$ values, and the data plot in Fig. 7.12 between the local meteoric line (relevant to recent groundwater) and the global meteoric line (which was apparently dominant in the region during past epochs). The water isotopes indicate the formation waters ori-

The Fossil Brine-Tagged Formation Waters

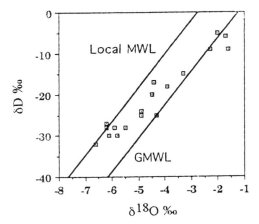

Fig. 7.12 The isotopic composition of the formation waters encountered in springs and wells around Lake Tiberias, Israel. The values, plot between the local meteoric water line and the global line, revealing an origin as meteoric waters. (Data from Mazor and Mero, 1969b.)

ginated as meteoric (i.e. continental) groundwaters, and entrapment of seawater or evaporitic brines is ruled out.

Local dilution effects. The composition diagrams reveal linear correlations between the concentrations of Cl, Br, Na, Ca, Mg, K, and SO_4 (Figs. 7.13 to 7.15). The correlation lines reach the zero points of the origin of the axes. Some dilution of saline formation waters by fresh shallow groundwater may occur in some of the springs and in poorly cased boreholes.

Diversity of fossil water types. Specific correlation lines are seen for the various geographic subregions, as demonstrated by the Ca–Mg and Cl–SO_4 graphs of Figs. 7.14 and 7.15. Thus, various saline end-members exist. The composition of the most saline water in each geographical subgroup has been plotted as a fingerprint diagram in Fig. 7.16. The variability of different water types is well seen.

External origin of Cl, Br, Na, K, Ca, Mg, and SO_4. The linear correlations seen as a function of Cl (Fig. 7.12), the Ca–Mg correlations seen in Fig. 7.13, and the Cl–SO_4 correlations seen in Fig. 7.14 indicate that Cl, Br, Na, K, Ca, Mg, and SO_4 are by and large allochthonous, i.e., they were brought in with the paleowater when the latter was recharged, and they do not stem from interaction of the water with host rocks. Different allochthonous sources supplied the salts to the various types of formation waters.

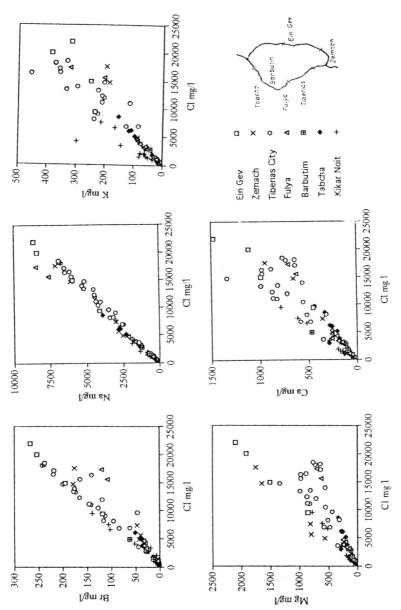

Fig. 7.13 Composition diagrams of formation waters encountered around Lake Tiberias. In general, positive correlations are observed between the concentrations of Cl and the other ions, with specific correlation lines in the geographical subgroups. (Data from Mazor and Mero, 1969a.)

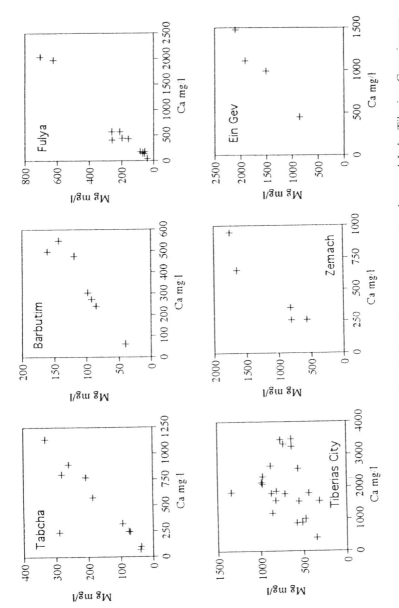

Fig. 7.14 Calcium–magnesium correlations in formation waters encountered around Lake Tiberias. Conspicuous positive correlation lines are observable for different regions, indicating presence of hydraulically separated subgroups and/or local dilution of saline formation waters. The Tiberias City region reveals no correlation, indicating presence of different water types. (Data from Mazor and Mero, 1969a.)

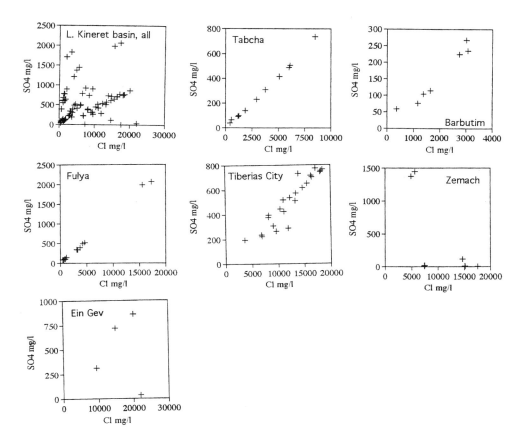

Fig. 7.15 Sulfate–chloride correlations in formation waters encounterd around Lake Tiberias. The data for the whole basin reveal no correlation, yet lines of the various geographical subareas are hinted. The data of the Tabcha region reveal a perfect line, reflecting different degrees of dilution and/or water tapped in separated rock-compartments, fed by the same type of paleo-brine-spray. The data from Zemach and Ein Gev clearly indicate presence of at least two distinct water types in each zone. (Data from Mazor and Mero, 1969a,b.)

Tagging by brine-spray. The Cl/Br ratio in most of the formation waters is in the range of 90 to 200 (Fig. 7.17), distinctly different from the seawater value (~293) or the value in halite (>3000). This value range is typical of evaporitic brines and rules out entrapment of seawater or halite dissolution.

The concentrations of Ca and Mg that balance part of the Cl are another indication of tagging by evaporitic brines.

The Fossil Brine-Tagged Formation Waters

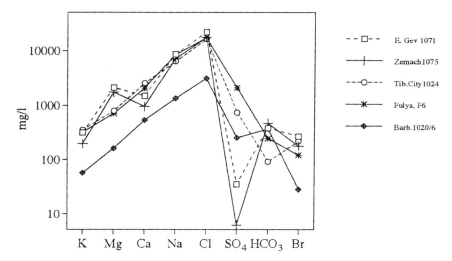

Fig. 7.16 Composition fingerprint diagrams of the most saline end-member at each geographical subregion around Lake Tiberias. The variability is well seen. Thus, a variety of types of formation waters is stored in distinct rock-compartments. (Data from Mazor and Mero, 1969a.)

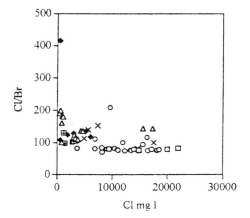

Fig. 7.17 The Cl/Br weight ratio as a function of the Cl concentration at the geographical subregions around Lake Tiberias. Values are in the range of 90 to 200, common in evaporitic brines. (Data from Mazor and Mero, 1969a.)

7.2.2 Deduced Working Hypothesis for the Lake Tiberias Basin

The saline formation waters encountered at different depths around Lake Tiberias are fossil waters that were formed in the subsiding rift valley that was dotted with evaporitic sabkhas and lagoons, as is also indicated by the gypsum and halite deposits. The formation waters were formed during the Neogene by (1) paleo-recharge of meteoric water that (2) washed down brine-spray and evaporitic dust, which (3) were concentrated by evapotranspiration that was effective in an arid climate. The different sediment beds subsided along with their groundwater and were confined by newly accumulating sediments that subsided in turn along with different types of groundwater.

The bulk of the dissolved Cl, Br, Na, K, Ca, Mg, and SO_4 are external, the great variability reflecting the variability of composition of the local evaporitic setups that varied geographically and over time.

7.2.3 Key Observations and Conclusions Related to Rift Valleys All Over the Globe

Thermal and saline groundwaters issue in springs and are frequently encountered in drill holes within rift valleys. The following observations and derivable conclusions are characteristic:

1. Elevated groundwater temperatures indicate storage at a depth of hundreds to several thousands of meters.
2. The salinity and chemical composition of the warm and saline waters differ from the currently formed groundwater at the respective locations, indicating formation under different environmental conditions; hence these waters are fossil.
3. The warm waters, and often also cold waters, lack tritium and ^{14}C and are enriched in ^4He, indicating they are old, thus supporting the conclusion that these waters are fossil.
4. The salinity, chemical composition, temperature, and isotopic composition vary significantly between adjacent springs and wells, disclosing the existence of numerous systems of the fossil waters, hence revealing storage within hydraulically separated rock-compartments.
5. The springs of fossil water issue in general along seismically active fault plains, disclosing ascent along hydraulically open pathways.
6. Drill hole data and results of geophysical surveys indicate rift valleys are filled with thousands of meters of geologically recent terrestrial sediments, including alternations of permeable and impermeable rocks.

7. Each stratum of these sediments was originally exposed at the land surface, where it acted as the active groundwater system, and subsequently it subsided along with its water content and was confined by the newer sediments that piled up in the subsiding basin.
8. In this way, rock-compartments that contain connate groundwater were formed, filling subsided rift valley basins.

7.3 Fossil Nonsaline Groundwaters Tagged by $CaCl_2$, Formed During the Messinian, at the Land Bordering the Dried-Up Mediterranean Sea

Wells drilled in the Mediterranean coastal plain of Israel tapped fresh water of the unconfined system. Over the years, heavy pumping lowered the local water table, and gradual salination of the pumped waters appeared, causing a management problem.

The intruding water contains around 700 mg/L Cl; it has a distinct Ca–Cl composition (Vengosh and Starinsky, 1992), and it has an isotopic composition of regular meteoric groundwater. This Ca–Cl groundwater must be brine-tagged ancient water, as no present-day recharge produces such groundwater. Neither is it an encroachment of seawater, as the occurrences are also away from the shoreline, and the chemical composition differs from seawater. The named Ca–Cl groundwater resides beneath the active unconfined coastal groundwater system (Fig. 7.18).

The dried-up Mediterranean Sea provided a mega-evaporitic environment. The above described observations reveal a link to the Messinian event of the drying up of the Mediterranean Sea, discussed in sections 5.8 and 5.9. The Mediterranean Sea dried up when the Strait of Gibraltar was temporarily closed, about 6 million years ago. This is borne out from thick gypsum and halite beds (Hsü, 1972a,b; Schreiber, 1988), encountered in research drillings into the sea bottom and by geophysical surveys. Thus the level of the Mediterranean Sea was lowered by a few thousand meters, as is also borne out by over 1-km-deep buried river channels encountered in the surrounding countries. The paleohydrology of the setup was as follows:

1. A significantly lower terminal base of drainage, followed by a much deeper level of the water table.
2. A drastic change from a sea spray regime to a residual evaporation brine-spray regime with a much drier climate, as the Mediterranean Sea was salty or even dry. As a result, saltier groundwater must have been formed for awhile inland, with a compositional brine imprint and a relatively deep water table.

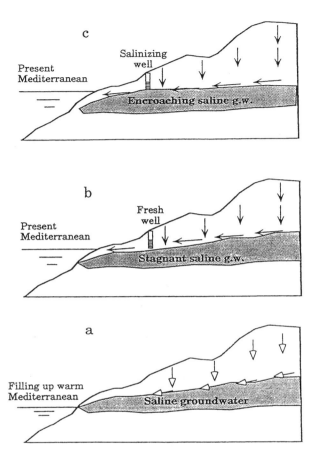

Fig. 7.18 Intrusion of fossil brine-tagged groundwater into overpumped coastal wells, Israel. (a) Deep groundwater, slightly saline and brine-tagged, formed during the refilling of the Mediterranean Sea. (b) The shallower recently formed groundwater that covers and confines the underlying fossil groundwater and a pumping well. (c) Overpumping lowered the water table to a degree that the underlying fossil water encroached, and the well was turned salty.

3. Once the Strait of Gibraltar reopened, Atlantic Ocean water refilled the Mediterranean basin, and as result

   The original sea level was restored.
   The sea spray regime was resumed.
   The current unconfined fresh water system was formed, overlying the ancient brine-like groundwater (Fig. 7.18).

The working hypothesis of the discussed formation of brine-spray-tagged groundwater during the Messinian episode leads to the prediction that similar fossil Ca–Cl groundwaters can be identified in other countries around the Mediterranean.

7.4 Some Physical Aspects of Formation Waters

7.4.1 Basin-Wide Groundwater Through-Flow is Ruled Out by First Principles Discussion and Observations

Erroneous models that claimed that groundwater actively flows through entire sedimentary basins, even at depths of 1000 m below sea level, are discussed in conventional textbooks. However, the lack of such flow is evident from first principles considerations, including (1) the lack of a source of energy that would be required to push the water down beneath the base of drainage and then force it up at the other end of the basin; (2) no flow takes place in the zone of zero hydraulic potential that extends beneath the ocean water level; and (3) uncountable layers of impermeable rocks block any cross-basin destination.

7.4.2 The Pressure Regime of Fluids Within Sedimentary Basins

The following is a brief overview of the different expressions of pressure acting on formation waters—all directly or indirectly caused by the terrestrial gravity pull downward.

1. *Hydrostatic pressure* is exercised by the groundwater column. It (1) increases with depth, (2) is proportional to the water density, and (3) is effective only within zones of hydraulic connectivity; it is cut off by effective hydraulic seals.
2. *Lithostatic pressure* exerted by the overlying rock column. It (1) increases with depth and (2) is proportional to the rocks' density.
3. *Pressure induced by volume increase due to petroleum genesis*, e.g., in the formation of gases and oil.
4. *Tectonically induced compression or dilatation.*

Hydrostatic pressure prevails within the zone of groundwater through-flow. It is implemented by the column of water within the rock system. Hence, it can be effective only in zones of hydraulic connectivity, e.g., the zone of groundwater through-flow. The latter is effective within the landscape, practically as deep as the sea level, which acts as the terminal base of drainage.

Lithostatic compaction is active everywhere. However, in the zone of active groundwater through-flow, dissolution constantly opens new flow paths that lead to the active base of drainage, e.g., the ocean surface. Hence, hydrostatic pressure is dominating in the through-flow zone.

Rock-compartments beneath the through-flow zone are lithostatically pressurized. Beneath the through-flow zone, within the no-flow static zone (of zero hydraulic potential) the structure of hydraulically isolated rock-compartments prevails. These closed compartments are hydraulically disconnected from the overlying through-flow zone, and hence no hydrostatic pressure from the overlying water column can be exercised on them. Thus the rocks of closed compartments are exposed to lithostatic pressure exerted by the overlying rock strata.

Fluids hosted inside closed rock-compartments can be under a range of pressures, up to full lithostatic. Within a hydraulically isolated rock-compartment built by noncompetent rocks, e.g., noncemented sandstone or conglomerate, the included fluids are under full lithostatic pressure. However, competent rocks like limestone, dolomite, or crystalline rocks, partially resist compaction-induced collapse, and the included fluids are in such cases "sheltered" and under a pressure that is lower than the full lithostatic one.

Highest possible pressure of entrapped fluids is lithostatic; above it fluid migration is activated. Once pressure of an entrapped fluid exceeds the respective lithological pressure, the rock-compartment lifts the overlying rocks, or ruptures in certain cases, wherein fluids are squeezed out until the lithostatic pressure reseals the compartment. Fluids migrate upward along preferred flow paths.

Compaction strengthens the sealing efficiency of the rock-compartment. Plastic rocks, like clay, are impermeable due to lack of interconnection between the water-containing pores. During subsidenc, pressure exerted by the overlying rocks closes microvoids in the clay beds, improving the impermeability, a process that is especially efficient thanks to the elevated temperature.

Pressure is lithostatic in closed compartments containing fossil water. Almost as a rule, artesian waters are fossil, as is indicated by their age (practically devoid of ^{14}C, containing measurable radiogenic ^4He); their warm temperature (reflecting their deep burial, beneath the through-flow zone); and their greater salinity compared to common recent recharged water, i.e., they were formed under different environmental conditions. Such fossil waters are disconnected from recent ongoing recharge, and hence they are entrapped within hydraulically closed rock-compartments.

As the density of rocks is more than twice the density of water, lithostatic pressure is always enough to push upward entrapped water from any depth to operate artesian wells, the wells providing the freeflow path.

A key conclusion: Pressure differences within the zero hydraulic potential zone disclose fossil water (or petroleum) entrapment in rock-compartments separated by effective impermeable rock barriers.

7.4.3 Included Water Prevents Collapse of Rocks as a Result of Compaction

Dry rocks would collapse under the load of overlying rocks, causing the volume of open interconnected voids to be significantly reduced and the permeability to be drastically reduced as well.

Observation: Many wells with high water or oil yields tap deep rock units. *Conclusion*: A high permeability has been preserved in all these cases. *Discussion and hypothesis*: The fluids stored in the rock voids supported the rock particles and prevented collapse. In these cases the included fluids are pressurized—up to full lithostatic pressure.

7.4.4 This Maintained the Hydraulic Connectivity that was Essential in Petroleum Migration into Traps that Make Up Exploitable Deposits

The preservation of water within interconnected voids, even in deeply buried rock units, provided the migration paths through which petroleum formed in a rock-compartment could migrate and be concentrated (this topic is discussed in Chapter 11).

7.5 The Fruitcake Structure of the Formation Waters and Petroleum-Containing Geosystem

Rock-compartments containing different types of formation waters are observed to be the rule in the confined systems that underlie the unconfined groundwater through flow zone. This is evident in the case studies discussed so far, and many more examples are discussed in Chapter 10. The useful criterion is the presence of formation waters that have different salinities, different relative ionic abundances, and different isotopic compositions.

Pressure compartments have been identified by petroleum explorers. So far we have identified hydraulically isolated rock-compartments by the intensive use of formation waters' chemical and isotopic compositions. Petroleum experts recognized compartmentalization by the use of an entirely different parameter—the pressure of entrapped fluids. The pioneering approach has been presented in a special book, *Basin Compartment and Seals* (Ortoleva, 1994b). This book contains 30 papers that deal with various aspects and case histories of formation water and petroleum setups that reveal pressure compartments of various scales, common in sedimentary basins.

The spatial arrangement of rock-compartments of different sizes, varying lithologies, and diverse fluid contents resembles, in a picturesque way, a fruitcake. The different fruit parcels are in our context different water types and oil and gas deposits that are hydraulically separated from each other.

A dense network of impermeable barriers is an integral element of the fruitcake structure. The impermeable rock barriers that engulf the permeable rock-compartments and separate them from each other make up a network that is an integral part of the observed fruitcake structure.

7.6 A Brief History of the Basic Concept of Connate Groundwater

The following are a few quotations from a paper by Lane, published in 1908, based on his observations of mine waters at Michigan:

> Downward working rain water percolates almost pure into the ground from the surface; but, starting as phreatic water practically pure or with but a few parts per million, it dissolves and gradually accumulates mineral content. Even in the first few inches of the soil it becomes much richer in gases, and its capacity to attack feldspars and other rock minerals increases.... Its content of carbonates, bicarbonates, perhaps also sulfates soon rises, especially in arid regions.

After defining recent meteoric water, Lane continued:

> Quite different may be the water buried with the beds in the first place, to which we may fitly apply the adjective connate. Most beds are laid down in water which may be either salt or fresh, and may indeed be quite fairly fresh close to the ocean. Such waters have to begin with some—perhaps much—mineral content.

And latter on he adds:

> Synonym of connate might be singenetic...Meteoric waters are dividable into (a) rain, vadose or pluvial waters, coming down from above, (b) buried or connate waters, which will be very different, according as they were originally marine or fresh.

Lane's almost century-old hypothesis was by and large ignored, mainly in light of theoretical and generalized mathematical water flow models that were detached from field observations. The term was occasionally applied to described old groundwaters in general. Since then our knowing and understanding the groundwater systems advanced substantially, and there is good evidence that a large portion of the deep waters encountered beneath the zone of active groundwater flow are indeed connate, i.e., entrapped in their host rocks since the latter were confined by overlying rocks. The hydraulically isolated rock-compartments contain connate formation waters.

7.7 The Bottom Line: Brine-Spray-Tagged Formation Waters Provide Markers of Paleo-Landscapes, Water Age, and Paleoclimate

The evidence for fossil brine-spray-tagged meteoric groundwaters is overwhelming, to a degree that "the quantity creates a quality." Recent publications rapidly expand the list of reported case studies whose data indicate existence of fossil brine-spray-fed groundwaters. In this way evidence is building up that during certain geological periods evaporitic conditions predominated over large sections of certain continents. This has, in turn, far-reaching implications in regard to the general geology; the respective terrains must have been large-scale flat lowlands, in the reach of frequent alternations of sea regressions and transgressions.

Low flatlands that border the ocean, and are characterized by a large-scale evaporitic environment, are scarce on the present Earth. Or, in other words, the present continents have a relatively elevated landscape. This is the cause of the general observation that all currently formed groundwaters are tagged by sea spray. In contrast, during long chapters of the geological record large parts of the continents had a low relief, and evaporitic conditions prevailed and the related groundwaters were brine-tagged. Hence it is quite safe to conclude that every groundwater that is found to bear a brine-spray imprint is fossil and old, most probably beyond the dating range of ^{14}C and often in the range of ^{36}Cl or ^{4}He dating, a topic further discussed in the next chapter.

The paleoclimate at which the preserved formation waters were formed must have been rather arid. This is borne out from the observation that these are meteoric groundwaters and yet they are saline, resembling the saline water described, for example, in the case study of the arid Wheatbelt of eastern Australia (section 4.5). This conclusion, that an arid climate prevailed, is in accord with the conclusion that evaporitic conditions prevailed during the evolution of the subsided sedimentary basins.

7.8 Solving a Great Puzzle: Why Are Recent Groundwaters Sea Tagged and Commonly Rather Fresh, Whereas Formation Waters Are By and Large Saline and Brine Tagged?

7.8.1 Sedimentary Rocks Accumulated Preferentially in Marginal Oceans

Observation: Our Earth experienced stages of orogenic mountain building that alternated with intermediate stages of orogenic tranquility. During the latter, erosion lowered the landscape, resulting in large-scale flat lowlands. *Second observation*: Sea sediments get quantitatively accumulated at the ocean part that is close to a continent, because the raw material from which marine sediments are formed is mainly of continental origin—the products of erosion. *Third observation*: Continental sediments, e.g., soils, calcretes, or conglomerates, are washed into the ocean; they rarely accumulate on the land.

Conclusion: The thousands of meters of sedimentary rocks, in which formation waters are encountered, were not accumulated on ancient continents but in marginal marine-subsiding basins.

7.8.2 The Thick Sedimentary Rock Assemblages Disclose Evaporitic Environments, Frequented by Sea Invasion and Retreat Over Large Paleo-Regions

Observation: The rocks filling the sedimentary basins are rich in clay, mud, sandstone, and dolomite, as well as gypsum and halite (Chapter 6). These are by and large rocks that are formed on flat evaporitic lowlands. Mud, sand, and evaporites are formed in flatlands that host sabkhas and lagoons and are frequented by sea transgressions and regressions. *Second observation*: The named rock types are stacked on each other, disclosing frequent and rather abrupt changes in the environment of the sediment-accumulating system, i.e., frequent changes in the depth of the sea and common alternations between subacrial exposure and a submarine environment. *Third observation*: Many of the sedimentary basins are of lateral extent of hundreds to thousands of kilometers.

Conclusion: Sedimentary rocks accumulated in sedimentary basins, which evolved in paleo-landscapes of large flat lowlands of evaporitic facies. Hence, practically all the preserved fossil groundwaters are brine tagged.

7.8.3 The Rocks Recorded the Transgressive Phases, and the Entrapped Formation Waters Recorded the Sea Regressions

The geological evidence provided by rocks is biased toward the marine facies, as marine rocks are well preserved. Phases of sea retreat are poorly recorded because hiatuses, or unconformities, are easily overlooked. At this point the meteoric brine-tagged formation waters are of great importance; they clearly disclose frequent phases of exposed land. In other words, the brine-tagged formation waters complement the geological record provided by the rocks.

7.8.4 The Answer to the Puzzle

At present Earth's landscape is rather mountainous, an echo of young orogenic activity, and flat evaporitic lowlands are scare. In such mountainous landscapes the open ocean provides the spray, transported by the atmosphere inland, and the result is the commonly observed sea-tagged recent groundwaters (Chapter 4). In elevated landscapes all the groundwater is drained and recycled into the oceans, and hence it is in most cases fresh. Such regimes were common during previous mountain building periods as well, but practically no terrestrial sediments were preserved of these periods, and hence also no related sea-tagged waters were confined and preserved.

In contrast, the thick sedimentary rock assemblages were accumulated and preserved along with their contained paleo-groundwaters in low flatlands of evaporitic environments, frequented by sea invasions and retreats over large paleo-regions, as concluded in the previous sections. Hence the preserved formation waters were as a rule saline and tagged by brine-spray.

7.9 Summary Exercises

Exercise 7.1: The following terms are applied in the text of the last chapters: *groundwater, formation water, interstitial water*, and *connate water*. Are they synonymous, or has each term its specific meaning? Try to define these terms.

Exercise 7.2: Compare the composition fingerprint diagrams of formation waters from the Upper Silesian Coal Basin in Poland (Fig. 7.10) to those from the Niagara Falls area, New York (Fig. 7.5). The waters reveal a variety of compositions, yet can you identify a general similarity, a general pattern? What can be learned from it?

Exercise 7.3: Let us compare the structure of the depth profile of the interstitial waters extracted from cores in Hole 533 of the Deep Sea Drilling Project (Fig. 5.1) and the depth profile of formation waters encountered at the Alberta Basin (Fig. 7.4). What is the commonality? What does it indicate?

Exercise 7.4: How were the remarkably different types of formation water introduced into the distinct rock-compartments?

Exercise 7.5: The following weight ratios of Cl/Br have been applied in the discussions: > 3000, 293, and 80 to 200. Which natural system does each describe?

Exercise 7.6: What care has to be taken in collecting formation water samples for laboratory analyses?

Exercise 7.7: Which parameters should be analyzed in water samples collected in drillings and wells reaching formation waters?

8
FOSSIL FORMATION WATERS RANGE IN AGE FROM TENS OF THOUSANDS TO HUNDREDS OF MILLIONS OF YEARS

8.1 Confinement Ages of Connate Waters and Criteria to Check Them

The age of connate water is the age of confinement, i.e., the age of the directly overlying rock. Thus the resolution by which the confinement age can be determined is a function of the stratigraphic resolution of the respective rocks. The stratigraphic column is well preserved and continuous in many of the basins of the world that host connate waters, and in these cases the confinement age of the connate waters is compatible with the age of the host rocks.

The confinement age of connate waters is semiquantitative, yet most important for the understanding of groundwater systems and the management of the resource.

A working hypothesis is sound if it includes criteria to check it in individual study sites. In the present case the check is provided by independent isotopic dating of the waters.

8.2 Isotopic Dating of Fossil Groundwaters

The dating of groundwater was briefly discussed in section 3.3.5, addressing the following methods: tritium, carbon-14, chlorine-36, helium-4, and argon-40. Let us now discuss the application of the isotopic dating methods to formation waters, which by their confinement ages turn out to have an age range from Cambrian to recent.

In contrast to the isotopic dating of rocks, the dating of groundwaters is semiquantitative, yet of prime importance. Isotopic dating is well established for whole rocks and minerals, being based on the measured concentrations of a radioactive parent isotope and the respective radiogenic daughter isotope in a given sample and knowledge of the half-life. The deduced age is the actual age if a set of conditions has been met by the studied system, e.g., the rock or mineral had no initial concentration of the daughter isotope and the system was all the time closed. Agreeing ages obtained for a rock sample by applying different parent–daughter pairs confirm the calculated age as well as the basic assumptions. Disagreeing ages provide a warning that the calculated ages are not applicable, and the pattern of the disagreements provides an insight into geological processes, e.g., heating events.

The arena of isotopic dating of groundwater is less clear. There is basically only one measured parameter—the concentration of the radioactive parent isotope left in the water in the case of the tritium, ^{14}C, and ^{36}Cl dating methods or the amount of accumulated daughter isotope in the case of the ^{4}He and ^{40}Ar age indicators. The initial concentration of tritium, ^{14}C, and ^{36}Cl produced by cosmic radiation interacting with the atmosphere is assessed in each case from general surveys of the concentration in precipitation. The concentration of U and Th in the host rocks, needed to calculate ^{4}He and ^{40}Ar ages, is measured in drilled rock samples or estimated for the inferred type of aquifer rock. The cosmic ray–produced nuclei were also introduced by the nuclear bomb tests, and ^{36}Cl is also formed in situ within the rock system.

Does the above portrayed picture convey the message that dating of groundwater is impossible? No! It is true that water cannot be dated by single parameter measurements, but dating can be meaningful if it is part of comprehensive studies that aim at the *understanding* of studied groundwater systems and include a wide range of physical, chemical, and isotopic measurements, along with geological and hydrological observations.

Isotopic dating establishes the age in terms of orders of magnitude. The mentioned isotopic methods provide *semiquantitative* water ages, expressible as orders of magnitude of years, i.e., 10^1, 10^2, 10^3, 10^4, 10^5, 10^6, 10^7, or 10^8 years. The hydrologically relevant orders of magnitude of ages were fully spelled out in the previous sentence in order to illustrate the vast range of water ages encountered in sedimentary basins. In every study area there occur groundwaters of a wide range of ages, the youngest ones being present in the unconfined through-flow system and the older ones within deeper static confined water systems. In certain cases knowledge of the isotopic age is the key information in the identification of the nature of a studied groundwater system, and in other cases it provides an essential check for conclusions based on other observations or measured parameters.

Each water dating method has its built-in uncertainties and intrinsic shortcomings, leading to definable practical dating ranges. The latter are semiquantitative yet reliable and extremely helpful in hydrological studies, topics that are briefly discussed in the following sections.

8.2.1 Highest He Concentration Measured in a Water System is Closest to the Indigenous Concentration

Several processes tend to interfere with ideal sampling for He determinations. Loss of gases as a result of pressure reduction during ascent in a pumped well is especially important in cases of water that is rich in other gases that escape and serve as carriers of the He. This process can be checked by simultaneous measurement of the atmospheric noble gases ^{36}Ar, Kr, and Xe. If they occur in concentrations expected for meteoric recharge water (e.g, at an ambient recharge temperature of 10°C), then little or no gases were lost (section 3.3.4). Often lower concentrations of the atmospheric noble gases are observed, indicating gas was lost, and a correction factor can be retrieved. This may be illustrated by an extreme example from well No. 358 in the Great Artesian Basin, Australia. The water in this self-flowing well has a temperature of 87°C and is rich in CO_2 and methane. Only 16% of the atmospheric noble gases were found in a collected sample, indicating that the He was lost at least at this rate, and the measured concentrations could be corrected accordingly (Mazor and Bosch, 1992).

Another common interference is intermixing with shallower young water within boreholes. This may be recognized by simultaneous measurements of different age indicators, of both short and long dating ranges.

Thus, ^{4}He measurements tend to be lower than the indigenous values. Therefore, when different results are obtained for duplicate samples, or samples collected from different wells tapping the same water system, the highest value will be closest to the indigenous one, not the average value.

8.2.2 Ages at the Limit of a Method Can Be Much Higher

Ages at the limit of an applied dating method warrant special caution, especially if no other age indicators have been measured. If the measured tritium is 1 TU, the age is not "a few decades" but more than a few decade, and that can mean any older age, even very old. The literature is full of examples of calculated ^{14}C ages of 30,000 to even 40,000 years. In a realistic way all these ages are essentially >30,000 years, and it has to be borne in mind that the real age may be greater by orders of magnitude. Hence, a result reported in the literature as a ^{14}C age of ≥30,000 years and a ^{4}He age of 7×10^7 years is absolutely possible, the old age being a minimal age of the old water end-member.

8.2.3 Waters Trapped in Subsided Host Rocks Have Ages Increasing with Depth, a Pattern Confirmed by ^{4}He Based Ages

Different water types could be introduced into each rock unit only when the latter was exposed at the land surface and acted as an unconfined groundwater through-flow system. The water of each rock unit was in many cases retained upon subsidence, as is disclosed by the "packing" of very different water types throughout sedimentary basins. Hence the age of the water is in such cases between the age of the host rock and the age of the confining rocks of the overlying rock strata. Thus the age of the groundwaters entrapped within sedimentary basins is by and large expected to increase with depth. This pattern has been confirmed by surveys of available He concentrations in water samples from different depths in a number of study areas (Mazor and Bosch, 1990, 1992).

8.3 Hydraulic Age Calculations—An Erroneous Approach to Confined Groundwaters, Which Are Static

8.3.1 Hydraulic Age Calculations Assume the Relevant Groundwater Is Flowing, But Formation Waters Are Entrapped and Static

Hydraulic calculations of groundwater ages are based on Darcian-type equations that are applied to calculate groundwater flow velocities, incorporating the hydraulic conductivity and suggested flow gradient of the modeled systems. This calculated flow velocity is applied to calculate travel times by dividing the distance between two studied points of a system by the calculated flow velocity. The travel time between a suggested zone of recharge and a studied well, situated along a suggested downflow direction, is applied to calculate groundwater ages.

Thus the hydraulic age calculation approach is based on the assumption that all groundwaters flow. Therefore, Darcian-type equations are intrinsically not applicable to confined groundwater systems with zero flow velocity.

Hence the application of hydraulic water age calculations to confined groundwater systems is erroneous and provides misleadingly young ages (Mazor and Nativ, 1992, 1994). In contrast, confined waters provide ideal samples for isotopic dating by ^{4}He and ^{40}Ar, the accumulation of these noble gases being directly proportional to the water age.

8.3.2 Crustal Helium Diffusion Has Been Recruited to Explain He Ages that Are Higher than Respective Modeled Hydraulic Ages

Groundwater He ages that were obtained in the pioneer studies soon revealed a repeated pattern: they were higher than respective hydraulic ages that had been previously computed on the assumption of basin-wide through-flow. Several of the pioneers of He dating assumed, in an axiomatic way, that the hydraulic ages are irrefutable and searched for a process that could perturb the He age calculations. The result was the hypothesis that groundwater gets significant contributions from He that is diffusing from the whole crust, swamping groundwater with much more He than that acquired from the host rocks. The reliance on the new theory was so great that crustal He diffusion fluxes were computed on the basis of the assumed hydraulic ages and the observed He concentrations (Andrews and Lee, 1979; Andrews et al., 1985, 1989; Heaton, 1984; Torgersen and Clarke, 1985).

Observations disproving that crustal He diffusion interferes with the dating of formation waters. The crustal He diffusion theory questioned the applicability of ^{4}He for groundwater dating, but the theory itself seems to be basically erroneous for the following reasons:

1. Water has a high capacity to dissolve He. The saturation value of He in groundwater has been calculated to be in the range of 0.2 to 1.0 cc STP/cc, for pressures prevailing in various basins. These values are two to eight orders of magnitude higher than the He concentrations observed in groundwaters, indicating that in no case could He losses occur due to oversaturation (Mazor and Bosch, 1990, 1992). Hence it is concluded that water acts as a He sink, well preserving the dissolved radiogenic helium.
2. Groundwater is encountered in all rocks so far probed by drill holes, i.e., down to over 7000 m. Thus the water in each rock layer dissolves all the radiogenic He produced by the associated rocks, shielding the water stored in the overlying rocks from any flux that would be formed by He diffusion.
3. Gas diffusion through water could be efficient over long geological periods, but the fruitcake structure of formation waters stored in hydraulically isolated rock-compartments reveals that hydraulic connectivity needed for the diffusion of mantle He does not exist.
4. Groundwater has been observed to have He concentrations from 5×10^{-8} up to $590,000 \times 10^{-8}$ cc STP/cc (Mazor and Bosch, 1990, 1992). This enormous range demonstrates that water has the ca-

pacity to preserve He as much as it dissolves from the associated rocks, the observed concentrations being correlated to the water age. The highest concentrations were observed in waters stored in Paleozoic rocks (Mazor and Bosch, 1992).
5. Helium concentrations are observed to vary abruptly between adjacent rock-compartments in sedimentary basins (Mazor and Bosch, 1992) and fracture-compartments within crystalline shields (Bottomley et al., 1990).
6. Hydraulic calculations of water age are misleading if applied to nonflowing water, e.g., connate water entrapped in a hydraulically isolated rock-compartment. After all, how can a flow velocity be applied to static water?!

Conclusion: ^4He *dating is valid and meaningful.* On the basis of the mentioned saturation limit calculations, and the listed observations, it is concluded that water acts as an efficient helium sink. Hence the ages calculated by the original equation, assuming all released radiogenic ^4He is dissolved in the water that is associated with the rocks, are valid and meaningful (Mazor, 2003). This point can in many cases be checked by comparison to the deduced confinement age, i.e., an age that equals the age of the immediately overlying confining rock formation, in cases that comply with a list of criteria identifying connate waters.

8.4 Radiogenic ^{40}Ar Dating

Potassium-40, the rare radioactive isotope of potassium, decays into ^{40}Ar with a half-life of 1.3×10^9 years, and an age equation, similar to the one given in section 3.3.5 for ^4He, can be formulated. The water age is a function of the measured concentration of radiogenic ^{40}Ar, the concentration of ^{40}K in the reservoir rocks (a few ppm), the emanation coefficient (~0.5), the rock density, and the water/rock ratio. The concentration of dissolved radiogenic ^{40}Ar is calculated via the measured $^{40}Ar/^{36}Ar$ ratio: The value of atmospheric argon is 295.6, and any excess over this figure is attributed to additions of radiogenic argon. In systems with significant contributions of mantle-derived gases a correction for mantle ^{40}Ar has to be applied, based on the measured $^3He/^4He$ ratio.

Because of the rather significant concentration of atmospheric argon, the presence of a radiogenic component is measurable only in waters of an age of 10^5 years and older. Thus the ^{40}Ar dating of groundwater is efficient in the range of 10^5 to 10^8 years. Water dating by ^{40}Ar is highly recommended for very old groundwaters. The presence of significant concentrations of this age indicator supports deduced high ^4He water ages.

The development of the ^{40}Ar water dating methodology is especially important in light of the mentioned controversy around the crustal diffusion of He, postulated by various investigators to interfere with ^{4}He dating of formation waters (section 8.3). Argon has a significantly larger atom than helium, making its diffusion through materials orders of magnitude smaller than that of helium. Thus if crustal He diffusion is important, observed ^{4}He water ages will have to be orders of magnitude larger than the calculated ^{40}Ar-based ages. On the other hand, agreement of ^{4}He and ^{40}Ar water ages provides independent evidence that crustal He is negligible and in no way interferes with ^{4}He water dating.

The ^{40}Ar dating method shares with the ^{4}He method the great advantage of being formed only by one process—the radioactive decay of ^{40}K in the host rocks, with no formation by cosmic radiation and with no bomb production. The ^{40}Ar dating method works as a simple cumulative clock.

8.5 Mixed Water Samples Are Commonly Encountered

8.5.1 The Problem of Mixed Samples

All isotopic water dating methods are based on the assumption that the measured values pertain to a single component of water. Whenever this is the case, agreeing ages are obtained by different methods that are simultaneously applied. However, in reality many of the studied water samples are mixtures of waters of different ages, and this point warrants discussion.

A borehole collects water from the rocks as it provides free space in which to flow, and pumping creates constantly new free space for flow toward the well hole. Whenever a borehole intercepts several aquifers, various water-carrying fractures, or different rock-compartments, the different waters enter the hole, and when sampled a mixture of different waters is obtained. Furthermore, short-circuiting by boreholes can cause migration of water between systems of different pressures. The problem exists to some extent even for samples collected from packered intervals of a borehole.

The common mixing of young and old water samples is borne out, for example, by tritium measurements. In many of the cases samples obtained from deep boreholes did contain measurable concentrations of tritium. Mixing is evident in such cases whenever the chemical composition of the water differs significantly from that of the recent local groundwater, indicating the presence of a fossil groundwater component. This situation emphasizes the absolute need to include tritium (and/or other anthropogenic contaminants) in the analyses made for every sample in groundwater studies.

Absence of measurable tritium and of other anthropogenic pollutants indicates no recent water or drilling fluids were introduced into the sample, but mixing of different older waters is still possible.

8.5.2 Pattern of Calculated Ages in the Case of Mixed Samples

Young water has high concentrations of short-range age indicators, e.g., high tritium and ^{14}C concentrations; old water is devoid of the named age indicators, but has a high concentration of long-range indicators such as ^{4}He. Hence a mixed sample of young and old waters contains measurable concentrations of both short-range and long-range age indicators. If the concentration of each isotopic age indicator is applied to calculate an age in the case of a mixed water sample, the following pattern will always emerge:

$$\text{tritium age} < {}^{14}C \text{ age} < {}^{36}Cl \text{ age} < {}^{4}He \text{ age}$$

That is, the age calculated using the measured tritium will be orders of magnitude younger than the age calculated by ^{4}He.

Thus disagreeing ages of the discussed pattern (1) provide a clear indication of mixing, and hence (2) none of the obtained ages has a straightforward meaning and (3) the conclusion that the sample is a mixture has to be taken into account in the interpretation of the results obtained for all the other measured parameters.

8.5.3 Natural Mixing of Waters of Different Ages in Certain Spring Systems

Springs emerging at fault planes may be fed by several aquifers. An example is one of the Hammat Gader springs in Israel that contained 6 TU, along with 7 pmc, clearly a mixture a post-1952 water and an old water. The measured ^{14}C was probably all brought in by the recent water along with the tritium, so that the old end-member contains no ^{14}C-14 and has an age beyond the limit of the ^{14}C method (Mazor et al., 1973a,b). This can be checked by the presence of radiogenic helium.

As this mixing in springs is a natural phenomenon, it is important to identify it as part of the endeavor to understand studied systems.

8.5.4 Retrieval of the Age of the Young and Old End-Members in Mixed Samples

In the most straightforward way, the age of the youngest water end-member in a mixed sample is younger than the age calculated from the concentration of the short-range age indicator, whereas the age of the old end-member is older than the age calculated from the concentration of the longest-range age indicator. In a second stage some mutual corrections can be made, including the following:

1. In a sample from the Lodgepole Warm Spring, Montana, 31.8 TU was found along with 28 pmc (Plummer et al., 1990), a clear case of mixing of recent and old waters, in this case naturally. The recent end-member contains ^{14}C along with the tritium, thus it is plausible that all the observed 28 pmc was brought into the mixture with the recent water, and the old end-member is devoid of ^{14}C. This assigns the old end-member an age of >30000 years, and, as mentioned before, this leaves open the possibility of a very high age.

   A second example of this nature is one of the Feshcha springs, on the Dead Sea shore, Israel, that was observed to contain 34 TU along with 33 pmc—again the old end-member seems to be devoid of ^{14}C, and hence of an age >30,000 years (Mazor and Molcho, 1972).

2. In a case of no measurable tritium, ^{14}C in a concentration of 30 pmc, and a significant concentration of ^{4}He, a water of an age of 10^3 years was mixed with a much older water.

These examples reveal the type of semiquantitative considerations that may help to estimate the concentration of the age indicators in the end-members of intermixed water samples.

8.6 Isotopic Dating of Very Old Groundwaters

8.6.1 Palo Duro Basin Case Study

Zaikowski et al. (1987) presented an excellent set of noble gas data for wells that penetrated to various Permian to Pennsylvanian rocks in the Palo Duro Basin, Texas. Their data, including $^{40}Ar/^{36}Ar$ ratios and radiogenic ^{40}Ar concentrations, are reproduced in Table 8.1, along with dates calculated from their data. The following pattern emerges:

1. The nine samples, from depths of 950 to 2500 m, revealed ^{4}He ages of 60 to 400 million years, i.e., in excellent accord (in terms of orders of magnitude) with the age of the rocks, namely, Pennsylvanian and Permian.
2. The high ages are well supported by very high $^{40}Ar/^{36}Ar$ ratios, of 350 to 2090 (compared to the atmospheric value of 295.6).

Age of Fossil Formation Waters

Table 8.1 Palo Duro Basin Radiogenic Helium and Argon and He Age

Well and formation (zone in parentheses)	Depth (m)[a]	He_{sat} $\times 10^{-5}$ (cc STP/cc)[b]	He $\times 10^{-5}$ (cc STP/cc)[c]	He age $\times 10^6$ (years)[d]	$^{40}Ar/^{36}Ar$	^{40}Ar $\times 10^{-5}$ (cc STP/cc)
Sawyer I						
Granite wash (4)	1,300	50,000	179	60	1,560	38
Wolfcamp (5)	950	40,000	223	200	430	19
Mansfield I						
Wolfcamp (2)			≥155	≥100	788	≤43
Zeeck I						
Pen. C. (1)	2,200	20,000	243	200	498	5
Wolfcamp (2)	1,700	70,000	≤92	80	353	4
Wolfcamp (3)	1,700	70,000	367	300	569	16
J. Friemel						
Granite wash (1)	2,500	100,000	514	200	2,090	61
Granite wash. (4)	2,400	100,000	589	200	1,870	89
Pen. C. (6)	2,200	20,000	423	400	1,123	47
Wolfcamp (7)	1,800	70,000	473	400	954	53

[a] Read from Fig. 4 of Zaikowski et al. (1987).
[b] Saturation concentration of helium in water at reservoir conditions (Mazor and Bosch, 1990).
[c] Data from Zaikowski et al. (1987).
[d] Applying 2.3 ppm U and 0.5 ppm Th for the Wolfcamp and Pennsylvanian carbonate rocks and 3 ppm U and 12 ppm Th for the Granite Wash rocks (Zaikowski et al., 1987); 6% porosity, and density of 2.6 g/cc.

3. The observed He concentrations were four orders of magnitude smaller than the calculated He saturation concentration, indicating there was no He escape or outward diffusion (section 8.3).

8.6.2 Paris Basin Case Study

Marty et al. (1988) presented a detailed study of helium concentrations measured in 35 low-enthalpy geothermal wells that tap different limestone beds of the Jurassic rock assemblage of the Paris Basin. The researchers calculated an age of 1.5×10^8 years for the well with the highest He concentration, but commented that it is much higher than a previously calculated hydraulic age, based on a basin-wide through-flow model. So they preferred to apply for the age calculation not the highest measured He value, but the average value, reducing the age to 7×10^7 years. In a later publication Marty et al. (1993) recalculated the age, assuming much of the observed He came from crustal diffusion, and reduced the age drastically to only 4×10^6 years.

This is a classic example of the above-discussed controversy that arose from applying hydraulic flow equations to static formation waters (section 8.3). In light of the preceding discussion (sections 8.2 and 8.3) it is suggested that the original age calculation made by Marty et al. (1988) provides the correct water age, namely, $\sim 1.5 \times 10^8$ years, based on the highest measured He concentration. This is borne out by ample observations that reveal a structure of water contained in hydraulically isolated rock-compartments, as is seen from the following points, taken from Marty et al. (1988):

1. The Paris Basin is large, about 350 km across.
2. The exploited rock sequence is complex and has a large number of thin producing limestone beds, separated by shales.
3. The waters are saline, the TDS concentrations vary from 6.4 to 35 g/L, and hence these are fossil waters.
4. These are deep waters, tapped at a depth interval of 1600 to 1850 m, i.e., deep in the zone of zero hydraulic potential, which hosts static formation waters enclosed in isolated rock-compartments.
5. The general chemistry is diversified. Hence, these waters are trapped in isolated rock-compartments and no through-flow takes place, making the calculation of a hydraulic age meaningless (Mazor and Nativ, 1992, 1994).

The bottom line: *the ^4He age agrees with the confinement age.* The He-based age, on the order of $\sim 1.5 \times 10^8$ years, is of the same order of magnitude as the connate water age, the host rocks having a Triassic age.

8.6.3 Milk River Formation Case Study

Groundwaters encountered in the Cretaceous Milk River Formation have been found to contain high ^{36}Cl concentrations in an outcrop region of Montana, indicating a post-bomb age within an active recharge system. However, significantly lower concentrations were found in the confined part of the Milk River Formation rocks in Alberta (Phillips et al., 1986). These lower concentrations have no geographical pattern that may indicate a flow direction, and the observed values may well be in the range of the in situ ^{36}Cl formation (section 3.3.5), i.e., beyond the limit of this dating method. Thus an age of $>0.6 \times 10^6$ years has been concluded (Mazor, 1992a), which leaves open the possibility that these are trapped waters of a significantly higher age.

8.7 Conclusions and Management Implications

1. The isotopic groundwater dating methods based on tritium, ^{14}C, ^{36}Cl, and ^{4}He are semiquantitative and provide reliable and meaningful orders of magnitude of ages.
2. A sample that is a mixture of water of different ages reveals the following age pattern:

$$\text{tritium age} < {}^{14}C \text{ age} < {}^{36}Cl \text{ age} < {}^{4}He \text{ age}$$

   In this way mixtures can be identified, natural in the case of springs, and manmade in the case of boreholes and wells.
   Hence, prior to any age computation, every set of data has to be checked for possible mixture of waters of different ages. This requires measurement of more than one age indicator and preferably all four. In the case of a mixture, the concentration of the age indicators in the end-members have to be determined in order to estimate the ages of the intermixed end-members.
3. Each dating method caters for a certain age range. Very small concentrations of an age indicator may stem from contamination or be at the limit of detection. Hence small concentrations mark the limit of a dating method.
4. Age determinations have to be regarded as an integral part of a wide approach to groundwater studies, which have to include hydrological, geological, physical, chemical, and stable isotope aspects.
5. Helium-4 dating is suitable for the dating of very old groundwaters, and it has the enormous advantage of He-4 being formed by a single process—the radioactive decay of uranium and thorium in the host rocks. Its measurement should always be accompanied by the measurement of all five noble gases in order to be able to correct the measured He concentration for air trapped during sample collection or gas losses that occurred prior to or during sample collection.
6. Isotopic age determinations are independent methods, and there is no theoretical and practical ground to "make them fit" to hydraulic ages.
7. Helium-4 ages that are on the same order of magnitude as the deduced confinement ages indicate the studied water is connate.
8. In general, isotopic ages are older in deeper rock units within sedimentary basins, providing independent evidence for the gradual accumulation and subsidence history of the water-containing rocks.

9. Isotopic ages are especially tailored for the dating of water confined in rock-compartments and fracture-compartments, i.e., systems of static water of uniform age. The latter provide groundwater samples "bottled" by nature and preserved in the geological record.
10. Determination of the age of groundwater is crucial to the management of the resource, as it indicates degrees of renewability. Recent through-flowing groundwater is renewed, and old fossil formation water is not renewed. The first system is vulnerable to anthropogenic pollution, whereas the latter is disconnected from the land surface and hence immune to pollution.
11. Nuclear waste repositories seem to be best sited in rock-compartments in which very old groundwater is trapped, indicating the respective systems are well sealed.

8.8 Summary Exercises

Exercise 8.1: The following results were obtained for a sample of water, collected from the 1111-m-deep Examination borehole: tritium age of less than 50 years, ^{14}C age of 31,000 years; ^{36}C age around 10^6 years; and ^4He age of around 1.8×10^7 years. Is this a logical set of results? What does it reveal? What is the age of this groundwater?

Exercise 8.2: Artesian flow in the Petroleum Hope drilling yielded water that contained only 20% of the expected atmospheric noble gases (water recharged at $10°C$; see section 3.3.5). Can the measured ^4He concentration be used directly to calculate the age of the water? What should be done?

Exercise 8.3: When multiple water samples of the same water source yield different ^4He concentrations, how do we select the value that will provide the most reliable water age?

Exercise 8.4: Spring A contained ^{14}C in a concentration that yielded an age of 32,500 years, and well B has a calculated ^{14}C age of 38,200 years. Is the water emerging in well B really older?

Exercise 8.5: Can the age of a fossil formation water be calaculated using the hydraulic flow equation? Discuss.

Exercise 8.6: Groundwater age is essential to understand groundwater systems. Has the age also management implications?

9
BRINE-TAGGED METEORIC FORMATION WATERS ARE ALSO COMMON IN CRYSTALLINE SHIELDS: GEOLOGICAL CONCLUSIONS AND RELEVANCE TO NUCLEAR WASTE REPOSITORIES

Formation waters are well known from sedimentary basins and are as a rule encountered in petroleum exploration and production boreholes. What do we know about groundwater in granitic and metamorphic shields? These are impermeable rocks in which water can move or be stored only in fractures.

The topic is practically ignored in textbooks, but we devote to it a special chapter because of the scientific and practical interest. The formation waters provide means to map interconnected fracture systems, including fracture-compartments. Understanding the formation waters sheds light on the paleo-environments to which the shields have been exposed, and the direction of fracture propagation from the land surface downward.

Crystalline rocks are considered as sites of nuclear waste depositories, thus understanding the water flow and entrapment in crystalline rocks is essential.

9.1 The Special Nature of Data Retrieved from Boreholes in Crystalline Rocks

Hydraulic connectivity of crystalline rocks is basically different from that to which we are accustomed from sedimentary rocks; there are no voids in the structure of igneous or metamorphic rocks; and in that sense they are impermeable. The movement and storage of groundwater within crystalline rocks is limited to fractures. This specific nature of crystalline rock hydrology places new challenges to researchers attempting to understand movement of water between fracture-compartments that are interconnected or short-circuited by our drillings.

9.1.1 The Filling of Fracture-Compartments with Saline Waters

Saline water is encountered in the crystalline terrains down to depths of several kilometers in mines and repository-related boreholes, and they have been observed down to depths of around 11 and 9 km, respectively, in the crustal research boreholes in the Kola Peninsula, Russia, and the KTB hole, Germany. Any hypothesis addressing the question of how the saline waters were inserted to these depths has to take into consideration three deduced boundary conditions, discussed later on in light of field observations:

1. The discussed waters are different from local recent recharge water and hence are fossil.
2. The saline waters originated at the land surface as meteoric water.
3. They are at present trapped within hydraulically isolated fracture-compartments, each hosting a distinct water type.

Fractures are dynamic and over geological periods they change shape, dimensions, and depth. Figure 9.1a portrays an irregular fracture open from the surface down to some depth. Slight shifting of the rocks in the directions marked in Fig. 9.1b causes compression and closure of the fracture at some segments and widening of others (similar to the formation of rhomb-shaped grabens in rift systems). Thus an initial fracture can be turned into a series of isolated fracture-compartments. An open fracture filled with water at the land surface (Fig. 9.1c) can later on be turned via this mode into a series of water-containing voids that are sealed from the land surface and from each other (Fig. 9.1d). Subsequent dilatation phases can cause gradual propagation of fracture segments to greater depths, coupled with repeated closure of shallower segments at compressional locations. The entrapped water will trickle down along with the step-wise downward propagation of the fractures.

In fact, the occurrence of saline water of land surface origin serves as a tracer indicating that fracture systems did evolve from the land surface downward to the discussed depths.

9.1.2 Short-Circuiting and Mixing

In the interpretation of data from waters encountered in crystalline rocks we have to bear in mind that every borehole can cause groundwater stored in one

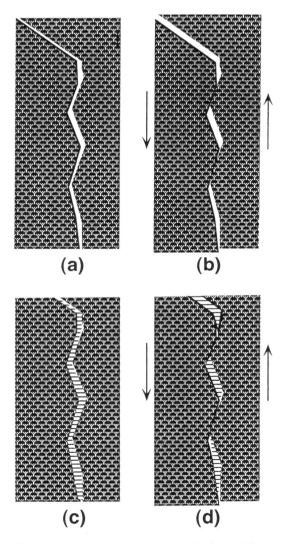

Fig. 9.1 Stages of opening and closing of fracture segments and entrapment of groundwater in isolated compartments: (a) Opening of a fracture at a stage of local dilatation. (b) Closing of segments of the fracture by sliding motion marked by the arrows. (c) A fracture filled with water from the surface. (d) Formation of hydraulically separated fracture-compartments capable of storing entrapped water over geological durations.

fracture-compartment to communicate with water stored in other compartments (Fig. 9.2). The amount of water in each system is probably small, and compaction-induced pressure may be noneffective inside open fractures, and hence it seems most likely that water may sink, or flow down, in a borehole, from shallow fracture systems to deeper ones (e.g., boreholes A, B, and C in Fig. 9.2). Occasionally the migration inside a borehole may be upward, in cases where deep waters are pressurized by lithostatic compaction or tectonic compression.

9.1.3 Occasional Exaggeration of the Depth of Groundwater Occurrence

In certain cases, a borehole may be terminated in a dry fracture-compartment, but downflow of water through the borehole may create the false impression of water being present at the greater depth (e.g., fracture-compartment 4 in Fig. 9.2). Or a borehole can be terminated at a region that has no significant

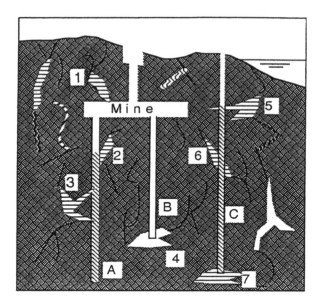

Fig. 9.2 Interception of fracture-compartments (FC) by boreholes (BH) and mines. FC 1 drains into the mine, possibly supplying a good sample; BH A gets filled by FC 2 and 3, giving the impression of presence of water down to its bottom; BH B reached empty FC 4, but the latter will soon be filled with mine drainage; FC 5, 6, and 7 are short-circuited by BH C, and samples collected in the latter will reveal a density-controlled depth profile.

fractures, but water from shallower compartments may flow down the borehole, creating the false impression of water being present at the greater depth (e.g., borehole A in Fig. 9.2).

9.1.4 Mixing with Recent Shallow Groundwater

Short-circuiting by a borehole can also introduce recent shallow groundwater into compartments hosting fossl water, so that a mixed sample is retrieved and analyzed. Such cases may be recognized by the presence of measurable tritium and/or other anthropogenic contaminants. Yet, the presence of an indigenous water end-member can be identified by specific compositional properties that significantly differ from the characteristics of the local recent groundwater.

9.1.5 Contamination by Drilling Fluids

This is a constant danger in groundwater research, but one has to be especially alert in the case of fracture-compartments, as the amount of originally contained water may be small. For this reason it seems essential that information of applied drilling fluids is provided in published case studies. Absence of drilling fluids may be assessed in various modes, e.g., whenever different water types are encountered within one borehole, at least some them are genuine. Mixing of the drilling fluid with a tracer may assist in estimating content of drilling fluid in collected samples, as reported by Bottomley et al. (1990).

9.1.6 Questions Related to Nuclear Waste Disposal Programs

Abandoned mines and specially quarried galleries in crystalline shields are considered in several countries as possible places for repositories of high-level nuclear waste. A major consideration in this regard was the observation that crystalline rocks are impermeable, in the sense that they have no interconnected voids that may serve for groundwater through-flow. Fractures in crystalline rocks were observed to be of limited extent and with poor hydraulic interconnections, so it was hoped that repositories placed within crystalline rocks would be preserved in an impermeable environment.

The saline groundwater encountered at various depths within crystalline rocks raised a number of new questions: (1) Does their presence indicate that the crystalline rocks were intruded by evaporitic brines and, if so, are they possibly open to through-flow of fluids also today? (2) Do the saline groundwaters present a danger of corroding containers, and to what extent will the corroding fluids remain buried at depth; or may they migrate in a mode that is dangerous to the environment? Thus the connectivity of fractures and the age of the encountered fluids are of prime interest, and the short-circuiting effects of boreholes and mine galleries have to be well understood. In addition, of interest is the possibility of washing-out of aline waters, prior the deposition of the wastes, in order to reduce corrosion. These topics are discussed in the following sections.

9.2 Observations Based on Data from the Fennoscandian and Canadian Shields and Deduced Boundary Conditions

The conclusions derived so far, based on a basic principles discussion, are examined herewith in light of regional data from the Fennoscandian Shield, Tables 9.1 and 9.2 (Nurmi et al., 1988), and data for the Canadian Shield, Table 9.3 (Bottomley et al., 1994). Local data for the Stripa mine, situated within the Fennoscandian Shield, are presented in Table 9.4 (Nordstrom et al., 1989a,b), and data for the Sudbury mine, situated in the Canadian Shield, are given in Table 9.5 (Frape et al., 1984). The data are plotted in Figures 9.3 to 9.8.

The reader is encouraged to take the time to have a close look at Tables 9.1–9.5 and Figures 9.3 to 9.8. Are the described formation waters really saline? Are they all the same? Can any general patterns be seen?

9.2.1 Near-Surface Intensified Rock Fracturing Related to an Active Through-Flow Zone

Nordstrom et al. (1989a) reported that at the Stripa site, Sweden, high fracture frequency and high hydraulic conductivity prevail in the granitic rocks down to a depth of 250 m, whereas at greater depths fracturing is significantly less developed. Dickin et al. (1984) reported that at the Atikokan site, the Canadian Shield, northwest Ontario, the upper 150 m of the granitic rocks reveal an interconnected network of fractures that provide secondary hydraulic conductivity in the range of 10^{-6} to 10^{-8} m/s, whereas below a depth of 300 m the hydraulic conductivity drops dramatically to less than 10^{-11} m/s. Raven et al. (1987) report that in the East Bull Lake Pluton, also in the Canadian Shield, rock fracturing is intense near the surface and decreases with depth, resulting in a deduced hydraulic conductivity of 10^{-8} m/s near the surface and less than 10^{-12} m/s at depths greater than 450 m.

This reported intensified near-surface fracturing is supportive evidence for the existence of a through-flow zone active in the shields within the landscape relief down to the terminal base of drainage at the ocean. The notable drop in fracture density at greater depths reflects the static nature of

Table 9.1 Chemical and Isotopic Composition of Waters from the Fennoscandian Shield

Name	Depth (m)	Cl (mg/L)	Br (mg/L)	Cl/Br	SO$_4$ (mg/L)	HCO$_3$ (mg/L)	Na (mg/L)	Ca (mg/L)	Mg (mg/L)	K (mg/L)	Sr (mg/L)	dD (‰)	d^{18}O (‰)	^{3}H (TU)
1. Outokumpu	270	3	0		22	51	10	14	3	3	1	−99	−14	70
	600	8,100	64	126	1	20	1,500	3,600	11	7	26	−102	−14	12
	770	8,400	67	125	1	20	1,500	3,700	13	7	27	−101	−14	14
	810	8,900	77	115	1	21	1,500	3,900	13	7	28			
	1,010	16,500	130	127	2	31	3,100	5,700	950	32	44	−75	−13	7
	1,070	16,800	140	120	2	34	3,200	5,700	1,100	32	43	−74	−13	1
2. Kerimaki	140	16	0		3	10	18	10	2	5	1	−94	−13	4
	540	2,400	23	104	0	7	1,100	340	23	13	9	−94	−13	19
	700	3,200	32	100	0	7	1,600	360	26	15	12	−95	−13	14
3. Parainen	280	70	0		41	20	71	26	9	4	1	−79	−11	53
	410	410	2	205	180	23	240	120	20	7	1	−76	−11	45
	490	3,700	20	175	370	12	1,500	720	130	37	5	−68	−10	43
4. Liminka	100	2,700	18	150	1,300	10	1,500	560	150	16	10	−106	−14	8
	400	7,400	8		2,700	17	4,000	1,600	240	24	21	−89	−13	8
	600	19,200	250	77	4,400	100	8,100	4,200	700	48	83	−84	−12	15
	720	22,100			4,100	98	8,200	5,500	760	51	110	−81	−12	11

Note: Boundaries between different water "layers" are marked by dashed lines.
Source: Data from Nurmi et al. (1988).

Table 9.2 Temperature–Depth Profiles in the Four Boreholes of the Fennoscandian Shield

Name	Depth (m)	Drilling year	Sampling year	Temperature (°C)	ΔT (°C/100 m)
Outokumpu	190	1981	1985	6.8	1.1
	365			8.5	
	530			10.3	
	700			12.5	
	860			14.7	
	1,060			16.1	
Kerimaki	190	1984	1985	6.2	1.1
	540			10.0	
	700			11.8	
Parainen	190	1980	1985	8.8	1.5
	370			11.6	
	500			13.4	
Liminka	190	1960	1985	9.0	2.1
	370			13.2	
	500			16.4	
	770			23.2	

Source: Data from Nurmi et al. (1988).

the groundwaters stored in isolated fracture-compartments, as depicted in Fig. 9.2.

A description of shallow through-flow zones that are underlied by saline water trapped in fracture-compartments is reflected in a report by Kelly et al. (1986). The investigators studied the hydrology and hydrochemistry of copper mines situated within crystalline rocks of the Keweenaw Peninsula, northern Michigan. They cite reports from the beginning of the century that in many mines "The upper 300 m were typically wet with fresh water that required much pumping. Below that the mining entered relatively dry ground in which the rocks were damp but yielded no appreciable flow. The working gradually became wetter again with increasing depth, and pockets of saline waters were locally breached in drilling or blasting. In some cases, the brines were encountered at shallower depths within as little as 100 m of the ground surface."

The saline waters within the discussed study area are similar to other shield groundwaters, as they have a TDS exceeding 200,000 mg/L, are Ca–Cl dominated, have Cl/Br ratios of around 130, and the water has a light isotopic composition of nonevaporated meteoric water.

Table 9.3 Chemical and Isotopic Composition of Waters from the Canadian Shield

Name and depth	Cl (mg/L)	Br (mg/L)	Cl/Br	SO₄ (mg/L)	HCO₃ (mg/L)	B (mg/L)	Na (mg/L)	Ca (mg/L)	Mg (mg/L)	K (mg/L)	Sr (mg/L)	δD (‰)	δ¹⁸O (‰)
1 E. Bull Lake (110 m)	3			3	123	0.70	60	1	3	1	<0.06	−94	−12.9
2 Doyon (500 m)	7	3		860	33	0.01	19	260	22	3	2	−97	−14.1
3 Creighton (380 m)	10			240	98	0.35	85	120	24	6	1	−70	−10.1
4 Campbell (640 m)	160	21	8	380	159	0.13	76	79	57	6	2	−91	−12.6
5 Macassa (915 m)	170			110		0.18	170	45	21	1	21	−92	−12.8
6 Kiena (610 m)	620	45	14	260	59	1.8	380	130	7	14	4	−89	−13.6
7 Birchtree (580)	740	10	74	250	120	0.69	320	100	24	7	3	−122	−16.3
8 Kidd (1,450 m)	1,000	6	170	3		0.42	270	140	21	2	5	−113	−15.5
9 Geco (1,220 m)	1,800	21	86	57		0.43	470	340	1	3	16	−71	−11.1
10 Portage (1,020 m)	2,000	42	48	10	110	0.03	870	450	97	13	9	−88	−11.6
11 Copper C. (730 m)	2,100	11	191	540		1.6	770	380	1	5	15	−94	−13.8
12 A. White (660 m)	2,300	20	115	2	44	0.84	420	310	110	4	15	−106	−14.7
13 Lupin (1,130 m)	7,700	66	117	290	40	0.16	2,700	1,500	170	37	55	−167	−21.3
14 Thayer L. (1,310 m)	8,300	67	124	9	15	1.7	1,800	2,200	9	17	39	−78	−11.8
15 Stanleigh (960 m)	29,000	200	145	142		12.3	8,730	5,750	700	103	250	−62	−10.1
16 Thomp. 1 (1,280 m)	42,000	360	117	830	40	0.91	5,000	13,000	880	110	300	−94	−13.4
17 Strathcona (580 m)	62,000	500	124	690	19	1.6	5,900	22,000	6	48	400	−41	−10.0
18 Stall Lake (870 m)	62,000	630	98	4	3	0.37	8,900	20,000	1,500	150	500	−63	−11.5
19 Thomp. 3 (1,070 m)	64,000	440	145	1	2	1.0	17,000	17,000	990	120	490	−106	−15.5

Note: Table arranged by ascending concentration of Cl.
Source: Data from Bottomley et al. (1994).

Table 9.4 Composition of Groundwaters Encountered in Crystalline Rocks at Stripa, Sweden

Depth (m)	Temperature (°C)	Cl (mg/L)	Br (mg/L)	Cl/Br	SO$_4$ (mg/L)	HCO$_3$ (mg/L)	Na (mg/L)	Ca (mg/L)	Mg (mg/L)	K (mg/L)	Sr (mg/L)
0–80	10.2	14			12	20	2.5	13	2.5	1.7	0.03
89–104	8.0	3.7			8.8	143	12.5	34	4.5	1.7	
355	10.4	50	0.64	78	0.5	63	30	26	0.1	0.2	0.19
765–861	10.6	630	6.5	97	102	9	277	172	0.2	1.2	1.7
908–969	8.6	460	4.5	102	57	18	218	94	0.03	0.4	0.89

Source: Data from Nordstrom et al. (1989).

Table 9.5 Composition of Groundwaters Encountered in Crystalline Rocks of the Canadian Shield at Sudbury, Canada

Depth (m)	Temperature (°C)	TDS (mg/L)	Cl (mg/L)	Br (mg/L)	Cl/Br	SO$_4$ (mg/L)	HCO$_3$ (mg/L)	Na (mg/L)	Ca (mg/L)	Mg (mg/L)	K (mg/L)	Sr (mg/L)	Tritium (TU)	δD (‰)	δ^{18}O (‰)
488		4,227	97			2,950	213	180	480	330	24	4	105	−82	−11.9
793		126	7			43	32	12	16	3	2	>0.1			
991	17.0	95,000	58,300	579	101	123	48	8,350	27,100	105	37	351	<3	−43	−10.4
1,006	19.5	18,800	11,300	78	149	530	15	2,300	4,470	2	23	72	<3	−76	−12.8
1,100		83,370	55,600	451		116	25	7,130	19,300	80	122	490	3	−73	−11.9
1,600	22.0	249,000	163,000	1,250	130	223	58	18,900	63,800	78	430	1,580	<3	−39	−10.9
1,650	21.5	255,800	168,000	1,530		213		22,000	62,000	48	320	1,700	<3	−45	−10.8

Source: Data from Frape et al. (1984).

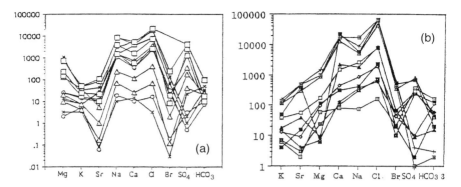

Fig. 9.3 Composition fingerprint diagrams of groundwaters from the Fennoscandian Shield (a) (Table 9.1) and the Canadian Shield (b) (Table 9.3). A pronounced variability in salinity and ionic abundance patterns is well seen.

Intense fracturing near the land surface is induced by pressure release of the overlying rocks that have been eroded away, and by diurnal and annual temperature changes coupled with ice formation within water-filled fractures.

The space between the landscape relief and the terminal base of drainage, which commonly is the sea surface, functions as an unconfined water system, and the through-flowing water removes material, thus developing a

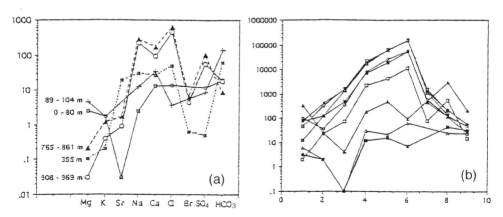

Fig. 9.4 Composition fingerprint diagrams for the crystalline rocks at the Stripa Site, Sweden (a) (Table 9.4) and the Sudbury site, the Canadian Shield (b) (Table 9.5). The large variability in salinity and ionic abundances is also conspicuous in such local scales.

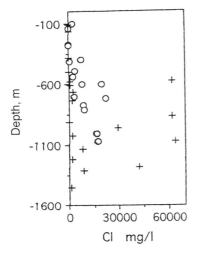

Fig. 9.5 Plot of Cl concentration versus sampling depth for the Fennoscandian Shield (O) and Canadian Shield (+). (Tables 9.1 and 9.3). No correlation is seen.

hydraulic connectivity. These processes are not active below sea-level depths, as is reflected by the reports of a drastic drop in fracture density and connectivity.

9.2.2 The Large Variety of Encountered Saline Waters Indicates Storage in Hydraulically Isolated Fracture-Compartments

Diversity of salinity and chemical and isotopic composition characterizes the saline groundwaters encountered in boreholes drilled in crystalline shields. This is seen on a large scale in Tables 9.1 and 9.3 and Figs. 9.3 and 9.5 of the Fennoscandian and Canadian shields, and on a small scale in Tables 9.4 and 9.5 and Figs. 9.3 and 9.7 of Stripa and Sudbury, two sites within the named shields. It is most interesting to compare the chemical composition patterns seen in Figs. 9.4 to 9.7; each set is made up of very different water types. Detailed examples of water diversity, even in the tens-of-meters scale, are reported by Nordstrom et al. (1989a,b,c) for the Stripa mine, a situation summed up by the researchers: "Any borehole can vary significantly and erratically in TDS for either a horizontal or vertical direction."

This diversity provides a clear indication that the groundwaters are stored in fracture-compartments that have been hydraulically isolated from each other since the time each water was entrapped.

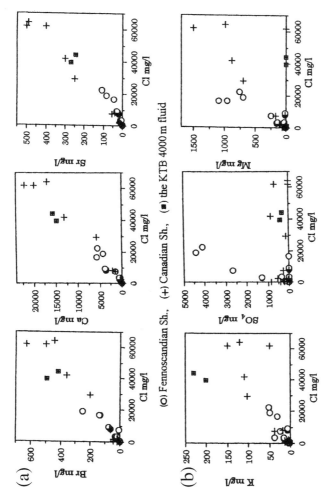

Fig. 9.6 Composition diagrams of the data from crystalline regions: (○) Fennoscandian Shield (Table 9.1); (+) Canadian Shield (Table 9.3), (■) KTB (Table 9.8). (a) Positive linear correlations are observed between Cl and Br, Ca, and Sr for data from the three regions, interpreted as indicating that the respective ions have a common source outside the rock systems, namely, brine-spray and brine-dust incorporated in groundwaters. (b) No correlation is observed between Cl and K, SO_4, or Mg, indicating variability of the brine-spray composition and/or modification by local water–rock interaction.

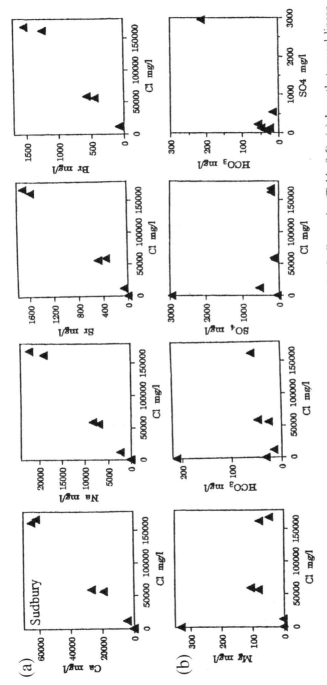

Fig. 9.7 Composition diagrams of groundwaters encountered in mines in the Sudbury site (Table 9.5). (a) Ions that reveal linear correlations between their concentrations, indicating a common allochthonous source. (b) Noncorrelated ions indicating site specificity and hydraulic separation of the various saline waters and ruling out dilution as the explanation for the linear correlation seen in the group of correlated ions.

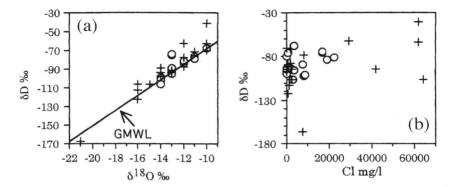

Fig. 9.8 Stable isotopes data in groundwaters (Tables 9.1 and 9.3): data from the Fennoscandian Shield (O) and the Canadian Shield (+): (a) Data plot along the present global meteoric water line (GMWL), indicating that the line was the same in the geological period of water entrapment and ruling out entrapment of residual evaporation brines, which would be located off the line at the upper right corner of the graph. (b) The δD and Cl data reveal no correlation, indicating separate origins.

9.2.3 The Large Variety of Encountered Saline Waters Precludes Ongoing Through-Flow in the Shields at Depths Greater than Sea-Level

Person (1996) and Toth and Sheng (1996) deduced from computer models that in all shields groundwater through-flow takes place down to depths of several kilometers. This hypothesis is not plausible, as through-flow of recently recharged water would result in homogenization of groundwaters and similarity to recent shallow groundwaters. The widely observed spatial variability of the deeper shield groundwaters indicates they are entrapped in hydraulically isolated compartments, and no intermixing has occurred since their emplacement.

9.2.4 Artesian Pressures or Downward Flow Into Mines?

Nordstrom et al. (1989a) state that in the Stripa mine, in the Fennoscandian Shield, "All mine water flows readily due to artesian condition." Kelly et al. (1986) state that in copper mines located in the Keweenaw Peninsula, within the Canadian Shield, when brines were encounterd they emerged as "seeps or pressurized flows that gradually ebbed and ceased altogether in periods ranging from days to years."

The described inflows may actually be a slow downflow of water stored in fracture compartments situated above the mine (e.g., fracture-compartment 1 in Fig. 9.2). Yet water may be compressed within some closed compartments due to tectonic compression, but this is entirely different from artesian flow in sedimentary basins.

9.2.5 Measurable Tritium Concentrations Reveals Mixing with Recent Shallow Groundwater—A Process Occurring Within Boreholes

The data in Table 9.1 reveal that samples collected at the Fennoscandian Shield contained tritium from below the limit of detection up to 70 TU. The latter value was observed in fresh water sampled at a depth of 270 m, and it represents recent through-flow water, either at its real flow depth or descended through the borehole. In the same borehole a sample collected at a depth of 1070 m contained only 1 TU, which is at the limit of detection, and thus the 16,800 mg/L of Cl observed at this depth is indigenous, and so are the Cl/Br ratio of 120 and the dominance of Ca among the cations—clear affinities to a residual evaporitic brine. The isotopic composition of this sample is $\delta D = -74‰$, and $\delta^{18}O = -13‰$, i.e., clearly reflecting that the indigenous water component is nonevaporated meteoric water.

Other samples in Table 9.1 reveal by their tritium concentrations intermediate degrees of dilution by recent fresh water, but their original nature is recognizable. No tritium data have been reported for the Canadian Shield samples in Table 9.3, yet their low Cl/Br ratios of <200 and the common presence of a Ca–Cl component clearly indicate that the indigenous water properties are well preserved, in spite of possible introduction of fresh shallow groundwater.

The data of the Sudbury mines (Table 9.5) indicate that the saline water samples were free of tritium, excluding mixing with recent shallow fresh water.

Much care was taken to retrieve well-defined water samples from different depths in two boreholes, up to 1270 m deep, in a pluton in the Atikokan area, the Canadian Shield. Yet significant tritium concentrations were encountered in all samples (Dickin et al., 1984). The data (Table 9.6) reveal a picture that is similar to the previous cases: tritium is detected in significant amounts in all the samples, down to the depth of 1050 m, and with no correlation to the depth. Thus recent shallow water has been introduced by the drilling operations. The salinity is here, too, seen to increase with depth, but this seems to be density-controlled rearrangement of different waters within the borehole that short-circuited several fracture-compartments. The characteristics of the indigenous waters involved are recognizable by the chemical compositions that differ from the composition of the fresh water used for the

Table 9.6 Measured Data in Two Boreholes in a Granitic Pluton in the Atikokan Area, Northwestern Ontario, the Canadian Shield

Borehole	Average depth (m)	Cl (mg/L)	Br (mg/L)	SO$_4$ (mg/L)	HCO$_3$ (mg/L)	Na (mg/L)	Ca (mg/L)	Mg (mg/L)	K (mg/L)	Sr (mg/L)	Tritium (TU)	δD (‰)	δ^{18}O (‰)
ATK 1	100	2	0.2	7	137	5	31	5	1	0.1	64	−82	−11.4
	210	31	<0.2	20	112	10	45	5	1	0.4	83	−84	−10.3
	430	400	3.4	337	12	63	293	1	1	5	74	−80	−9.5
	570	365	3.5	240	18	66	332	1	2	6	66	−82	−9.2
	680	895	7.4	89	6	59	480	2	2	9	101	−79	−9.1
	790	2,380	18	80	13	613	978	2	3	24	106	−77	−9.2
	880	25,470	141	430	45	1,830	11,900	4	10	308	66	−72	−9.8
ATK 5	100	2	0.1	3	92	3	32	4	2	>0.1	52	−85	−11.8
	320	194	1.2	53	25	34	102	2	1	1.6	69	−85	−10.2
	380	170	1.6	126	29	67	110	2	3	1.5	27		
	420	428	3.3	184	16	84	297	1	1	4	41	−89	−11.1
	680	1,860	17	91	>2	86	940	1	2	24	112	−89	−9.3
	700	2,560	19	114	12	116	1,240	1	3	31	86	−77	−9.3
	980	18,100	137	263	2	816	8,930	1	7	220	62	−70	−9.5
	1,050	17,700	119	986	21	779	7,460	1	9	205	72	−84	−9.1
Drilling water		>0.5		2.5	>10	0.6	1.8	0.6	0.5		129	−69	−7.8

Source: Data from Dickin et al. (1984).

drilling (Table 9.6) or the local shallow recent fresh water. Also, the water isotopes are lighter than the composition reported for the drilling water. Thus mixing with recent shallow water must be limited to about a 1:1 ratio, and the different types of nonmixed indigenous groundwaters are in fact more diverse than reflected from the analyzed samples reported in Table 9.6.

In regard to the presence of tritium in the majority of water samples, a limited amount of tritium could enter by exchange with the atmosphere when the water was standing in a borehole for a long time prior to sample collection. In such cases the degree of dilution by drill water, or shallow recent groundwater, is less than estimated by assuming that all the observed tritium came from intermixing recent water.

Of crucial importance is the conclusion that the introduction of recent water is a malfunction of the boreholes and not an ongoing natural process. This conclusion can be checked by improving future sampling techniques, e.g., by utilizing packers and by pumping away enough water to clean the borehole prior to sample collection.

9.2.6 Downflow of Waters Within Boreholes is Reflected in Temperature Profiles

The conclusion that some shallow groundwater trickled down in the boreholes reported in Table 9.1 is supported by auxiliary data provided for the same boreholes in Table 9.2 (from Nurmi et al., 1988). The boreholes were drilled between 1960 and 1984, and temperature profiles were conducted in 1985. The geothermal gradient, ΔT, calculated from the data, reveals a suspiciously low value of only $1.1\,°C/100$ m in the boreholes drilled 1 and 4 years prior to the measurement, $1.5\,°C/100$ m in the borehole drilled 5 years prior the measurement, and $2.1\,°C/100$ m in the borehole drilled 25 years prior to the temperature measurement. The fluids accumulated in the last borehole had the longest time for temperature equilibration, but even they reflect some deviation from equilibrium. Thus cold water from various fracture systems flowed down in all the wells, and partial equilibration with rock temperatures took place, a process that was slow and time dependent. The temperature gradients seen in each borehole indicate that after completion of drilling, each borehole was filled with water from short-circuited fractures, resulting in an artificial profile.

Data from the Stripa mine, Sweden, disclose a similar picture (Table 9.4): water temperatures revealed no depth–dependent pattern and were in the range of 8.0 to $10.6\,°C$. The depth interval of 908 to 969 m revealed $8.6\,°C$—nearly $20\,°C$ lower than the equilibrium temperature expected at this depth. Thus the boreholes triggered flow of cold shallow water into deeper fracture-compartments.

9.2.7 Density-Controlled "Layering" Observed in Boreholes is Sometimes Artificial

Table 9.1 reports the results of water samples collected at different depths in the four boreholes discussed in the previous section. It is readily seen in the table that in each borehole the salinity increased downward, thus giving the impression that the groundwaters within the shield are layered, saltier waters prevailing at greater depths. However, this type of layering seems to be manmade. Different waters cannot coexist in the same space for a long time as they will be mixed and be homogenized by diffusion. Thus the variability of water types observed in a borehole indicates that the latter has been lately filled by water from distinct reservoirs, i.e., different fracture systems (Fig. 9.2). The borehole provided free space for flow that short-circuited several fracture-compartments, and as a result different types of water trickled into the borehole and were rearranged by density. The rearrangement whithin the boreholes was not of a single type of saline water diluted to different degrees by a shallow fresh water, but rather several water types were involved. For example, three different water types are seen in borehole 1 (marked by dashed lines in Table 9.1), as revealed, for example, by Mg being equal to K in the upper layer, nearly twice than K in the middle layer, and about 30 times more in the bottom layer. Similarly, several water types are seen in the other boreholes reported in Table 9.1.

Salinity or density layering is reported also from two boreholes in the Atikokan area, in the Canadian Shield (Table 9.6), as reported by Frape et al. (1984). All the samples were found to contain significant concentrations of tritium, e.g., 72 TU at the depth of 1050 m. Thus downflow of waters took place within the boreholes, and the rearrangement of fluids was density controlled.

9.3 What Typifies Formation Waters Within Crystalline Rocks?

9.3.1 Lack of Regional Salinity–Depth Correlation

Another side of the same coin—there is no depth–salinity correlation. A glance at the four boreholes described in Table 9.1 reveals that there is no overall correlation between salinity and depth, e.g., boreholes 1, 2, and 4 each have a sample at around 700 m depth, yet the respective Cl concentrations are 8400, 3200, and 22,100 mg/L. Thus each water containing fracture-compartments reflects an individual history of evolution.

Conclusion: boreholes (and mines) tend to short-circuit fracture-compartments, each compartment containing one type of water. First principles consideration leads us to the conclusion that water entrapped in a fracture-compartment is homogenized, and thus each fracture-compartment can contain only one type of groundwater. The data of Table 9.2, published by Nuri et al. (1988), provide a most important insight into the hydraulics of boreholes in an environment of fracture-compartments in crystalline shields. Regular boreholes short-circuit waters encountered in distinct fracture-compartments and cause fluid rearrangement. This nature has to be taken into account when interpreting data, e.g., possible dilution with shallow recent water, possible intermixing of different fossil waters, and possible placement of waters to greater depths (Fig. 9.2). Similar lessons have to be learned for possible short-circuiting and mobilization of waste products from repositories in crystalline rocks.

9.3.2 Low Cl/Br Ratios Indicate an Affinity to Residual Evaporitic Brines and an Origin Outside the Rock System

The bulk of the groundwater samples reported from crystalline shields reveal Cl/Br ratios that are distinctly lower than the seawater value of ~293. The values reported in Tables 9.1 and 9.3 range between 80 and 200, indicating an affinity to a variety of evaporitic brines, i.e., residual brines of a range of evaporitic bodies that reached different degrees of precipitation of halite and other salts. These observations indicate an external origin of these two ions, i.e., they were washed in by the original recharge water.

9.3.3 Presence of a Ca–Cl Component Provides Additional Evidence of an Affinity to Evaporitic Brines and an Origin Outside the Rock System

Chloride is the dominant anion in the shield samples (Tables 9.1 and 9.3–9.6), and the dominant cation in the medium saline waters is Na, but it never suffices to balance all the Cl. A Ca–Cl component is always present in significant amounts, and in part of the more saline samples Ca is the dominant cation. The general and pronounced presence of a Ca–Cl component is another distinct marker of a compositional affinity to residual evaporitic brines. The Ca–Cl component thus must have an origin from outside the host rocks, i.e., it was brought in with the original recharge water.

9.3.4 The Saline Waters and Brines Are Not Associated with Any Specific Rock Type, Supporting Salt Supply from Outside the Rock System

As pointed out by Dickin et al. (1984), and as is revealed from reading other publications, no correlation is observed between the general composition of the encountered deep groundwaters and the nature of the host rocks. Thus the majority of the dissolved ions stems from outside the rock systems, supporting the conclusion that the bulk of the ions were brought in by the recharged water.

9.3.5 Positive Linear Correlation Between Cl, Br, Na, Ca, and Sr Concentrations Indicates Water–Rock Interactions Were Minute

The data of the two shields, as well as data from the KTB hole, are plotted in Fig. 9.6a in composition diagrams that reveal a high linear correlation between the concentrations of Cl, Ca, Br, and Sr, and the same holds true for Na. These ionic correlations indicate that these ions all have an origin from outside the rocks, i.e., by and large they were brought in with the original recharge waters. Thus in the discussed shield data the bulk of the Cl, Br, Ca, Sr, and Na are allochthonous, and modification of the concentrations by water–rock interactions are minute or negligible.

9.3.6 Lack of Correlation Between the Concentrations of Cl, K, Mg, SO_4, and HCO_3 Reveal Storage in Separated Fracture-Compartments

In contrast to the discussed correlated ions, there is a second group of ions that reveal no correlation between their concentrations. Examples, seen in Fig. 9.6b, include Cl, K, Mg, and SO_4, and the same holds true for HCO_3. The relative abundances of these ions are site-specific, determined by local variations in the composition of the ancient brine-spray and/or resulting from water–rock interactions. This observation indicates, together with the large range of salinity, that the saline groundwaters encountered in the crystalline shields differ from one another and hence must be stored in distinct compartments that are hydraulically separated.

The noncorrelated ions are instrumental in clarifying that the observed linear correlations, seen in Fig. 9.6a are not the outcome of dilution of a brine by different amounts of fresh groundwater, as in such a case all ions are expected to be correlated.

9.3.7 A Closer Look at Correlated and Noncorrelated Ion Groups

A division into correlated and noncorrelated ions is also observable on a local scale, as revealed by the data from the Sudbury site in the Canadian Shield (Table 9.5, data from Frape et al., 1984). In spite of a very wide range of salinity, linear correlations are seen between the concentrations of Cl, Ca, Na,

Sr, and Br (Fig. 9.6a), and a complete lack of correlation is seen between Cl, Mg, SO_4 (Fig. 9.6b), as well as HCO_3. Thus the first group stems from a common alochthonous source, and the second group of ions is site specific, reflecting local conditions that prevailed at the time that each water type entered a fracture system and/or a variety of water–rock interactions. The pattern of the ion abundances seen in Fig. 9.6b rules out dilution by fresh water as a quantitative explanation of the linear correlations seen in Fig. 9.6a.

9.3.8 The Isotopic Composition Discloses an Origin from Terrestrial Meteoric Groundwaters

The water isotopes reveal in the Fennoscandian Shield ranges of $\delta D = -106$ to $-68‰$ and $\delta^{18}O = -14$ to $-10‰$ (Table 9.1), and the values for the Canadian Shield are $\delta D = -167$ to $-41‰$ and $\delta^{18}O = -21$ to $-10‰$ (Table 9.3). These are very light isotopic compositions, revealing a clear affinity to nonevaporated meteoric groundwaters. Data of both shields are seen in Fig. 9.8 to plot along the global meteoric water line, providing further evidence that these are nonevaporated meteoric waters. This characteristic indicates a crucial boundary condition; the discussed deep saline waters encountered in the crystalline rocks of the shields are meteoric groundwaters by origin.

Tritium is observed in part of the collected shield water samples (Tables 9.1, 9.5, and 9.6), reflecting presence of a recent water component in these cases. This is a severe problem of sample collection in crystalline rocks, as discussed above. The chemical compositions of the shield waters, which are distinctly different from present local recharge water, reveal that we deal with indigenous fossil waters, and the chemical characteristics are not masked by additions of recent water during sample collection. Similarly, the isotopic composition affinity to meteoric water and the distinct variance from residual evaporitic brines are borne out by the following observations:

1. Residual evaporitic brines plot outside the respective meteoric water line, yet the data from the shield groundwaters plot along the global meteoric line (Fig. 9.8a), indicating no evaporitic water component is involved.
2. Among the most saline waters are cases with an extremely light isotopic composition, e.g., the sample from a depth of 1070 m at the Thompson Well 3 (Table 9.3) contains 64,000 mg/L Cl and has a δD value of $-106‰$ and a $\delta^{18}O$ value of $-15.5‰$ (more examples are seen in the tables).
3. There is no correlation between the Cl concentration and the stable isotopes' compositions (Fig. 9.8b). Thus some addition of local recent groundwater to part of the samples did not mask the original

light isotopic composition and the indigenous affinity to non-evaporated meteoric water.

9.4 Results from the KTB Deep Research Boreholes

Two KTB research boreholes were drilled into the continental crust in southeastern Germany. A pilot borehole reached 4000 m, and a 200-m-away main hole reached 9101 m, at a temperature of ~265°C. The wells were drilled into paragneiss, metabasanite, and other crystalline rocks varieties. The following observations were made in regard to the formation waters (Pekdeger et al., 1994; Emmermann and Lauterjung, 1997; Möller et al., 1997; Haak et al., 1997):

1. *Diversity of water compositions.* Waters of different salinities and a variety of relative ionic abundance combinations were encountered at distinct depths. The main types were as follows:

Hole	Depth (m)	Water composition
Pilot	Down to 650	Fresh
	3447	Saline
	3850–4000	Saline (Table 9.7)
Main	Down to 1500	Minor salinity
	2000–3000	Salinity increasing with depth
	3110–3300	Saline
	3600	Saline
	3800	Saline
	4800	Saline
	5300	Saline

2. *Evaporitic composition.* At the bottom of the pilot borehole, at a depth of 4000 m, a large water sample was pumped, and the composition is given in Table 9.7. The Cl concentration is 44 g/L; the ions' abundance pattern is

   $Ca > Na \gg Sr = K > Li$ and $Cl \gg Br = SO_4 > HCO_3$

   $Ca\,Cl_2$ predominates and the Cl/Br ratio is ~90. Thus the chemical composition reveals a similarity to a residual evaporitic brine.

3. *Resemblance to formation waters in sedimentary basins.* The chemical composition resembles the composition of waters reported from

Table 9.7 Chemical Composition of Water Samples Obtained at a Large-Scale Pumping Test at a Depth of 4000 m in the KTB Borehole

Depth (m)	Date	Na (mg/L)	Ca (mg/L)	Mg (mg/L)	K (mg/L)	Sr (mg/L)	Li (mg/L)	Cl (mg/L)	Br (mg/L)	CL/Br (mg/L)	SO_4 (mg/L)	HCO_2 (mg/L)
4,000	April 1990	5,500	14,700	6.8	200	270	5.4	39,500	493	79	390	98
4,000	December 1991	7,160	15,700	2.2	231	244	2.4	44,100	417	105	307	45

Source: Data from Möller et al. (1997).

sedimentary basins adjacent to the Fennoscandian and Canadian shields, discussed in Chapter 7.
4. *Light isotopic composition of meteoric water.* The isotopic composition of the water is $\delta D = -10‰$ and $\delta^{18}O = -5‰$, reflecting a terrestrial meteoric origin, probably somewhat evaporated.
5. *Lack of hydraulic connectivity between the 200-m-apart boreholes.* The hydraulic tests indicated just one zone at which the fractures crossed by the 200-m-apart wells did communicate hydraulically, whereas all other fracture systems were found to be hydraulically isolated!

9.5 Isotopic Dating of the Fossil Groundwaters Within Shields

The resemblance of the formation waters encountered at the crystalline shields to the brine-tagged formation waters encountered within the bordering sedimentary basins is overwhelming. This leads to the conclusion that the respective shields bordered the vast evaporitic lowlands that were subject to alternating sea transgressions and regressions, which supplied the brine-spray to the meteoric groundwaters, as deduced from the observations addressed in Chapter 7. The confinement ages of the formation water in the sedimentary basins go back to the Paleozoic, and this was confirmed by isotopic dating, addressed in Chapter 8.

Similarly, isotopic dating provides an indispensable tool to understand the anatomy of the formation waters encountered at the crystalline shields. The obtained water ages provide an independent criterion to check the suggested origin of the crystalline shields formation waters.

9.5.1 Old Age Deduced from the Evaporitic Composition

The common occurrence of salt and gypsum deposits as well as predominance of saline formation waters of evaporitic composition indicate that large-scale evaporitic conditions prevailed around the Canadian Shield during the Lower Paleozoic and around the Fennoscandian Shield from the Permian to the Jurassic (Chapters 6 and 7) and to some extent later on in the Neogene. It seems that during the respective geological periods the crystalline rocks of the shields were exposed and received brine-spray from the large neighboring evaporitic provinces; then the airborne salts were washed down into rock fractures by rain and snowmelt within the through-flow zone, i.e., within the landscape relief that protruded above the sea surface that acted as the base of drainage.

The trapping of the fossil groundwaters in deep fractures at shields had to be enhanced by large-scale fracturing that caused shallow groundwater to move down in the fractures as the latter propagated to greater depths, coupled with sealing of fracture segments by lateral and oblique displacements (Fig. 9.1). With the termination of the large-scale fracturing, new recharge into the crystalline rocks must have been restricted to the through-flow zone, similar to the present situation.

On the basis of the evaporitic affinity of the fossil shield groundwaters, it is thus concluded that the Canadian Shield waters are by and large of Lower Paleozoic age. The age of the waters in the Fennoscandian Shield may be somewhat younger, between Permian and Jurassic or even Neogene.

9.5.2 Helium-4 Ages Deduced for the Saline End-Members of Mixed Samples Studied in Two Sites Within the Canadian Shield

A good example of dating groundwater encountered in crystalline rocks is presented in a detailed study conducted by Bottomley et al. (1990) at the East Bull Lake pluton, in the Canadian Shield. The study included, among many other parameters, the measurement of tritium, ^{14}C, and helium (Table 9.8). The $^3He/^4He$ ratios were measured as well, and no contribution from mantle helium was concluded. Also, Ne was measured and applied for small corrections of the He data to accommodate for free air trapped in sampling vessels. The data and ages deduced in the frame of the present communication are given in Table 9.8. Carbon-14 was assumed to be only little diluted by dead carbon, as one deals with crystalline rocks and as was reflected by one sample of well P-10 that revealed a nondiluted post-bomb ^{14}C concentration. The concentration of ^4He in recent recharge water is around 5×10^{-8} cc STP/g water, and thus any higher concentration indicates presence of radiogenic He. The ^4He age was computed for the case that in a steady-state fossil water dissolves most of the helium produced in the host rocks. Needed auxiliary data were reported by Bottomley et al. (1990), e.g., concentration of U and Th in the local rocks and the water/rock ratio. The outcome of the East Bull Lake pluton data (Table 9.8) reflects indeed a mixture of young and old waters:

$$\text{tritium age} < {}^{14}C \text{ age} < \text{helium age}$$

Thus old groundwaters are present in the studied system. Interestingly, the low Cl concentrations, given in Table 9.8, indicate that in the study area old fresh groundwater was also entrapped.

Bottomley et al. (1990) reported data obtained from a carefully pumped sample of the EBL-4 borehole, which yielded progressively higher ^4He concentrations as pumping went on. The last collected sample contained 5.2×10^{-3} cc STP/g, for which the investigators computed an age of 1.2×10^8 years. This is to be regarded as a minimum age, as some intermixing with recent water could still have taken place. Thus the age is $> 1.2 \times 10^8$ years,

Table 9.8 Concentration of Age Indicators in Water Samples from the East Bull Lake Pluton, Canadian Shield

Borehole	Depth (m)	Tritium (TU)	^{14}C (pmc)	He (cc/g)	Tritium age (years)	^{14}C age (years)	He age (years)	Cl (g/L)
P-3	0–16	44	14	6.3×10^{-8}	10^1	10^3	10^3	1.0
P-5	88–98	26	33	2.0×10^{-6}	10^1	10^3	5×10^4	1.2
P-10	0–34	69	132	1.7×10^{-7}	10^1	10^1	5×10^3	2.8
P-10	82–99	22	63	3.5×10^{-7}	10^1	10^2	10^4	1.1
EBL-1	147–298	72	38	4.3×10^{-6}	10^1	10^3	10^5	17.9
EBL-2	111–126		12	1.7×10^{-5}		10^4	5×10^5	6.8
EBL-4	105–117		23	3.6×10^{-4}		10^3	10^7	3.1

Note: Deduced ages reflect mixing of recent and old waters.
Source: Data from Bottomley et al. (1990).

leaving room for older waters to be sampled in the future, possibly in the range of Lower to Middle Paleozoic, as suggested by the prevalence of continent-wide evaporitic provinces during that time.

A second case study, reported by Bottomley et al. (1984), dealt with formation water samples collected at boreholes in a Precambrian granitic batholith in Pinawa, Manitoba. The findings included a spectrum of fresh to saline waters (up to 9570 mg/L Cl), the more saline waters having a pronounced Ca–Cl component, and theses samples contained tritium (0–34 TU) and ^{14}C (26–70 pmc), along with ^4He (10^4 to 10^{-1} cc STP/g water), and in addition radiogenic ^{21}Ne was detected. Bottomley et al. (1984) concluded: "Helium and ^{21}Ne ages of these groundwaters, calculated on the basis of known crustal production rates of ^4He and ^{21}Ne, are unreasonably high up to 2×10^8 years and incompatible with the ^{14}C ages."

Mixing of recent and old waters seems most plausible in this case, and the calculated He age is probably close to the real age of the nonmixed oldest formation water end-member.

9.5.3 High ^4He and ^{40}Ar Concentrations Observed Along with Tritium and ^{14}C Indicate Mixed Water Samples Collected at the Stripa Mine, Sweden

Andrews et al. (1989) reported high ^4He concentrations in waters collected at research boreholes in granite at the Stripa mine site (up to 4×10^{-3} cc/cc water in the 457–862 m sampling interval), accompanied by high ^{40}Ar concentrations ($^{40}Ar/^{36}Ar$ up to 377, as compared to the atmospheric value of 296).

The researchers indicated that a straightforward He age calculation with the local auxiliary parameters "yielded a very old groundwater age of

840 ka." They added: "The estimated brine ages are, however, more than an order of magnitude greater than hydrological estimates," and further on they comment that the ^{4}He age is "much greater than the ^{14}C-derived ages." This lead the researchers to conclude "that this discrepancy could be due to diffusive movement of ^{4}He into the aquifer from other geological horizons."

It is here suggested that the discrepancy in the computed ages stemmed from the fact that the Stripa samples were mixtures of different proportions of shallow recent water and fossil old water. The conclusion that the Stripa mine water samples were mixtures of shallow recent water with deeper saline water was here reached based on tritium measurements; Nordstrom et al. (1989a,b) reached the same conclusion based on the water chemistry; and Mazor (1992a) reached it based on ^{36}Cl data.

The high ^{4}He concentrations at Stripa are accompanied by a significant concentration of radiogenic ^{40}Ar, strongly supporting an age of $>> 10^6$ years.

9.5.4 Helium-4 and ^{40}Ar Ages Deduced for the 4000-m-Deep KTB Water

The KTB 4000-m-deep water was dated by Möller et al. (1997), who stated:

> Using the radiogenic fraction of ^{4}He and ^{40}Ar, apparent accumulation times of fluids are derived under "closed system" conditions, radioactive equilibrium, and nondestructive gas-composition sampling. The low variabilities of He contents, as well as rather constant He isotope ratios, support the assumption of a large-scale, homogeneous, fluid reservoir with respect to these parameters.
>
> Accumulation times of 1.5×10^7 to 8×10^7 years were obtained from the ^{4}He data and 3×10^7 to 3×10^8 years from the ^{40}Ar data. Möller et al. (1997) concluded that these water ages support a fluid exchange after the Cretaceous denudation, in agreement with their measured He and Ar loss rates from the host rocks.

9.5.5 Conclusions Regarding Mixed Waters in Crystalline Shields

Mixing of recent and old waters is observed in a number of case studies within crystalline shields, borne out by the coexistence of significant concentrations of short-range and long-range age indicators. The old water end-members have, at least in some cases, an age in the range of 10^7 to 10^8 years, in accord with the prevalence of large-scale evaporitic environments in the neighboring regions during ancient geological periods. The application of age indicators is highly recommended for routine investigations connected to nuclear waste repository siting.

9.6 Working Hypothesis: Tectonic "Fracture Pumps" Introduced Meteoric Groundwater to Great Depths

The long list of discussed observations leads to a rather simple conclusion related to saline groundwaters encountered in crystalline rocks in shields beneath the ongoing groundwater through-flow system:

1. These are fossil groundwaters that washed down brine-spray that arrived from vast evaporitic provinces that surrounded the shields during certain geological periods.
2. Under arid climatic conditions, water of small rain events dried up leaving the salts at the surface. Once a year or once in several years powerful rains dissolved the accumulated salts and washed them down into the groundwater through-flow system. In the crystalline rocks the saline groundwaters filled shallow open fractures within the groundwater through-flow zone, as is observed today in crystalline rocks exposed in arid climates (section 4.5).
3. Part of the shallow fractures gradually propagated to greater depths as a result of local or regional tectonic dilatation phases, allowing the trapped water to trickle down to greater depths.
4. Fractures were commonly bent and had rough walls, and, as a result of small shifting of their walls or cessation of the dilatation regime, they became pressure sealed at narrow parts, whereas the water was trapped at wider sections.
5. The process of opening and closing of fracture segments was repeated many times, allowing the water to percolate to greater depths whenever deeper routes were temporarily opened. Thus fracture-compartments were formed at various depths within crystalline rocks, some of them containing groundwater and others being dry.
6. Compositional modifications by interaction with rocks were limited, as is indicated by the preserved evaporitic composition.

9.7 The Saline Waters in Shields Serve as a Geological Record

1. *Downward propagation of fractures.* The occurrence of fossil waters in closed fracture-compartments implies that certain fractures developed first at the surface and then gradually propagated, or migrated, to significant depths.
2. *Large-scale fracture propagation coincided with the large-scale evaporitic conditions and formation of the subsiding basins.* The pre-

dominance of brine-spray-tagged groundwaters in the deeper parts of explored crystalline rock systems implies that fracture formation and downward propagation activity went on simultaneously with the extensive evaporitic processes. Thus the ages of the fossil groundwaters of the shields provide ages of periods of intensive tectonic fracturing.

3. *Exposure of the shields at the time of saline water entrapment.* The saline paleo-waters could be recharged into the crystalline rocks of the shields only through exposed fracture systems. Thus the age deduced by the prevalence of large-scale evaporitic systems, or the concluded isotopic age of the contained saline waters, marks in each study case a geological time of exposure and lack of cover by sedimentary rocks.

4. *Results indicate a relatively arid paleoclimate.* The Canadian and Fennoscandian shields have at present humid climates, and presently formed groundwaters are fresh. However, in crystalline rock regions that are at present arid zones, saline groundwaters are formed, e.g., in northeast Brazil, vast areas of the Wheatbelt, Australia (section 4.5), and in many of the world's deserts. It seems that at the time that the fossil saline groundwaters were entrapped, the shields had semi-arid to arid climatic conditions.

9.8 Nuclear Waste Disposal Implications

A long list of observations supports the conclusion that at depths greater than sea level no groundwater through-flow takes place within crystalline rocks. Fossil groundwaters are trapped in fracture-compartments for very long geological durations, often on the order of 10^7 to 10^8 years. Adjacent compartments are observed to host distinct water types, and hence they are hydraulically disconnected. On this basis it seems that crystalline rocks in shields do provide a promising environment for nuclear waste repositories.

The distribution of water-containing fractures is uneven, and zones with practically no—or very few—fractures have been identified besides more fractured zones. Also, fossil saline groundwaters are observed to be mobilized by short-circuiting between fracture-compartments, caused by mining operations and borehole drilling. Hence special guidelines have to be worked out for safe and optimal siting and operation of nuclear waste repositories within crystalline rocks. The following are a few tentative suggestions:

1. *Identification of the local active through-flow zone.* The local groundwater regime has to be explored and understood in terms of the active through-flow system, and identification of its terminal base

of drainage must be made. Potential siting of repositories should be done at safe depths beneath the through-flow zone.

2. *Proper sampling of studied water samples and complete analyses.* Groundwaters encountered at a potential site have to be sampled with no intermixing of shallow water and be fully analyzed, including dating with short-range and long-range methods. Groundwater with a composition distinctly different from the local recent groundwater can be regarded as fossil, an observation that has to be checked in light of determined ages. Presence of fossil groundwater indicates hydraulic separation from the land surface.

3. *Sites with minimal fracturing.* Regional and local exploration results should be applied to locate sites with the least fracturing, so that the quantity of fossil groundwater that may be traversed by the repository site will be minimal.

4. *Sites with least corrosive formation water.* Regional and local exploration results should also be applied to locate sites with fossil groundwaters that are the least saline and the least corrosive to waste containers. Possible pumping of fossil water from fracture-compartments should be considered.

5. *Eventually all boreholes have to be safely plugged.* Construction of selected repository sites should include plugging of exploration and research boreholes and sealing of observable open fractures, even dry ones.

9.9 Summary Exercises

Exercise 9.1: Study in detail Fig. 9.2. How could formation water enter the at present isolated fractures 1 to 7? Borehole C causes mixing of which fracture compartments? Which other disturbance may happen?

Exercise 9.2: Why were crystalline rocks considered appropriate for depositing high-level nuclear waste? What are the requirements of a safe repository?

Exercise 9.3: Study Table 9.1. Which patterns can you see? The Limnika borehole encountered at a depth of 600 m water with how much Cl? Can this be entrapped seawater? Look for the answer in the information provided by the other ions analyzed in this sample.

Exercise 9.4: Which observations lead to the conclusion that an active groundwater through-flow system also operates in crystalline rocks, and beneath the active base of drainage formation waters are stored in hydraulically isolated fracture-compartments?

Exercise 9.5: Let us have a look at the 1650-m-deep water sample in Table 9.5: Is this fossil groundwater? Is the light isotopic composition indigenous, or does it reflect filling of the borehole with recent shallow groundwater?

Exercise 9.6: The chemical composition of formation waters encountered at great depth within crystalline shields reveals a large variety of compositions, but a general trend is observable—what is it? Is it unique to formation waters within shields, or has it a wider occurrence?

Exercise 9.7: Can old formation waters be identified even in samples that have been mixed upon sampling with recent groundwater? How?

PART IV
PETROLEUM HYDROLOGY

~~~~~~~~~~~~~~~~~~~~

Formation water
recorded the
formation of petroleum
within
closed rock-compartments

~~~~~~~~~~~~~~~~~~~~

10
ANATOMY OF SEDIMENTARY BASINS AND PETROLEUM FIELDS HIGHLIGHTED BY FORMATION WATERS

So far we have become familiar with the vast geological information encoded in formation waters. Let us now apply this information to gain boundary conditions to the processes that led to the formation of associated oil and gas deposits.

10.1 Petroleum and Associated Formation Waters Are Complementing Sources of Information

As strange as it may seem, the vast literature dealing with petroleum genesis and exploitation remained detached from the characteristics of the associated formation waters. Field work ignored the mapping of the associated water facies and neglected the meaning of their detailed chemical and isotopic composition as well as isotopic dating. The research articles and textbooks that deal with petroleum occurrences and genesis do not address topics that are related to the nature of formation waters. Hence a special part of the present book is devoted to it. In a way, all our knowledge and understanding of hydrology and hydrochemistry comes to full use when dealing with oil and gas occurrences.

In the last chapters we became acquainted with the brine-tagged facies of meteoric formation waters that are encountered all over the globe in small and huge sedimentary basins, within rift valleys, and throughout shields built of crystalline rocks. The evidence from the chemical and isotopic composition of these fossil formation waters leads to two major conclusions:

1. The formation waters originated mainly at large-scale evaporitic flatlands, frequented by sea invasions and retreats.
2. The formation waters are entrapped in isolated rock-compartments.

In the present chapter we take a closer look at formation waters that are directly associated with petroleum deposits. The topic will by now sound familiar to the reader, and it may almost seem superfluous to address it further. However, the emphasis is on case studies of formation waters that are *directly associated with petroleum*.

10.2 Petroleum-Associated Formation Waters in the Western Canada Sedimentary Basin

The topic of connate waters in this basin is discussed in light of data published by Hitchon and Friedman (1969) and Hitchon et al. (1971), and representative examples are compiled in Table 10.1.

The water samples were separated from oil and gas at the production sites, thus representing the deep water found at known exploitation depths. This is of much value, as dilution by drilling fluids or shallow short-circuited fresh groundwater, common in exploration boreholes or pumped water wells, were avoided.

Pronounced spatial variability of water characteristics indicates confinement in hydraulically separated rock-compartments. The compositional fingerprint diagrams seen in Fig. 10.1 reflect an array of very different water types, varying from one another both in salinity and in the relative ionic abundances. Different compositions of the formation waters are seen for the Cretaceous, Triassic-Jurassic, Carboniferous, and Devonian rock formations. Practically each sample is seen in Fig. 10.1 to have a distinct salinity/composition combination. The number of encountered water types is to a large extent dependent on the spatial resolution of the sample collection density, and many more water types are likely to be found as the sample collection resolution is improved. The large salinity and composition variability is also reflected in the composition diagrams of Fig. 10.2, the isotopic composition diagrams of Fig. 10.3, and the depth profiles depicted in Fig. 10.4.

The discussed waters are fossil, as concluded on the basis of the composition that drastically differs from the composition of the ongoing recharge. Hence the different formation water types must have been efficiently isolated from each other for the geological durations that passed since these waters were introduced into their host rocks.

Anatomy of Sedimentary Basins

Table 10.1 Chemical and Isotopic Composition of Formation Waters Separated from Oil and Gas Samples, Western Canada Sedimentary Basin

Age	Field	No.	Na (mg/L)	K (mg/L)	Li (mg/L)	Ca (mg/L)	Mg (mg/L)	Sr (mg/L)	Cl (mg/L)	Br (mg/L)	Cl/Br	SO_4	HCO_3	δD (‰)	$δ^{18}O$ (‰)
Upp. Cret.		82	6,360	25	0.4	55	87	13.3	8,650	92	94	4	1,220	−104	−9.2
		28	3,750	14	1.6	41	43	17.4	5,370		153	3	784	−110	−9.3
L. Cret.	L. Colorado	27	8,290	50	2.2	249	79	38	13,100	33	397	5	1,030	−115	−8.9
		85	1,510	10	0.3	17	8	0.52	994	2	497	5	1,870	−98	−8.9
		11	22,800	98	4.2	1,230	409	219	38,200	169	226	11	761	−109	−7.6
L. Cret.	Mannville	58	16,800	440	14	489	104	11.4	25,400	74	343	901	2,080	−8.1	−0.8
		10	34,000	840	21.6	4,540	1,000	276	61,100	208	294	441	509	−98	−5.8
		30	42,900	1,820	42	9,740	1,760	481	95,800	375	255	466	473	−85	−3
Jurassic		88	730	68	0.8	55	43	3	496	0.5	101	20	1,090	−126	−15.2
		93	1,850	62	1.6	43	19	1.1	912	9	473	10	3,700	−130	−14.6
Triassic		20	53,700	1,660	30	3,070	827	152	91,400	193	473	558	784	−82	0.8
Carbonif.		51	26,300	800	31	2,220	388	539	46,900	128	366	27	728	−80	0.4
		87	510	70	0.8	96	58	6.1	320	0.5		9	1,070	−132	−15.9
		69	10,600	108	6.4	21	40	15	12,100	58	209	1,290	1,820	−103	−5.1
U. Devonian	Wabamum	17	73,800	5,800	72	23,500	2,490	900	171,000	466	376	276	474	−57	3
U. Devonian	Winterburn	14	33,000	1,980	36	7,770	1,620	222	69,100	363	190	1,050	581	−85	−4.2
		41	41,300	4,600	44	12,800	2,280	392	98,000	642	153	778	157	−71	−0.2
		73	9,360	740	9.8	1,380	428	36	15,800	20	790	3,910	1,030	−134	−14.1
		90	65,000	2,580	40	25,200	3,380	820	158,000	516	306	512	138	−53	−0.3
		16	92,800	5,200	74	24,300	2,160	900	157,000	396	396	228	767	−41	3.1
		45	48,800	8,800	76	22,800	2,590	945	126,000	961	131	367	767	−18	7.8
U. Devonian	Beaverhill L.	92	73,600	1,560	32	10,900	1,140	460	139,000	267	520	588	251	−54	2.7
Granite Wash		23	85,300	2,380	30	24,900	1,430	735	183,000	252	726	248	216	−64	−2.7

Source: Data published by Hitchon and Friedman, (1969) and Hitchon et al. (1971).

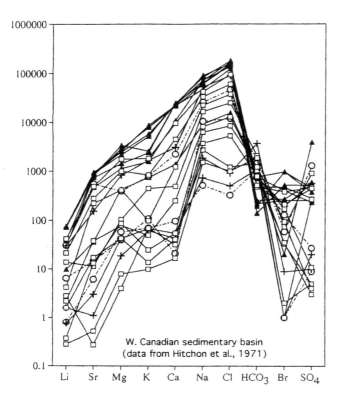

Fig. 10.1 Composition fingerprint diagrams of formation waters separated from oil and gas encountered in Upper Devonian to Upper Cretaceous formations in the west Canadian sedimentary basin (Table 10.1). (□) Cretaceous, (+) Triassic–Jurassic, (○) Carboniferous, (▲) Devonian. A remarkable variability in salinity and relative abundances is observed—almost each well has its own distinct composition. (Data from Hitchon et al., 1971.)

An allochthonous source of Cl, Br, and Na. The composition diagrams (Fig. 10.2) reveal a positive correlation between the concentration of Cl in the Cretaceous and Carboniferous formation waters and the concentration of Na and Br. This indicates that not only did Cl have an external origin, as deduced above, but Br and Na did also. The composition diagrams of the Devonian formation waters reveal distinctly lower correlations, reflecting a variability of the allochthonous sources, e.g., different facies of evaporitic systems.

The concentration of Cl and the Cl/Br ratio exclude an origin from halite dissolution. The Cl concentration in the waters encountered in the Cretaceous

Anatomy of Sedimentary Basins

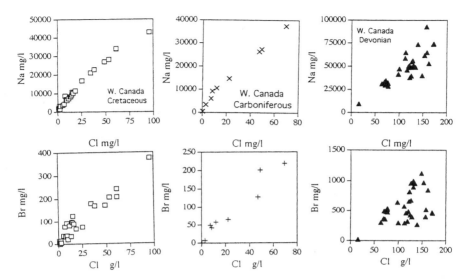

Fig. 10.2 Composition diagrams of the formation waters separated from oil and gas in the western Canada sedimentary basin. Positive linear correlations are observed between the concentrations of Cl and Na and Br in the Cretaceous and Carboniferous waters, indicating a common external origin. In contrast, almost no correlation is seen in the Devonian waters, indicating high variability of the respective paleo-environments at the time the formation waters were formed and/or in situ water–rock interactions. (Data from Hitchon et al., 1971.)

to Carboniferous formations varies from 0.3 to 96 g/L, i.e., less than the concentration of saturation with respect to halite, which is ~180 g/L. Thus these formation waters are undersaturated with respect to halite, indicating the latter is absent from the hosting rocks. The Devonian formation waters vary in the range of 16 to 170 g/L Cl, indicating that in most cases the respective formation waters are undersaturated with respect to halite as well. The Cl is, thus, external and came in along with the water.

The weight ratio of Cl/Br varies in all the studied waters in the range of 60 to 790, i.e., significantly lower than the ratio in halite, which is >3000, thus independently excluding halite dissolution.

A prominent Ca–Cl component indicates brine-tagged formation waters. The Cl in the more saline waters is balanced by a significant component of Ca, which amounts up to half the weight of the Na component, reflecting an origin from evaporitic brine-spray.

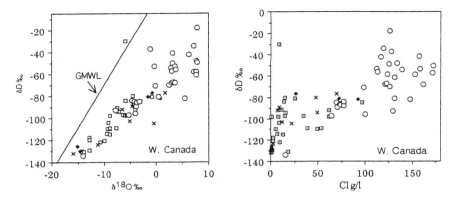

Fig. 10.3 Isotopic composition and Cl concentration of formation waters separated from oil and gas in the western Canada sedimentary basin. All the δD values resemble meteoric water, whereas most of the $\delta^{18}O$ values reveal an ^{18}O shift. As observed in many systems discussed so far, δD and Cl are not correlated, revealing different intake histories. (Data from Hitchon and Friedman, 1969; and Hitchon et al., 1971.)

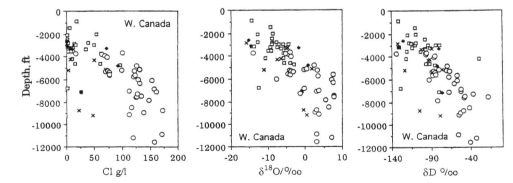

Fig. 10.4 Depth profiles of formation waters separated from oil and gas in the western Canada sedimentary basin. The data are rather scattered, revealing a wide range of Cl concentrations for each depth, and a wide range of isotopic compositions, reflecting the fruitcake structure of separated rock-compartments filled with different water facies and different petroleum facies. (Data from Hitchon and Friedman, 1969; and Hitchon et al., 1971.)

The δD values are meteoric. The δD values of the studied formation waters vary in the range of −140 to −20‰ (Fig. 10.3). These values indicate that these are meteoric groundwaters, and an origin from entrapped seawater or entrapped evaporitic brines can be ruled out. Thus the formation waters encountered in the Western Canada sedimentary basin have a continental origin.

Formation at an arid paleo-climate. The high salinity indicates that the formation waters were formed in an arid climate, at which the airborne brine-spray and dust accumulated on the ground and were washed into the active groundwater system only by exceptionally heavy rain events that preserved their meteoric δD values (section 6.13).

The $\delta^{18}O$ values are in part meteoric and in part reveal a temperature-induced ^{18}O shift. The $\delta^{18}O$ values vary between −16 to +8‰. The negative values reflect the meteoric origin, and the positive values reflect an ^{18}O shift, often observed in geothermal waters as a result of water-rock interaction at elevated temperatures, as was originally pointed out by Craig (1963). The Devonian formation waters reveal the most positive $\delta^{18}O$ values, as is seen in Fig. 10.3, probably reflecting the largest ^{18}O shift. This conclusion is plausible as the Devonian rocks are the deepest and hence must have experienced relatively high burial-induced heating.

The discussed formation waters reside at the zone of zero hydraulic potential and are static. The discussed formation water samples were retrieved at the depth range of 2 to nearly 12 km (Fig. 10.4), i.e., deep below sea level. The sea level is the plain of zero hydraulic potential, and all deeper groundwaters reside in the zone of zero hydraulic potential, which is another way to describe the lack of drainage (section 2.13). Thus, from a first principles discussion, the petroleum associated formation waters in the Western Canada sedimentary basin are static. This is well supported by the boundary condition, derived above, that these are fossil waters entrapped within hydraulically isolated rock-compartments.

These observations and conclusions rule out basin-wide groundwater through-flow. Many of the West Canada Basin researchers assumed that recent recharged groundwater flows through the basin even at depths of thousands of meters. These notions stem from mathematical models, but the field observations and first principles considerations reveal that beneath the unconfined groundwater system, i.e., deeper than sea level, no through-flow of fluids occurred in the geological past and none happens at present.

The preservation of deposits of oil, gas, and halite are additional indicators that no fluids flow through the sedimentary basin. The preservation of thick halite beds, of different geological ages, indicates the lack of water through-flow. The same is evidenced by the preservation of even very old oil and gas deposits.

10.3 Petroleum-Associated Formation Waters Within Ordovician Host Rocks, Ontario, Canada

Dollar et al. (1991) published an extensive set of data on the composition of formation waters associated with petroleum hydrocarbons from the regions of Ontario and Michigan. Wells in Ontario tapped the shallow margin of the huge sedimentary basin that attains great depth in Michigan. The studied Late Cambrian rocks are mainly sandstone and dolostone, and the reader is referred to the mentioned publication for the geological background. Data from 13 wells reveal the following pattern:

1. The very high salinity, e.g., Cl concentrations of 100 to 180 g/L (Fig. 10.5), reveals that possible intermixing with fresh recent groundwater during sample retrieval was negligible.
2. The waters have a composition that differs drastically from the local recently formed fresh groundwater, indicating these are fossil waters.
3. The Cl/Br ratios of the reported waters are in the range of 85 to 150, ruling out an autochthonous origin from halite dissolution. The Cl/Br ratios reveal an affinity to residual evaporitic brines, thus indicating that Cl and Br are allochthonous.

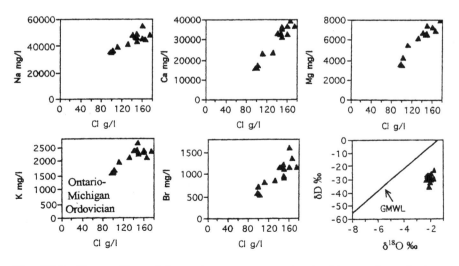

Fig. 10.5 Composition of formation waters associated with petroleum hydrocarbons from the regions of Ontario and Michigan. The familiar pattern is again seen: positive linear correlations between Cl and other ions, prominence of $CaCl_2$, light meteoric δD values, and ^{18}O enrichment. (Data from Dollar et al., 1991.)

4. The waters are rich in Ca, which matches the concentration of the Na (Fig. 10.5), i.e., a Ca–Cl component is dominant, indicating an affinity to a residual evaporitic brine.
5. The positive correlation of the concentration of Cl to those of Na, Ca, and Mg reveals that these cations are of the same external evaporitic source.
6. The waters are seen to have rather light δD values, in the range of -3 to -20‰, indicating the studied water are nonevaporated meteoric groundwaters. The $\delta^{18}O$ values plot in Fig. 10.5 to the right of the global meteoric water line, a trend that is common in formation waters and is interpreted as reflecting an ^{18}O shift due to exchange with rocks during the burial to the depth of the petroleum formation temperature window.
7. The compositions of the studied waters vary between adjacent wells, indicating storage in hydraulically separated rock-compartments, a feature that is also borne out from the occurrence of associated gas and oil.

10.4 Kettleman Dome Formation Waters Associated with Petroleum—Key Observations and Concluded Boundary Conditions

Kharaka et al. (1973) published data of their intensive study of formation waters separated at the head of oil wells in the Kettleman Dome oil field, California. The data are here processed in the mode the reader is already familiar with, providing a detailed insight into the anatomy of an oil field.

Diversity of Cl concentrations discloses superimposed compartments containing different formation waters. Depth profiles of formation waterss separated from oil reveal a juxtaposition of water types of different ion concentrations, with abrupt changes along the profile. The changes of Cl concentration in depth profiles are plotted in Fig. 10.6, indicating that the formation waters (like the petroleum deposits) are contained in hydraulically isolated rock-compartments.

The formation waters are also diverse in the relative ionic abundances. The differences between the various water types are not only by salinity but also by the relative ion abundances (Fig. 10.7). The large diversity indicates that exchange of fluids between individual compartments must have been negligible. If waters had flowed through the individual compartments, the composition would be homogenized. In other words, if some fluid exchange took place, then the original composition diversity was even more accentuated than observed today, further supporting the concept of different water types originally entrapped in separated rock-compartments.

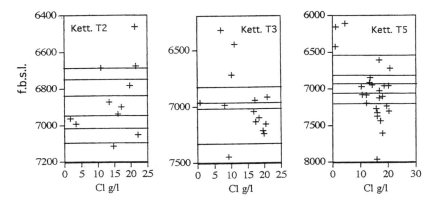

Fig. 10.6 Depth profiles of Cl concentration in wells that penetrated three of the Miocene Temblor Formation subunits (T2, T3, T5), Kettleman Dome. Abrupt changes in the Cl concentration (marked by horizontal lines) indicate that the formation waters and the associated oil are stored in distinct hydraulically separated rock-compartments. (Data from Kharaka and Berry, 1974.)

This observation might provide a means to map the distribution of rock-compartments, i.e., different waters indicate different compartments, whereas similar waters *may* stem from the same compartment. The spatial mapping resolution of rock-compartments is higher as more ions are applied for this analysis and as more samples are available from different locations and different depths.

Positive correlations between the concentration of Cl, Br, Na, K, Li, and Sr disclose an external origin. The composition diagrams (Fig. 10.8) reveal linear positive correlations between the concentration of Cl and a list of other ions, the correlation being better defined within each of the Temblor subunits.

Dilution by fresh water as an artifact, e.g., by inclusion of drilling water or by short-circuiting between shallow and deep waters in poorly cased boreholes, is ruled out by the following arguments:

1. Dilution of a saline end-member would equally dilute all ions, and the result would be parallel lines in the fingerprint diagrams, but this is far from being the case, as is well seen in Fig. 10.7.
2. Dilution of a saline end-member by a local fresh water would produce a positive linear correlation between the δD values and the Cl concentration, but this is not the case—in Fig. 10.9 these two components are seen to be noncorrelated.
3. Local fresh water has δD and $\delta^{18}O$ values that are light and plot along the global or local meteoric water line. Hence on a δD–$\delta^{18}O$

Anatomy of Sedimentary Basins 215

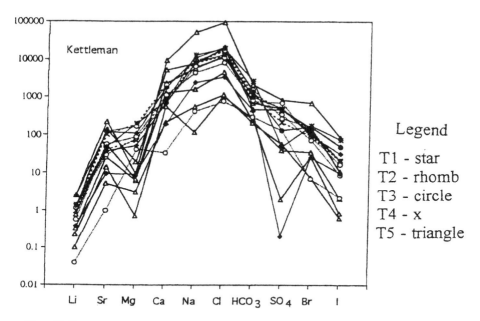

Fig. 10.7 A compositional fingerprint diagram of representative water samples, Kettleman Dome, all Temblor wells. A pronounced diversity is seen in ion concentrations and relative abundances. This is another reflection of the presence of distinctly different water types within the Miocene Temblor section of the Kettleman Dome. (Data from Kharaka and Berry, 1974.)

diagram dilution products of a saline end-member and fresh water would result in a linear distribution along a mixing line. No such line is seen in the Kettleman data (Fig. 10.9).

It is thus concluded that the differences in ion concentrations observed in the discussed Kettleman Dome formation waters are genuine. A further check can be tritium and ^{14}C analyses. If some tritium and ^{14}C are found in a sample, intermixing with recent water is established, and its extent might be estimated. No detection of ^{14}C assures that no recent water was intermixed.

These are meteoric groundwaters, formed at ancient exposed land. The δD values of the Temblor Formation waters are in the range of -50 to -15‰, with most of the waters having values around -30‰, i.e., well within the range of meteoric groundwaters. These values exclude entrapment of residual evaporitic brines or seawater.

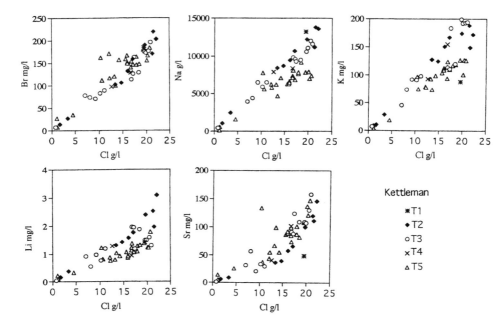

Fig. 10.8 Composition diagrams of water samples of the five Temblor Formation subunits, Kettleman Dome. The positive correlation lines disclose a common external origin. (Data from Kharaka and Berry, 1974.)

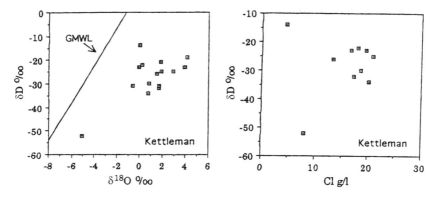

Fig. 10.9 Isotopic composition of the waters of the Kettleman Dome. The δD values are well in the range of nonmodified meteoric waters, indicating the formation waters are entrapped (connate) groundwaters. An ^{18}O shift is observed, indicating the waters subsided along with their host rocks and reached the temperature–depth window of petroleum formation. The lack of Cl–δD correlation indicates the origin of the water and that the solutes were decoupled. (Data from Kharaka et al., 1973.)

The meteoric formation waters subsided into the petroleum temperature window. The $\delta^{18}O$ values plot in Fig. 10.9 to the right of the global meteoric water line, but not along a mixing line. Thus an oxygen isotope shift is seen, indicating the formation waters were buried with their host rocks when the latter entered the petroleum formation depth–temperature window.

The δD values reveal no correlation with the Cl concentration (Fig. 10.9), indicating the origins of the water and the dissolved salts were decoupled, in good agreement with a model of meteoric water that washed down brine-spray salts.

10.5 Shallow Formation Water and Petroleum in Devonian Rocks, Eastern Margin of the Michigan Basin

Saline groundwaters were encountered at depths of 90 to 140 m in Devonian oil-bearing strata in the shallow eastern part of the Michigan Basin, southwest Ontario. The groundwaters were tapped in the same formations at localities that are 10 to 20 km apart. Data reported by Weaver et al. (1995), compiled in Table 10.2, lead to the following observations and deducible scenario:

Variable compositions, different from recent groundwater, disclose these are fossil formation waters stored in isolated rock-compartments. A wide range of ion concentrations has been seen, e.g., Cl occurs in the range of 8 to 31 g/L—highly different from the local recent groundwater, indicating that we deal with fossil brines. The large differences in the concentrations of the various dissolved ions (Fig. 10.10), reported from wells sometimes only a few hundreds of meters apart, indicate storage in separated rock-compartments. The waters differ in salinity and the relative abundance combinations, mainly of Br, HCO_3, and SO_4.

Brine-spray tagging. The Cl/Br ratio is around 130 (Fig. 10.11), ruling out halite dissolution and indicating that both ions originated from residual evaporitic brines, a conclusion supported by the prominence of $CaCl_2$. This origin was not from direct entrapment of residual brines, but from brine-spray, as is indicated by the next observation.

Meteoric origin. The isotopic composition of the waters is distinctly light: $\delta D = -83$ to -36‰ and $\delta^{18}O = -9.5$ to -5.9‰ (Fig. 10.12), disclosing nonevaporated meteoric water. The last characteristic is emphasized by the location of the data along the global meteoric water line. Thus entrapment of a residual evaporitic brine is ruled out, and the discussed fossil groundwaters must have been formed as nonevaporated groundwater that washed down brine-spray and evaporitic dust.

No ^{18}O shift. The present case study is unique as it depicts no isotopic oxygen shift, of the kind we have seen to be typical to formation waters

Table 10.2 Composition of Some Formation Waters in the Shallow Eastern Part of the Michigan Basin, Southwest Ontario

Field	Well	Na (mg/L)	Ca (mg/L)	Mg (mg/L)	K (mg/L)	Sr (mg/L)	Cl (mg/L)	Br (mg/L)	SO$_4$ (mg/L)	HCO$_3$ (mg/L)	δD (‰)	δ^{18}O (‰)
Petrolia	LAI-2	3,470	1,180	715	91	23	8,650	58	1,700	264		
	LAI-3	4,830	1,580	898	115	29	12,000	91	2,200	325		
	WB-11	4,390	1,580	908	117	30	10,900	78	2,000	258		
	WB-2	3,870	1,400	793	105	26	9,920	82	1,940	258		
	WB-7	3,420	1,140	675	85	24	8,870	62	1,380	248		
N. Oil Sp.	LBO-2	4,030	1,530	940	132	34	11,300	99	448	40		
	LBO-3	3,130	1,640	920	125	35	10,400	69	630	113		
	CFN-14	2,390	1,310	608	85	29	7,990	66	796	219		
S. Oil Sp.	CFN-161	10,900	5,990	2,750	445	100	31,400	277	1,240	328		
	CFN-C	8,690	3,830	2,430	307	69	27,400	200	7,600	221		
	CFN-E	8,090	4,020	2,270	288	67	26,800	216	1,220	328		
Bothwell	LBH-1	4,450	1,700	1,170	152	32	13,200	47	1,400			
	LBH-2	4,950	2,000	1,320	156	33	14,300	86	2,340			
	LBH-3	4,900	1,940	1,340	169	36	14,100	97	2,140			

Source: Data from Weaver et al. (1995).

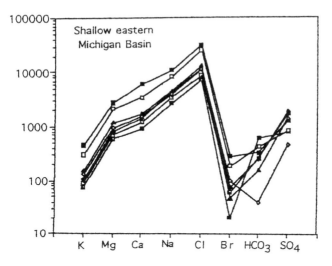

Fig. 10.10 Fingerprint diagrams of formation waters encountered in Devonian rocks at depths of 90 to 140 m at the eastern shallow part of the east Michigan Basin. The compositional variability, mainly revealed by the concentrations of SO_4 and HCO_3, indicates entrapment in hydraulically isolated rock-compartments. The salinity differences provide another indication of entrapment in separated compartments. (Data from Weaver et al., 1995.)

associated with petroleum. Two explanations are possible; (1) the host rocks provided no exchangeable ^{18}O or (2) the associated petroleum was formed at a temperature that was somewhat lower than usual.

External ions rule out a list of water–rock interactions. Distinct linear positive correlations are observed in Fig. 10.11 among the concentrations of Cl, Br, Na, Ca, Mg, K, and Sr. This indicates that not only are Cl and Br allochthonous and originated from brine-spray, but the bulk of Na, Ca, Mg, K, and Sr have the same origin. The named positive correlations rule out in the present case study in situ water–rock interactions such as dolomitization and de-dolomitization (which would result in a negative Ca–Mg correlation), as well as absorption, ion exchange, or membrane filtration.

Conservative ions. The positive linear correlations observed in Fig. 10.11 also indicate that the original concentrations of Cl, Br, Na, Ca, Mg, K, and Sr were preserved in the aquifers, and these ions were not involved in secondary water–rock interactions to any noticeable degree.

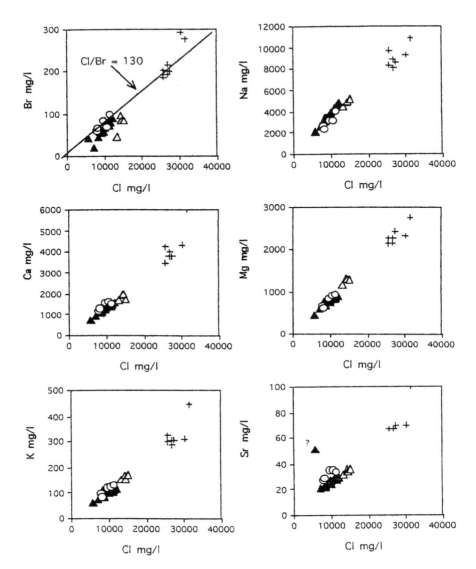

Fig. 10.11 Composition diagrams of formation waters from the eastern shallow part of the Michigan Basin. (▲) Petrolia, (○) N. Oil Springs, (+) S. Oil Springs, (△) Bothwell. The already familiar linear positive correlations indicate that Cl, Br, Na, Ca, Mg, K, and Sr had a common allochthonous source, suggested by the Cl/Br ratio to have been brine–spray. The linear correlation further indicates that these ions were not involved in secondary water–rock interactions. (Data from Weaver et al., 1995.)

Anatomy of Sedimentary Basins

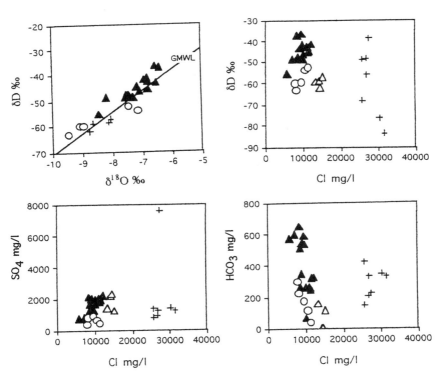

Fig. 10.12 Isotopic and chemical composition of formation waters of different oilfields in the shallow eastern Michigan Basin. (▲) Petrolia, (○) N. Oil Springs, (+) S. Oil Springs, (△) Bothwell. The isotopic values are distinctly light and plot along the global meteoric water line, clearly indicating an origin from nonevaporated meteoric water. The isotopic composition reveals no correlation to the Cl concentration, indicating the origin of the solutes was decoupled from the origin of the water. The concentrations of SO_4 and HCO_3 are not correlated to Cl, indicating independent origins. (Data from Weaver et al., 1995.)

The SO_4 and HCO_3 values reveal a marked variability in the original local paleo-environmental conditions. The concentrations of SO_4 and HCO_3 reveal a marked variability, as can be seen in Fig. 10.12. This may reflect variability of the evaporitic landscape units that fed the water that was recharged into each rock-compartment at the time of its formation.

The rock-compartments remained locked even at the shallow depth of 90 to 140 m! The discussed fossil brines and associated petroleum, located at the eastern margin of the Michigan Basin, manifest the efficient hydraulic barriers that engulf each of these compartments. The local unconfined active groundwater through-flow zone must be shallower, and no open fractures were formed during the uplift.

Conclusions: The discussed formation waters are connate waters, of Devonian confinement age. The water, organic raw materials, and hosting rocks subsided as a closed unit to a depth at which the petroleum was formed; and subsequent uplifting and erosive removal of overlying rocks placed the studied units at their present shallow depth.

10.6 Petroleum-Associated Brines in Paleozoic Sandstone, Eastern Ohio

Breen and Masters (1985) published a comprehensive set of chemical and isotopic data of formation waters that were separated from producing oil and gas wells in eastern Ohio. The pertinent geology, described in that paper, encompasses three formations rich in sandstones, subdivided and separated by shales:

- Rose Run sandstone, probably Upper Cambrian
- Clinton sandstone, Silurian
- Berea sandstone, Mississippian

The data (Table 10.3, please note the use of mmol/kg) lead to the following pattern of formation water occurrences, with direct bearing on the occurrence of the associated oil and gas:

1. A wide range of ion concentrations and different ion abundance ratios are seen in Table 10.3 and in the compositional fingerprint diagrams of Fig. 10.13. These brines are distinctly different from the local groundwaters, which are rather fresh. Hence the discussed brines were formed in different paleo-environments, and they are fossil.
2. The δD values are -48 to $-13‰$, the bulk of the data being around $-35‰$ (Fig. 10.14), i.e., well in the range of meteoric water. Hence, the brines originated as meteoric groundwaters, and entrapment of residual evaporitic brines (bitterns) or seawater can be ruled out.
3. The $\delta^{18}O$ values plot to the right of the global meteoric water line in Fig. 10.14, reflecting a temperature-induced isotope shift. This indicates that the studied formation waters subsided into the pe-

Table 10.3 Composition of Some Petroleum-Associated Waters, Eastern Ohio

	No.	Depth (ft)	Na (mmol/kg)	Ca (mmol/kg)	Mg (mmol/kg)	K (mmol/kg)	Sr (mmol/kg)	Cl (mmol/kg)	Br (mmol/kg)	HCO$_3$ (mmol/kg)	SO$_4$ (mmol/kg)	δD (‰)	^{18}O (‰)
Berea ss.	9BBEL	1,740	1,300	210	80			1,900		0.3	nd		
	3BLIC	735	1,000	86	43	0.3	1.6	1,200	1.3	2	0.15	−34.5	−5.5
	13BMAH	620	1,300	210	100	0.7		1,900	5.5	1.2	0.6		
	14BMED	460	1,100	110	58	5	nd	1,400	3.7	0.3	0.3		
	5BMEI	1,816	1,700	220	120	3	nd	2,400	5	nd	nd		
	8BMON	2,162	1,100	170	46			1,500		0.9	nd		
	1BMOR	1,422	2,100	270	120	0.5	4.1	2,800	20	1.8	0.06	−13.5	−2.2
	2BMOR	1,560	1,800	250	110	2.3	4.1	2,600	12	2.1	1.8	−35	−2.9
	6BNOB	1,760	650	65	33	12		840		5.6	nd		
	7BNOB	1,592	790	65	38	12		1,000		0.8	nd		
	4BSCI	315	1,000	280	110	5		1,800	6	0.3	0.53		
	11BSTA	920	1,000	110	60			1,300	4.8	0.9	0.15		
	12BSTA	835	1,100	120	60	2.1		1,200	2	1.1	0.09		
	10BTUS	880	1,300	180	100			1,800	3.1	0.6			
Clinton ss.	50CMUS	4,690	1,600	430	66	34	6.7	2,500	21	nd	7.8		
	51CMUS	4,580	2,200	700	120	39	12	3,800	19	nd	9.1		
	52CMUS	4,600	2,100	700	140	40	11.6	3,700	19	nd	8.2		
	53CMUS	4,030	2,100	540	110	22	9.2	3,300	17	0.5	9.3		
	54CMUS	4,030	2,000	520	110	29	8.4	3,200	20	1.3	10		
	55CMUS	4,070	2,900	440	170	52	5.8	4,100	16	0.7	7.6		
	56CMUS	4,000	1,800	530	67	27	8.7	3,000	15		7.4		
	57CMUS	4,636	2,200	750	130	37	12.8	4,000	21		9		
	58CMUS	4,370	2,000	650	100	37	11.3	3,700	18		7.4		
	59CMUS	4,129	2,200	810	160	32	13.6	4,200	22		7.4		
	60CMUS	3,641	1,700	580	130	27	9.1	3,100	16		10		
Rose Run ss.	1RCOS	6,225	2,000	980	270	90	3.4	7,400	41	0.86	4.6	−33	−1.6
	2RCOS	6,620	2,000	880	260	92	3.4	5,500	40	1.7		−33	−1.4
	3RCOS	6,550	2,100	900	260	88	8.2	6,200	37	0.8		−32.5	−1.3
	4RCOS	6,140	2,800	920	220	88	7.5	4,400	62	1.6	1.2	−32.5	−1.5
	5RCOS	6,145	2,800	1,000	2,100	81	0.9	4,800	63	1	1.2	−40.5	−4.2
	6RCOS	6,155	3,000	940	190	84	7.5	4,800	66	2.1	1.5	−30	−1.3
	7RCOS	6,490	3,600	940	280	82	nd	4600	22	0.2	8.5	−31	−3.7

Source: Data from Breen and Masters (1985).

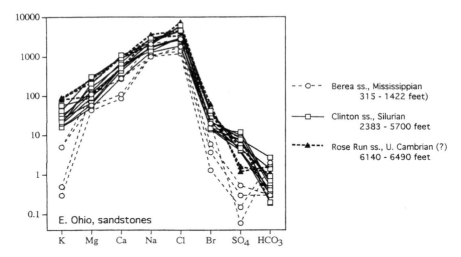

Fig. 10.13 Composition fingerprint diagrams (data in mmol/kg) of petroleum-associated brines from three Paleozoic sandstone assemblages, eastern Ohio (Table 10.3). A distinct variability is observed in the salinity and the relative ion abundances, indicating existence of a multitude of distinct brine types, all very different from the local recent groundwater. An overall pattern is Cl > Na > Ca > Mg > K > Br ≥ SO$_4$ ≃ HCO$_3$.

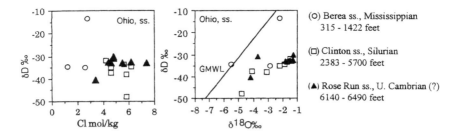

Fig. 10.14 Isotopic composition of petroleum-associated brines from three Paleozoic sandstone assemblages, eastern Ohio (Table 10.2). The δD and Cl concentrations are noncorrelated, indicating their origins were decoupled. The δD values are in the range of meteoric waters, indicating the brines originated as saline groundwaters and excluding entrapment of evaporitic brines or seawater. The δ^{18}O values reveal an isotopic shift, reflecting the deep paleo-burial into the petroleum production temperature window.

troleum formation window and have been associated with the petroleum since its formation.
4. The plot of δD as a function of the Cl concentration reveals that the two components are not correlated, indicating their origins were decoupled.
5. Figure 10.15 depicts compiled depth profiles for ion concentrations in wells that tap the Clinton and Berea sandstones. Distinct variations are seen between samples from different depths. In Fig. 10.15a at least nine water types are identifiable, and in Fig. 10.15b around five water types are seen. This indicates that different water types are contained within distinct strata that are hydraulically separated. Containment in separated rock-compartments is a common observation in producing oil and gas fields, evidenced by chemical variability as well as by fluid pressure differences (Ortoleva, 1994a,b).
6. Rather low Cl/Br weight ratios typify the Ohio formation waters: a range of 32 to 95 in the Rose Run sandstones, 32 to 116 in the Clinton sandstones, and 62 to 266 (and a single value of 409) in the Berea sandstones. The variability of the Cl/Br ratios is another reflection of the presence of distinct types of formation waters that were tagged by different brine-sprays.
7. A pronounced Ca–Cl component is present, disclosing tagging by brines that passed the stage at which halite started to precipitate and dolomitization went on.
8. Composition diagrams of several ions are presented in Fig. 10.16 for water samples from the Berea sandstone and samples from the Clinton sandstone. In both cases clear linear correlations are seen among the Cl, Na, Ca, and Mg concentrations, indicating these ions have a common allochthonous source and originated from the same brine-spray and evaporitic dust.
9. Of special interest are the Sr data, available for the Clinton sandstone brines—they plot on a perfect linear correlation line with Ca, as is seen in Fig. 10.16. This indicates that the Sr is allochthonous as well and did not stem from interactions with the host rocks.
10. No correlations are observed in the Ohio brines between the concentration of Cl and HCO_3 or SO_4, reflecting variability of the composition of the tagging brines and/or local water–rock interactions.

These observations and deduced conclusions clearly disclose that these are connate groundwaters, introduced into the host rocks when the latter were subaerially exposed in a vast evaporitic paleo-environment.

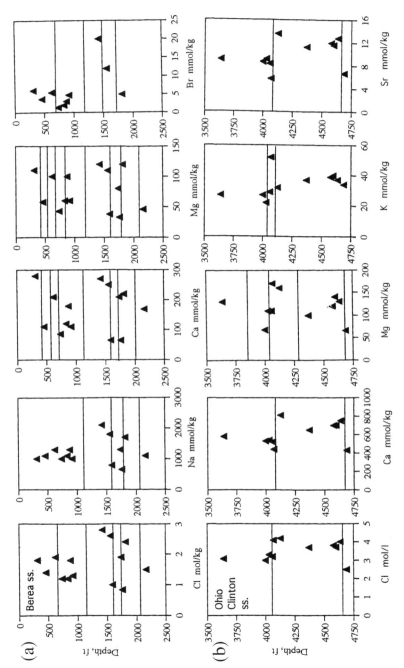

Fig. 10.15 Compositional depth profiles of petroleum-associated brines from Paleozoic sandstone assemblages, eastern Ohio (Table 10.2): (a) Berea sandstone; (b) Clinton sandstone. Abrupt changes in concentration (marked by horizontal lines) reveal that the brines are stored within hydraulically isolated rock-compartments. Measurements of many ions are required to gain high resolution of rock-compartment identification. About ten compartments are defined in the Berea sandstone data and at least five in the Clinton sandstone data.

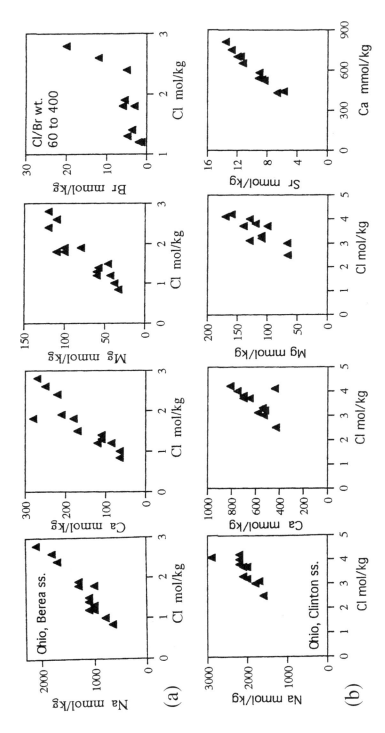

Fig. 10.16 Chemical composition diagrams of petroleum-associated brines from Paleozoic sandstone assemblages, eastern Ohio (Table 10.2). (a) Brines within the Berea sandstone (Mississippian, 315 to 1422 ft deep); (b) brines from the Clinton sandstone (Silurian, 2383 to 5700 ft deep). The linear correlations indicate a common allochthonous origin. (Data from Breen and Masters, 1985.)

10.7 Formation Waters of the Mississippi Salt Dome Basin Disclose Detailed Stages of Petroleum Formation

A comprehensive study by Kharaka et al. (1987) provides a wide spectrum of data of brines, separated at 1900 to 4000 m-deep wells that tapped Upper Jurassic to Upper Cretaceous rocks at six oilfields within the Mississippi Valley. This extensive set of data suits as a sort of summary case study, covering the list of subtopics already discussed for so many other case studies.

A perfect temperature-depth profile indicates the studied samples were well collected, ruling out dilution or mixing within the wells. The temperature measured in the sampling points is plotted as a depth profile in Fig. 10.17, revealing a gradient of 2.2°C/100 m. The constant gradient reveals that the sampling points represented single fluid systems, with no short-circuiting between fluids of different depths.

The formation waters differ from local recent groundwater and, hence, are fossil. The composition of the studied formation waters is drastically different from the composition of currently recharged groundwater in the study area, indicating the formation waters were formed at entirely different environmental conditions, and hence these waters are all fossil.

Diversity of the formation waters discloses entrapment of the water and associated oil in many separated rock-compartments. The compositional

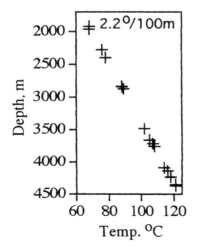

Fig. 10.17 A depth–temperature profile of sample collection points, the Mississippi Salt Dome case study. The excellent correlation confirms that the reported data are of high quality, as the samples were obtained from single fluid systems. (Data from Kharaka et al., 1987.)

fingerprint diagrams (Fig. 10.18) reveal a complex of many formation water types that differ from one another by their composition, disclosing entrapment within distinct rock-compartments that are separated from one another by a network of impermeable barriers.

Compartmentalization is conspicuous. Ionic depth profiles reveal frequent abrupt changes in the concentration, reflecting that different water types are contained in separated host strata (Fig. 10.19). Multiparametric measurements and high sampling resolution are needed to identify the whole list of different water types enclosed in hydraulically separated rock-compart-

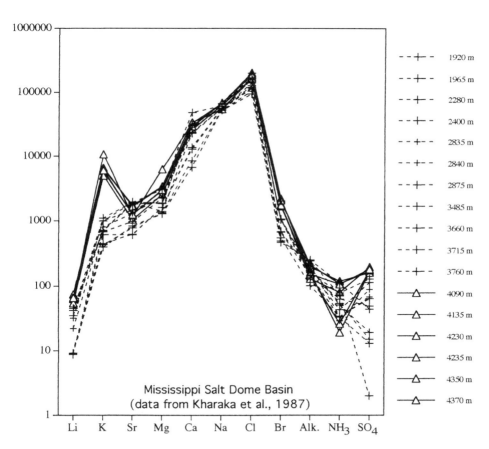

Fig. 10.18 Composition fingerprint diagrams of water samples collected at oil-producing wells, the Mississippi Salt Dome, from Upper Jurassic (△) and Upper Cretaceous (+) rocks. A large composition diversity is seen, indicating different water types are present within the studied area, encountered at a depth interval of 1920 to 4370 m. (Data from Kharaka et al., 1987.)

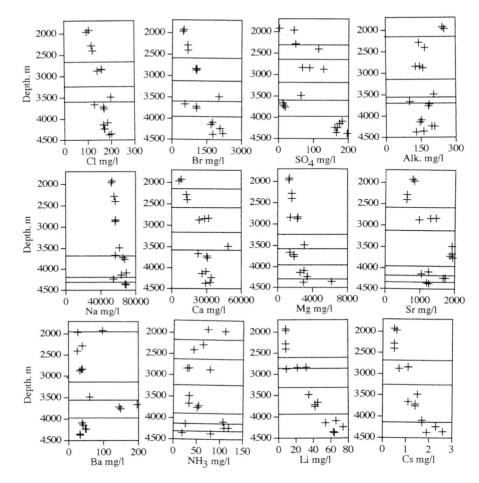

Fig. 10.19 Composition depth profiles, the Mississippi Salt Dome. Different water types are seen to reside at different depths, as is revealed by abrupt concentration changes (marked by horizontal lines). Different ions reveal discontinuities at different depths, indicating that for a good resolution of the water types analyses of many parameters are required. Together at least seven different formation water strata are identifiable. (Data from Kharaka et al., 1987.)

ments. With Cl alone one could get the impression of a steady salinity increase with depth, but the overall picture seen in Fig. 10.19 is of many distinct layers, with no depth-related general trend. At least seven distinct "layers" are distinguishable.

Linear correlations of ionic concentrations reveal an allochthonous origin. Besides the variability of the formation waters, revealed in Fig. 10.18, there is also an observable common relative abundance pattern:

$$Na \geq Ca > Mg > Sr \geq K > Li \text{ and } Cl \gg Br > Alk.$$

The same family of formation waters encountered in different host rocks discloses external origin. The mentioned general chemical similarity (Fig. 10.18) and the positive concentration correlations (Fig. 10.20) are observed irrespective of the lithology of the individual host strata. The rocks hosting the studied formation waters include different kinds of marl, shale, chalk, limestone, dolomite, cemented sandstone, sandstone, glauconitic sandstone, mudstone, siltstone, and conglomerate. This is another proof that the bulk of the dissolved Cl, Br, Na, Ca, Mg, Li, and Sr ions are allochthonous, stemming from outside the rock systems, i.e., brought in with the water when the latter was recharged into the host rocks.

The various formation waters are all tagged by brine-spray. The different formation water types, encountered in the discussed complex of oil fields, have in common (1) the Cl/Br ratios vary in the range of 90 to 200 and (2) $CaCl_2$ is prominent. Thus all these formation waters are tagged by brine-spray.

The low Cl/Br weight ratios typify evaporitic systems that reached the halite precipitation stage, at which the more soluble Br salts became enriched in the residual brine. The presence of $CaCl_2$ discloses dolomitization that resulted in exchange of Ca for Mg.

Entrapment of seawater is ruled out by the high salinity and chemical composition. The encountered saline groundwaters cannot be entrapped seawater because the concentration of the Cl is six to ten times the concentration in seawater. The relative abundance of the other ions are much different from the marine values as well, e.g., the Cl/Br weight ratio is significantly lower than the seawater value, and $CaCl_2$ is prominent.

Diffusion of water that dissolved halite is ruled out by the low Cl/Br ratios. Water that dissolved halite has Cl/Br weight ratios of >3000, whereas the discussed formation waters have Cl/Br ratios in the range of 90 to 200. This and the significant Ca–Cl component rule out in the Mississippi Salt Dome case an origin by upward diffusion of brines formed by dissolution of the large amount of halite present in the underlying Louann Salt Formation.

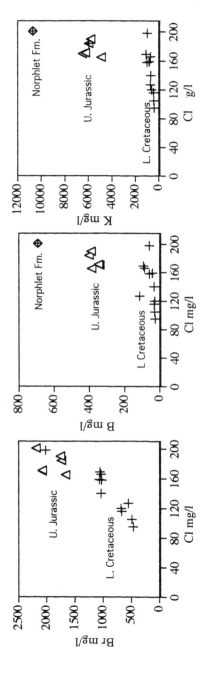

Fig. 10.20 Composition diagrams of Br, B, and K as a function of the Cl concentration, Mississippi Salt Dome. Three distinct composition groups of different ages reveal that the local environmental conditions changed over the respective geological periods. (Data from Kharaka et al., 1987.)

The evaporitic environment varied over the geological periods. Within the mentioned general composition pattern, there is a distinction between the Upper Cretaceous and Upper Jurassic water, as is seen in the fingerprint diagram (Fig. 10.18). The B and K plots as a function of Cl (Fig. 10.20) reveal a distinction also between the Norphlet Formation (lowest in the Upper Jurassic formations) and the rest of the Jurassic formations. Thus three very distinct water groups are seen to have evolved over the geological time in the Mississippi Salt Dome Basin.

The concentration of K, Ca, and SO_4 reveal differences in local paleo-environments and/or water–rock interactions. Besides the common compositional characteristics discussed above, the fingerprints of Fig. 10.18 reveal marked differences between the studied water samples, not only in the total salinity but also in the relative abundances of the different ions, especially K, Ca, and SO_4. This variability is another indication that the studied formation waters are stored within hydraulically isolated rock-compartments, and the observed salinity variations are indigenous and not at all the result of sampling-induced dilution or mixing.

These differences between the formation waters stored in adjacent rock-compartments could result from (1) differences in the local evaporitic paleo-environment at the time and location when the individual paleo-groundwaters were formed and entrapped and/or (2) local water–rock interactions.

A continental meteoric origin of the waters is disclosed by the light δD values. The water stored in the Upper Cretaceous rocks has δD values of around $-12‰$, i.e., distinctly lighter than seawater and residual brines. Hence a meteoric origin is identifiable. Some evaporation losses may be seen, as formation waters elsewhere have more negative δD values. The values of the water in the Upper Jurassic rocks is around $-2‰$, reflecting a different paleo-environment, possibly a higher degree of evaporation.

Temperature-induced ^{18}O enrichment discloses that the formation water and associated organic raw materials subsided together into the petroleum formation window. Two clusters are seen in the δD–$\delta^{18}O$ diagram (Fig. 10.21) of the Upper Cretaceous and the Upper Jurassic waters. Both groups are located right of the global meteoric water line (GMWL), a pattern suggested to reflect a temperature-induced ^{18}O shift. The shift is more pronounced in the deeper Jurassic water group, reflecting its deeper burial temperature or a higher degree of evaporation prior to entrapment.

The answer to the question of which of these two possibilities is the right one may come from a study of the associated petroleum deposits: Was the oil stored within the Jurassic rocks exposed to a higher temperature?

The water and salts had independent origins. The δD–Cl graph (Fig. 10.21) reveals distinct Jurassic and Cretaceous groups, with no δD–Cl correlation within each group. Thus the origins of the water and of the

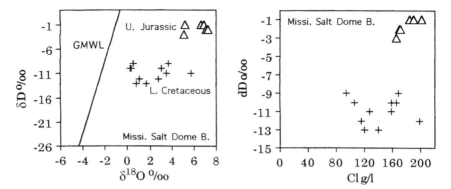

Fig. 10.21 The δD–$\delta^{18}O$ and δD–Cl diagrams of formation waters encountered within the Mississippi Salt Dome. A pronounced ^{18}O enrichment is seen, indicating a temperature-induced isotopic exchange with the host rocks occurred. This discloses, in turn, that the formation water subsided into the temperature window of petroleum formation. In the δD–Cl diagram no correlation is observed, indicating separate origins. (Data from Kharaka et al., 1987.)

dissolved ions were decoupled. The recharged water came from extra-strong rain events that caused water to infiltrate efficiently and reach the local paleo-groundwater systems. The salts were airborne and accumulated on the ground until the recharge water washed them into the rock systems.

Conclusion: These are connate groundwaters. The indicated structure of hydraulically isolated rock-compartments that host a variety of fossil water types implies that each compartment was separately filled with its distinct type of fossil groundwater. This could happen only during the time that each host stratum was exposed on the continent and got filled with continental groundwater. Upon subsidence and coverage by overlying confining rocks, the paleo-groundwater was entrapped and continued to subside together with the host rocks. The distinct compositions that are often observed to differ between adjacent wells or between different depths within the same borehole indicate the fossil groundwaters were not intermixed. Thus they were preserved within their host rocks, representing genuine connate waters.

The confinement age of the formation waters is of Upper Jurassic to Upper Cretaceous. The stratigraphic sequence at the study area is rather continuous, in spite of the common interphases of sea regressions. Thus the confinement age of the formation waters is close to the age of the host rocks, and this is also the age of the organic raw material.

The age of the petroleum deposits is somewhat younger, defined by the time of subsidence to the depth of high-enough temperature.

Anatomy of Sedimentary Basins   235

Dominance of evaporitic rock facies. Kharaka et al. (1987) provide the following summary of the rock facies:

> The Mississippi Salt basin contains a maximum of 8000 m of siliciclastic and carbonate sedimentary rocks and evaporites. The bulk of the sedimentary rocks are Jurassic and Cretaceous in age; however, up to 2000 m of Paleogene deltaic, nonmarine and continental sandstones and shales are present in this basin. The Mississippi Salt Dome basin contains 50 piercement type salt domes which, together with their associated faults, provide the most important entrapment mechanisms for petroleum in the basin. The Louann Salt, of Jurassic age, provided the salt present in these domes. This formation, consisting of up to 2000 m of massive, coarsely recystallized halite with minor amounts of anhydrite, is the oldest formation encountered in the Mississippi Salt Dome basin.

Conclusion: The initial rock-compartments included all the ingredients that were necessary for the formation of oil deposits. These included organic raw material, water, and porous rocks in which the formed petroleum could concentrate as a deposit.

10.8 Norwegian Shelf: Petroleum-Associated Formation Waters, Upper Triassic to Upper Cretaceous

10.8.1 Upper Triassic–Jurassic

Egeberg and Aagaard (1989) presented data of petroleum-associated formation waters from Upper Triassic to Jurassic sandstones, interbedded with shales and mudstone, located in the Norwegian shelf. The data, plotted in Fig. 10.22, reveal the following pattern:

1. A good temperature–depth correlation is observed, indicating that each well tapped only one rock unit, ruling out intermixing of water from different depths, which is often so troublesome in studies of this nature.
2. A large variability of ion concentrations is observed, indicating entrapment in hydraulically isolated rock-compartments.
3. The Cl/Br ratios are in the range of 142 to 247, well in the limits of evaporite brines and thus indicating these two anions are allochthonous and of an evaporitic origin.
4. The concentration of Na, Ca, Mg, K, and Sr are positively correlated with the Cl concentrations, indicating that these cations are of the same external source of evaporitic origin.

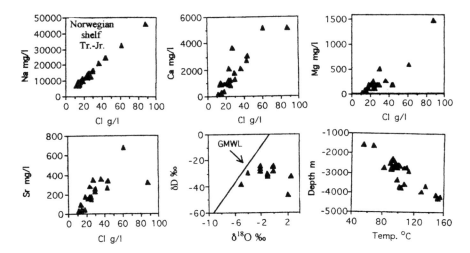

Fig. 10.22 Composition of petroleum-associated formation waters from the Norwegian Shelf, hosted in Upper Triassic–Jurassic sandstones, interbedded with shales and mudstones. The familiar positive linear correlation lines are seen between the concentrations of Cl, Ca, and Sr, revealing an external origin; the δD reveals we deal with meteoric waters; and the $\delta^{18}O$ reveals deep subsidence. (Data from Egeberg and Aagaard, 1989.)

5. The δD values, around $-26‰$, indicate an origin from non-evaporated meteoric water.
6. The $\delta^{18}O$ values disclose an ^{18}O shift (Fig. 10.22), indicating the water subsided into the petroleum formation temperature window.

Conclusion: These observations indicate that the discussed formation waters are meteoric, brine-tagged, and of an Upper Triassic–Jurassic confinement age.

10.8.2 Upper Cretaceous Host Rocks

A set of petroleum-associated formation waters, encountered in Upper Cretaceous chalk reservoirs, in the shelf offshore Norway has been presented by Egeberg and Aagaard (1989). The data, plotted in Fig. 10.23, reveal the following pattern:

1. The samples, retrieved from a depth range of 3.2 to 4.6 km, revealed a remarkable depth–temperature correlation (Fig. 10.23). This is taken as an indication that the wells were well cased so that no short-

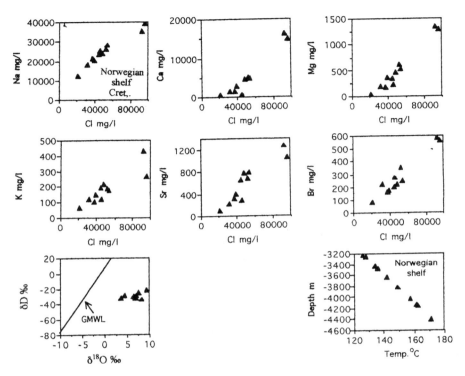

Fig. 10.23 Composition of formation waters from the Norwegian Shelf hosted in Upper Cretaceous rocks. The classic patterns are observable. (Data from Egeberg and Aagaard, 1989.)

circuiting of water from different depths occurred, and each water sample represents a specific host rock unit. The samples are thus confined and fossil.
2. The Cl/Br ratios are in the range of 140 to 260, ruling out halite dissolution and revealing an affinity to residual evaporitic brines. Thus the Cl and Br are allochthonous.
3. The concentrations of Na, Ca, Mg, K, and Sr are seen to be linearly correlated to the concentration of Cl, indicating these cations are of the same external origin related to evaporitic brines.
4. The δD values are light, in the range of -34 to $-22‰$, i.e., resembling nonevaporated meteoric water.
5. The $\delta^{18}O$ values reveal the ^{18}O shift.

Conclusion: These features indicate that the discussed deep formation waters are brine-spray-tagged fossil meteoric groundwaters of an Upper Cretaceous confinement age.

10.9 Lithostratigraphic Controls of Compartmentalization Were Effective from the Initial Stage of Subsidence and Further Evolved Under Subsidence-Induced Compaction

Formation waters, which were all formed at the land surface, provide evidence that compartmentalization started with confinement during the initial stages of subsidence. The lithostratigraphic sequence of all sedimentary basins is composed of hundreds of alternations of rock strata that are capable of hosting fluids, e.g., sandstone and fractured limestone or dolomite, and impermeable rocks, e.g., clay, shale, marl, or mudstone. The rocks of the second group provide efficient hydraulic sealing at shallow depths, as is evidenced by relatively young shallow pressurized water occurrences, e.g., at depths of 50 to 300 m in the Dead Sea rift valley (Mazor et al., 1995), or uplifted petroleum-containing strata that preserve their fluids at depths of 140 to 200 m in the eastern margin of the Michigan Basin (Weaver et al., 1995).

At greater depths the hydraulic barriers have to withstand greater pressures, induced by the overlying rocks, but at the same time compaction is increasing the sealing efficiency of the low-permeability rocks. Clay, shale, and marl are plastic, a property that is enhanced with temperature increase, and compaction is gradually closing in these rocks' minifractures and interconnected pores, thus increasing their sealing ability.

Furthermore, the plastic sealing rocks were squeezed into fractures developed in the competent rocks, along fault planes and along bedding thrust planes. The results were twofold: Large water-containing rock-compartments were subdivided into smaller compartments, and the wrapping of competent water-containing rock units by hydraulic barriers was improved.

Martinsen (1997) reached the following conclusion in regard to fluid-pressure compartments: "Studies in the Powder River Basin of Wyoming indicate that the boundaries of individual fluid *pressure-compartments* commonly correspond to the boundaries of various stratigraphic elements (e.g., lithofacies and unconformities)."

Thus independent types of evidence, of formation waters with varying compositions and of fluids with varying pressures, lead to the conclusion that lithostratigraphic controls of compartmentalization were effective, and they were formed at the initial stages of subsidence and confinement.

10.10 Summary Exercises

Exercise 10.1: Let us study Table 10.1. Rearrange the table in some logical order. What can be seen? Discuss.

Exercise 10.2: Certain observations are common among the case studies presented in this chapter and portrayed in all its figures. What are these common observations? To which key conclusions do they lead?

Exercise 10.3: Give a synonym for the following terms: *allochthonous ions* and *autochthonous ions*. Discuss these terms.

Exercise 10.4: Presence of a significant concentration of $CaCl_2$ is interpreted as indicating tagging by brine-spray. On which observations is this based?

Exercise 10.5: There are cases where formation water and associated oil are encountered at a depth that is distinctly shallower than the deduced depth of ^{18}O exchange. What may be concluded?

Exercise 10.6: Can Tables 10.1 and 10.2 provide a clear answer as to whether the respective formation waters are entrapped seawater, residual brines, or meteoric water? Discuss.

Exercise 10.7: From where arrived the salts dissolved in the formation waters described in Tables 10.1 and 10.2?

Exercise 10.8: The depth profiles of petroleum-associated formation waters of the Kettleman Dome reveal in Fig. 10.6 a significant composition variability. Can this be the result of different water–rock interactions?

11
EVOLUTION OF SEDIMENTARY BASINS AND PETROLEUM HIGHLIGHTED BY THE FACIES OF THE HOST ROCKS AND COAL

So far we have applied the characteristics of formation waters to understand the history, compartment structure, and hydrology of sedimentary basins—topics that have direct bearing also on the understanding of petroleum systems. In the present chapter let us have a glimpse at the more traditional tool of petroleum research and exploration, namely, the composition and facies of the rocks that fill the basins.

11.1 Sediments Formed in Large-Scale Sea–Land Contact Zones

Scanning the petroleum-related literature it is obvious that a majority of systems, studied all over the globe, is in one way or another connected to a sea–land transition zone, or large flatlands that were subjected to frequent sea transgressions and regressions. The following are a few case study examples in which the geological history was deciphered via the characteristics of the involved rocks.

11.1.1 Transgressive–Regressive (T–R) Cycles Disclosed by Mississippian Petroleum-Containing Rock Formations, Central Kansas

The following are quotations extracted from a paper by Watney et al. (2001):

"The initial mudstones to sponge-spicule... were deposited in transgressive–regressive (T–R) cycles on a shelf margin setting, resulting in a series of shallowing-upward cycles."

"After early silicification, inter- and post-Mississippian subaerial exposure resulted in further diagenesis, including sponge-spicule dissolution. ... *Meteoric water infiltration* is limited in depth below the exposure surface and in distance downdip into unaltered, cherty Cowley Formation facies."

"From bottom to top in a complete cycle seven lithofacies are present: (1) argillaceous dolomite mudstone, (2) argillaceous dolomite mudstone that has chert nodules, (3) clean dolomite mudstone that has chert nodules, (4) nodular to bedded chert, (5) autoclastic chert, (6) autoclastic chert that has clay infill, and (7) bioclastic wacke-grainstone. The uppermost cycle was terminated by another lithofacies, a chert conglomerate of Mississippian and/or Pennsylvanian age."

"Investigations in surrounding areas suggest that the Cowly Formation consists of Transgressive–Regressive (T–R) cycles that have complex stratal architecture exhibiting lithofacies interpreted as shallowing-upward successions... To the north in Iowa, in a high to mid-shelf position, three high frequency T–R cycles are recognized in the Osagean."

"The four cycles noted in the core and the recognition of regional T–R cycles would indicate numerous periods of possible subaerial exposure during the Mississippian....."

"Post-Mississippian subaerial exposure and erosion along the Mississippian subcrop was substantial."

Discussion: The paper by Watney et al. (2001), quoted above, brings up lithological evidence that the studied region was during the Mississippian and Pennsylvanian a flat lowland frequented by sea invasions and retreats. This far-reaching conclusion can be well checked by analyzing the associated formation waters. In any case, this type of paleo-landscape agrees very well with the general conclusion derived in the previous chapters that dealt with formation waters, namely, that the sediment collection was by and large in flat sea–land lowlands of evaporitic facies.

11.1.2 A Thick Sequence of Marginal Marine and Continental Sediments: Petroliferous Ghadames Basin, North Africa

Acheche et al. (2001) described and analyzed the rock facies of the Ghadames Basin. It is over 1000 km across and is located in parts of Tunisia and Algeria. The sedimentary column reaches up to 7000 m, from the Cambrian to the Senonian. The following sequence of stratigraphic units is observed in the many petroleum exploration and exploitation drill holes. Five "stratigraphic packages" were recognized, separated by four regional unconformities:

1. Cambrian: 550 m of continental mainly fluvial sandstone, covered by 400 m of Ordovician transgressive marine shales, that are truncated by the Taconian unconformity.
2. Early Silurian represented by 550 m of marine shales and black organic-rich muds. Above are 700 m of marine sandstones and shales, truncated by the Caledonian unconformity.
3. The Devonian is represented by 250 m of continental shallow marine sandstones; overlied by 200 m of shales. These are overlied by 200 m of argillaceous marine sediments. Above are around 2000 m of Late Devonian and Carboniferous shallow marine and deltaic sediments. These sediments are truncated by the Hercynian unconformity.
4. The following sediments are 150 m of lower Triassic fluvial-deltaic deposits, overlied by 500 m of alternations of sandstones and shales of lacustrine/lagoonal/evaporitic sediments, which are covered by 1150 m of fluvial to lacustrine sediments, followed by 400 m of continental sediments that are truncated by the Austrian unconformity.
5. Cenomanian and Senonian marine dolomites are covered by lagoonal sediments, attaining a thickness of 850 m.

Discussion: The listed sedimentary section of the Ghadames Basin provides a highly generalized overview. Local detailed studies revealed a much more variegated pattern, due to lateral facies changes and tectonic structures. Thus the basin seems to be highly compartmentalized. Second, it seems that over long geological periods the active landscape was a flat lowland, with frequent transgressions and regressions. However, the information attainable from the lithology has its limitations. For example, the evidene is biased in favor of marine phases of which marine sediments are well preserved, but during continental phases little or no sediments were accumulated and preserved. The common tool to identify continental stages are discontinuities, but these are easily overlooked in field sections and drilled cores.

In this context it will be most interesting to study the associated formation waters. In the previous chapters we have seen how well the identification of meteoric formation waters discloses episodes of subaerial exposure.

11.1.3 Frequent Alternations of Shallow Sea–Land Facies in Upper Paleocene Strata: Kopet-Dagh Basin, Iran

The following text and Fig. 11.1 are extracts from a paper by Mahabubi et al. (2001):

"The intracontinental Kopet-Dagh basin was formed after the Middle Triassic orogeny in northeastern Iran and southwestern Turkemenistan."

"Six stratigraphic sections in the central and eastern parts of the basin have been used to divide the Upper Plaeocene (Thanetian) carbonate supersequence into *four major carbonate lithofacies, each having multiple subfacies*. These lithofacies units represent open-marine, shoal, semi-restricted lagoon, upper tidal, and tidal-flat subenvironments that formed on a shallow carbonate rump. In addition there are two silisiclastic lithofacies consisting of calcareous shale (marl) and calcite-cemented sandstone."

"Upper Paleocene strata were deposited in about 4 m.y... We estimate that sea level fluctuations in the study area were between 5 and 11 m during development of parasequences."

"The upper Paleocene carbonates in the Kopet-Dagh basin constitute one of the potential petroleum-producing intervals that warrant exploration consideration."

"Thickness of this unit ranges from 103 m in the east to 234 m in the west... The Chehel-Kaman Formation is dominantly carbonate rocks, with silisiclastic intervals and minor amounts of evaporite."

Discussion: A number of detailed rock sections in the studied basin, described by Mahabubi et al. (2001), reveal frequent facies changes that were connected to sea level fluctuations. The latter were, in turn, translated in low flatlands to significant invasions and retreats of the sea. Analysis of formation waters that will in future studies be sampled from such sections will add a lot to our information. Based on the observations in suboceanic drilling in the Mediterranean Sea (Chapter 5) and in the many cases of meteoric saline water encountered in sedimentary basins (Chapter 7), it is anticipated that saline meteoric waters are common in the Kopet-Dagh basin as well. The study of such waters will provide an additional tool to identify continental stages in the history of this basin. Common abrupt changes of lithology may turn out to be the result of regression–transgression alternations that occurred more often than is recorded in the observable rock sequence.

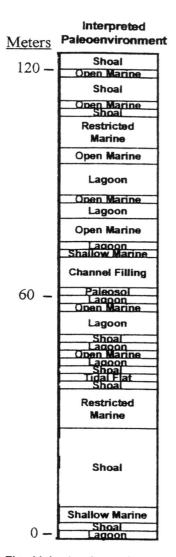

Fig. 11.1 A paleo-environment section of the Upper Paleocene in the Kopet-Dagh Basin, Iran. The frequent sea–continent shifts are evident. Over 60 different rock beds have been identified in this 120-m section. The beds include alternations of shale, a variety of carbonate rocks with different micro- and macrofossils, sandstones, gypsum, conglomerate, sandy dolomite, etc. (From Mahabubi et al., 2001.)

11.1.4 Petroleum-Hosting Chert Formed by Silicification of Anhydrite and Dolomite

Chert is a hard and dense rock, white or colored (the brown variety is also called flint). It is composed of microcrystalline to amorphous quartz. Chert has been formed by direct precipitation in a marine environment (especially so the flint), but it has been often formed diagenetically. Examples for the latter case were reported in a paper by Lubeking et al. (2001):

"Wolf Springs fields...located at the Yellowstone County, Montana...have produced more than 5.7 million bbl oil from the Pennsylvanian Amsden Formation reservoir rocks that occur in several laterally persistent and mappable zones."

"The Amsden was deposited in a peritidal to sabkha setting where evaporitic minerals, mainly anhydrite, were once common. These evaporites were partly replaced by silica (chalcedony and chert) soon after deposition. Later dissolution of the remaining evaporites soon after the silicification event, or during the pre-Middle Jurassic unconformity...produced the solution-collapse chert breccias that now serve as the best reservoir facies in the field."

"The lower part of the Amsden interval seen in cores consists of light-gray to medium-gray to reddish-gray dolomite mudstones that have some intermixed quartz sandstones and thin shale sequences."

"Textures of these samples suggest that deposition occurred in a peritidal environment (near tide level, i.e., shallow subtidal, tidal flat, and sabkha settings). Interclasts . . ., peloids, "soil" textures, fenestral vugs (now filled with anhydrite), and windblown quartz silt grains all support this interpretation. Anhydrite is faily common in this facies and occurs both as fairly early and finely crystalline masses and as later stage pore-filling and fracture-filling cements. The abundance of anhydrite and windblown silt indicates that *deposition occurred under arid conditions.*"

"Chert has always been recognized as being a constituent in the Amsden, but it was not until cores and thin sections were viewed in the Wolf Springs fields that chert as a factor in forming a major component of the reservoir became obvious. The chert, much of which formed during silicification of anhydrite...and to a lesser degree dolomite, became a brittle and easily broadcasted component of the interval in certain zones. Thus, where relatively fresh waters invaded the Amsden after precipitation of the chert, dissolution of the remaining beds of anhydrite created highly porous solution-collapse breccias."

"The Pennsylvanian was a time of relatively warm, arid climate in what is now central Montana. Thus, the shallow-water carbonate muds associated with the sabkha anhydrides were probably dominated by aragonite brines, left behind by the precipitation of the anhydrite, were undoubtedly rich in magnesium ions, and these fluids helped form dolomite in the aragonitic sediments they encountered. Thus, one of the earliest and most important diagenetic processes to affect the Amsden carbonates was penecontemporaneous dolomitization...of peritidal deposits."

"Another important diagenetic event soon after the early dolomitization was partial silicification of the evaporites...particularly of the anhydrite nodules. The source of the silica for the silicification is not apparent, but it may have been derived from salinity-tolerant salacious organisms such as diatoms, which are common in many shallow settings today. Alternatively, silica may have been derived from the associated scholastics within the Amsden. In any case, there was enough silica to partly replace former anhydrite nodules and the associated sediments with chert and length-slow chalcedony."

"Following silicification, much of the remaining evaporite in the Amsden was dissolved to create solution-collapse breccias."

Discussion: The lithological observations lead to the conclusion that (1) the chert in certain regions originated by silicification of anhydrite and hence (2) the close relation to an evaporitic paleo-environment; (3) much of the originally formed anhydrite was dissolved; (4) an arid climate prevailed on the evaporitic lowlands; and (5) the residual evaporitic brines were Mg rich, after the anhydrite precipitation, and this Mg was consumed in a dolomitization process, explaining the common occurrence of dolomite in evaporitic rock facies.

These conclusions complement and support the conclusions reached in the preceding chapters, based on the study of formation waters. It seems clear that study of both, formation waters and rocks, in petroleum zones seems very promising.

11.1.5 Deltas as a Petroleum-Forming Environment

Tissot and Welte (1984) provide the following summary (page 615):

"Large young delta areas represent a special setting for the origin, migration and accumulation of petroleum. The essential features and processes for the occurrence of petroleum are among the characteristics of a delta, provided that there is an organic rich source formation. The geologic setting is such that the marine shales of the delta front and the prodelta area, receiving mainly land-derived organic matter, can act as source rocks. The interlayered mixed sands and shales of the peralic sequence of the delta plan provide reservoir opportunities and sealing cap rocks in the direct neighborhood of the delta front. Furthermore, young deltas during their build-up and sediment accumulation undergo comparatively strong subsidence without uplifting. This is in turn essential for source rock maturation, generation of hydrocarbons, and migration. Thus the main geologic features and processes required for petroleum occurrences are suitably combined in space and time in delta areas; i.e., they offer a compact scenario of petroleum generation, migration and accumulation."

11.2 Lithological Evidence of Subaerial Exposure Phases

Unconformities observed between marine rocks are an indication of a continental phase of which no deposit was preserved. In rather rare cases other features disclose a subaerial exposure phase.

11.2.1 Meteoric Water and Karstification: Mississippian Oil and Gas Chert Reservoirs, Oklahoma

Rogers (2002) stated the following observations and conclusions:

"Study of well logs and completion records of more than 6000 wells in north-central Oklahoma shows that the chat [chert] is widespread but not continuous and that chat reservoirs are very heterogeneous."
"Oil and gas produced from such zones are generally *accompanied by salt water*."
"Eventually, a change in the conditions, such as continued uplift and a resulting *infiltration of meteoric water*, would result in flushing of the rock, dissolution of remaining calcite by low pH fluids, and no new silica precipitation... The dissolution of remaining calcite by leaching could have resulted where *meteoric water*, undersaturated with respect to calcite, flowed through the zone."
"The model suggested from this study develops chat from weathered and/or eroded Mississippian limestone that was deposited as debris flows or collapse breccias during an episode of *karst* development. The material was than subjected to silicification."

Discussion: The conclusions reached by Rogers (2002) in this section emphasize the important role of the meteoric groundwater in opening flow paths within the through-flow zone, for example, dissolution of carbonate rocks and karstification. This is a common process of erosion, but its features are rarely preserved in buried rock strata.

11.2.2 Dolines and Karstic Caves Delineate a Mid-Ordovician Unconformity, Newfoundland

Cooper et al. (2001) provide interesting descriptions of specific paleo-karst features:

"The St. George unconformity marks the end of the stable platform and base of flexural bulge mega-sequence. This unconformity represents an hiatus of 3–4 m.y."

"Dolines above and below the unconformity and peritidal dolo-laminites above the unconformity..."

"The Table Head Group represents progressive deepening from shallow to deep subtidal limestones infilling the karsted paleorelief at the St. George unconformity."

"Karst features such as cave-fill and solution pipes have been described... Discordant and bedding parallel matrix breccias support cave development in the Aguathuna Formation, and shale-filled caves are also common in the upper part of the Catoche Formation."

Discussion: These paleo-karst features are very convincing, but their geological occurrence is rare. At this point the formation waters provide an independent and much more readily available tool to identify continental exposure phases (Chapter 7).

11.2.3 Observations at the Sabkha Near Abu Dhabi, Along the Arabian Gulf

The following is a quotation from an abstract of a paper by Patterson and Kinsman (1981):

"During the last 4000 to 5000 years, sedimentary offlap and a relative fall in sea level have resulted in the development of broad, gently seaward sloping (about 1:3000) planar areas, or sabkhas, along much of the south shore of the Arabic (Persian) Gulf. The depth of the water table below the sabkha surface increases with distance from the sea until it reaches 1.0 to 1.5 m, after which it remains constant. When the water table depth is in the range of 1.0 to 1.5 m, the sabkha surface has reached a state of deflational equilibrium. The water table beneath the sabkha always slopes seaward indicating that the direction of groundwater flow is toward the sea. The subsurface flow system beneath the sabkha is part of a regional seaward flowing groundwater regime."

"The flow of sabkha brines beneath the lagoons in the Abu Dhabi area indicates a mechanism for preserving anhydrite during marine transgressions, thus permitting the development of stacked sabkha successions which are common in the rock record."

Discussion: Sabkhas developed at flat areas that became subaerially exposed as a result of a slight drop in the sea level, followed by a sea retreat. At the geological periods during which the large sediment accumulations in subsided basins took place, large-scale flatlands existed—the result of long orogenic tranquility.

11.2.4 Evidence of Subaerial Exposure and Fresh Water Involvement in Diagenesis of Devonian Cherts, Dollarhide Field, West Texas

Saller et al. (2001) provided the following information:

"Approximately 70 million bbl of oil have been produced from the chert-dominated Thirtyone Formation at Dollarhide field. The Thirtyone Formation is Devonian in age, and contains two reservoir intervals, an upper dolomite and a lower chert, separated by non-porous limestone. The chert reservoir... consists of two different facies—laminated microporous chert and burrowed chert."

"Following deposition, Devonian strata at Dollarhide underwent two periods of subaerial exposure and freshwater diagenesis."

"The upper reservoir zone includes 1 to 3 different intervals of porous dolomite that have a total net pay thickness of 0–90 ft (0–27 m)."

"Petrographic and isotopic relationships suggest that most of the chert diagenesis occurred during shallow burial in either marine or meteoric waters."

"Ultimate oil production from Dollarhide Devonian field is closely related to *chert reservoir facies and compartmentalization.*"

Discussion: The compartmentalization observed in the chert formation seems to indicate that the diagenetic process of chert formation followed the structure of already existing rock-compartments, formed at the initial stage of sediment formation and subsidence.

11.3 The Lithological Record of Inland Basins and Rift Valleys

In several rift valleys across the globe brine-tagged meteoric formation waters are encountered (section 7.2). It is of special interest to compare the type of information retrievable from rock studies to those reachable from the respective formation waters.

11.3.1 Oilfields Within a Saline Inland Basin: Qaidam, Northwest China

Let us follow the description of this interesting system from a paper by Hanson et al. (2001):

"The Qaidam basin is a non-marine, petroliferous basin that lies in the northeastern corner of the Tibetan Plateau in northwest China. The basin floor sits approximately 3–3.5 km above sea level, and the basin is surrounded on all sides by high mountains."

"Palynological studies of Cenozoic strata of Qaidam indicate that flora were principally species that were tolerant to extremely dry and saline habitats and suggest that, during most of the Tertiary, the basin was characterized by a dry climate. The dry climate and lack of significant higher plants have had a direct impact on the type of organic matter available for preservation in source rock materials in the basin."

"In simplest terms, basin fill is characterized by *stacked* coarse alluvial fan and braided stream deposits near the active basin margins that interfinger and grade laterally basinward into lacustrine deposits... Small modern lakes in the basin center continue to be highly saline in the present arid climate and are currently the site of deposition of unusual potash evaporites."

"Where thickest, Oligocene–Miocene strata exceed 3 km, whereas Pliocene and Quaternary sections exceed 5.5 and 2.8 km, respectively."

"Previous geochemical studies of produced oils from Qaidam basin have indicated the presence of two distinct genetic *groups* of oils."

"In addition...the Qaidam basin has a genetically distinct *group of freshwater lacustrine oils* that have been correlated with Middle Jurassic source rocks."

"Currently, producing *reservoirs* consist of sandstone deposited by meandering and braided fluvial systems, alluvial fan deposits, and fractured siltstones, shales, and carbonates. These reservoirs range in age from Oligocene to Pliocene."

Discussion: The Qaidam basin is an example of a large inland evaporitic system that hosts a multitude of oil bodies of different composition, entrapped in rock-compartments that were locally deposited. The concluded sedimentary history reached by studying the rocks is similar to the history deduced in inland rift valleys based on formation water studies. It seems highly recommendable that both research approaches are combined in future studies.

11.3.2 Large-Scale Tertiary Lake Sediments: The Green River Petroleum System, West-Central United States

The following are a few quotations from a paper by Ruble et al. (2001) describing the lithology of deeply buried lake sediments:

"The Tertiary Green River petroleum system in the Uinta basin generated about 500 million bbl of recoverable, high pour-point, paraffinic crude oil from lacustrine source rocks. A prolific complex of marginal and open-lacustrine source rocks, dominated by carbonate oil shales containing up to 60 wt% type I kerogen, occur within distinct stratigraphic units in the basin. Petroleum generation is interpreted to originate from source pods in the basal Green River Formation buried to depths greater than 3000 m along the steeply dipping northern margin of the basin."

"The Green River depositional system involves two early Tertiary lakes: Lake Uinta in the Uinta-Piceance basin of northwestern Utah and Lake Gosiute in the greater Green River basin of northwestern Colorado and southwestern Wyoming."

"The Uinta basin encompasses an area of approximately 24,000 km^2, and is both a topographic and structural trough filled by as much as 5000 m of late Cretaceous to early Oligocene lacustrine and fluvial sedimentary rocks..."

"When the sea withdrew because of uplifts to the west, thick Cretaceous marine shales...gradually gave way to inland flood-plain deposits of shale, sandstone, and *coal* (Mesaverde Formation), overlain by coarse clastic continental facies...at the beginning of the Tertiary."

"In its final stage, Lake Uinta regressed into the west-central region of the basin and became hypersaline, resulting in the deposition of an unusual assemblage of evaporitic minerals... Eventually the lake was filled in with coarse-grained sandstones and conglomerates during fluvial and alluvial deposition in the late Eocene to Oligocene or Miocene."

Discussion: The sediments of this highly complex and large inland system contained water when they subsided into the zero hydraulic potential zone, deep beneath their inland base of drainage. Hence they are anticipated to contain connate waters. It seems highly recommendable that formation waters of the Green River rock system be analyzed for their chemical and isotopic composition and undergo isotopic dating. The related research strategies include (1) mapping of hydraulically isolated rock-compartments and (2) distinguishing between sediments formed in fresh water lakes and those formed in identifiable evaporitic environments.

11.4 Rock-Compartment Structures and Their Evolution

In the previous chapters we came across observations, many times, that formation waters varied in their composition over rather short distances, indicating storage in hydraulically isolated rock-compartments. The following section presents lithological observations that lead to the same structural picture.

11.4.1 Petroleum in Isolated Compartments of Devonian Cherts, West Texas, and New Mexico

The following are quotations from a paper by Ruppel and Barnbay (2001):

"The lower Devonian Thirtyone Formation of west Texas and New Mexico is one of the largest chert reservoir *successions* in the world, having accounted for more than 750 million bbl of oil production."

"Proximal reservoirs, represented by Three Bar field, are composed of a single, thick, sheet like chert unit, which extends for hundreds of square miles. Heterogeneity in these reservoirs, which were formed by strike-parallel *deposition on a gently sloping outer platform during regional transgression*, is primarily a function of faulting, fracturing, and dissolution of associated carbonate along unconformities. *Small-scale (bed-scale)* heterogeneity also exits within the tabular chert body, resulting from variations in silica deposition and diagenesis between and among beds."

"By contrast, distal reservoir distribution, typified by University Waddell field, comprises *thin, vertically stacked and laterally discontinuous chert intervals* whose origin is a function of transport and deposition of siliceous sediments as debris flows and turbidites. *Flow units in these reservoirs are thin [10–20 ft (3–6 m)] and separated vertically and laterally from one another by low-permeability mud-rich, siliceous sediments and hemipelagic deposits*. The distribution of flow units is the result of both paleotopography and sea level cyclicity. *Chert units are most abundant in transgressive and early high stand legs of sea level rise–fall cycles and display offset stacking suggestive of topographically controlled sedimentation*."

"Both the lower and upper porous chert zones…contain as many as seven interbeds of nonporous chert and up to three interbeds of limestone, each more than 10 cm thick."

Discussion: The observations described for the Thirtyone Formation, west Texas and New Mexico, lead to the conclusion that during the Lower

Devonian the region was a flat lowland, of a near-shore evaporitic environment. Intensive compartmentalization is exhibited by the chert formations. Study of accompanying formation waters may provide independent clues to the occurrence of isolated rock-compartments and increase substantially the resolution of their identification and mapping.

11.5 Compartmentalization Was Effective from the Initial Stage of Subsidence and Further Evolved Under Compaction

Petroleum experts discuss pressurized compartments as entities that developed at depth, recruiting hydraulic barriers made of mineralized seals, mainly silica, or low-permeability rocks that contain multiple fluid phases (Martinsen, 1997). The reason for this notion—that compartments were formed at great depths—stems probably from the interest in oil and gas entrapment, and these fluids were formed at depth.

However, formation waters, which were all formed at the land surface, provide evidence that compartmentalization started with confinement during the initial stages of subsidence. The lithostratigraphic sequence of all sedimentary basins is composed of hundreds of alternations of (1) rock strata that are capable of hosting fluids, e.g., sandstone and fractured limestone or dolomite, and (2) impermeable rocks that act as seals, e.g., clay, shale, marl, or mudstone. The rocks of the second group provide efficient hydraulic sealing at shallow depths, as is evidenced by relatively young shallow pressurized water occurrences, e.g., at depths of 50 to 300 m in the Dead Sea rift valley (Mazor et al., 1995), or uplifted petroleum-containing strata that preserve their fluids at depths of 140 to 200 m in the eastern margin of the Michigan Basin (Weaver et al., 1995).

At greater depths the hydraulic barriers have to withstand greater pressures, induced by the overlying rocks, but at the same time compaction is increasing the sealing efficiency of the low-permeability rocks. Clay, shale, and marl are plastic, a property that is enhanced with temperature increase, and compaction is gradually closing in these rocks minifractures and interconnected pores, thus increasing their sealing ability. Furthermore, the plastic sealing rocks were squeezed into fractures developed in the competent rocks along fault planes and along bedding thrust planes, a process that caused subdivision into smaller compartments.

11.6 Summary Exercises

Exercise 11.1: Take the time to review a dozen monthly issues of the American Association of Petroleum Geologists (AAPG) Bulletin. Name at

least a dozen facies types of rocks that are discussed in the petroleum-related papers. Try to "get the flavor" of this type of case study presentation.
Exercise 11.2: Is there a geochemical interconnection between dolomitization and evaporitic processes?
Exercise 11.3: What is a diagenetic process? Give some examples.
Exercise 11.4: What type of lithological properties indicate a subsurface structure of rock-compartments?
Exercise 11.5: Do we have the means to conclude that certain shallow rock units were once buried substantially deeper?

12
PETROLEUM AND COAL FORMATION IN CLOSED COMPARTMENTS—THE PRESSURE-COOKER MODEL

The many discussed case studies and included observations and conclusions lead to the conclusion that petroleum was formed in rock-compartments that contained the needed ingredients, namely, organic materials, porous rocks, and formation water. In the present chapter this conclusion is presented under the pressure-cooker model of petroleum formation.

The previously discussed fruitcake structure of isolated rock-compartments, that host various fluids, contradicts the traditional notion of large-scale petroleum migration, so let us first discuss this paradigm.

12.1 Did Petroleum Migrate Tens and Even Hundreds of Kilometers?

Extended migration of petroleum is assumed in the literature for many case studies, and it is somehow taken as self-evident. However, the fruitcake structure of isolated rock-compartments separated by a network of impermeable barriers indicates migration could occur only within the hydraulically interconnected space of the individual rock-cells. Let us have a look at the petroleum migration notion.

12.1.1 Buoyancy is Recruited to Explain Lateral Migration over Hundreds of Kilometers

A book, *Petroleum Migration*, edited by England and Fleet (1991) includes 16 articles that model petroleum migration. The introduction includes on page 3 the following statement:

> Secondary migration of petroleum through a carrier bed from a source rock, or source-interval containing mudrock, to a reservoir is driven mainly by the buoyancy which results from the density contrast between the petroleum and the water which saturates the carrier bed... If there is hydrodynamic flow, it may act with buoyancy or against it. The main resistive force opposing buoyancy is capillary pressure which is a function of the pore-throat radii of the carrier bed, petroleum–water interfacial tension and wettability.

On page 5 of the named publication the Canadian Basin case study is summed up: "Piggot & Lines review the vast petroleum system of the West Canada Basin. They describe three discrete 'hydrocarbon cells' made up of Devonian, Mississippian–Neocomian and mid-upper Cretaceous strata each of which were sourced from intervals within the cell. *Regionally lateral migration over hundreds of kilometers dominated petroleum accumulation.*"

Discussion. This cited section is amazing: buoyancy is recruited to explain lateral migration over hundreds of kilometers! Can buoyancy drive fluids laterally? And for such extremely long distances? These views are discussed throughout *Petroleum Migration* without even mentioning the associated formation waters and their testimony of the prevalence of hydraulically isolated rock-compartments. The pressure-cooker model, presented in section 12.3, is freed from the illogical large-scale migration notion.

12.1.2 On what Arguments is Extended Petroleum Migration Based?

Tissot and Welte (1984) give the following description (page 293):

> It is highly improbable that the huge quantities of petroleum found in these rocks could have originated from solid organic matter of which no trace remains. Rather... it appears that fluid petroleum compounds are generated in appreciable quantities only through geothermal action on high molecular weight organic kerogen, usually found in abundance only in fine-grained sedimentary rocks, and usually some insoluble organic residue remains in the rock at least through the oil-generating stage. Hence, it can be concluded that the place of origin of oil and gas is normally not identical with

the locations where it is found in economically producible conditions, and that it had to migrate to its present reservoirs from its place of origin.

Discussion. This brief section presents three suggested, and widely quoted, paradigms:

1. Insoluble petroleum leftovers remained in the source rocks.
2. Only a specific portion of the organic raw material was turned into petroleum compounds.
3. Only certain rocks contained the raw material, e.g., not sandstones.

Based on these assumed—and unproven—paradigms, large-scale petroleum migration is deduced.

12.1.3 Petroleum is Assumed to be Formed in Low-Permeability Rocks and Migrate to High-Permeability Rocks

Tissot and Welte (1984) describe (page 243):

> The release of petroleum compounds from solid organic particles (kerogen) in source beds and their transport within and through the capillaries and narrow pores of a fine-grained source bed, has been termed *primary migration* by numerous workers. The oil expelled from a source bed passes through wider pores of more permeable porous rock units. This is called *secondary migration.* Petroleum compounds may migrate through one or more carrier beds with similar permeability and porosities as reservoir rocks before being trapped by an impermeable or a very low permeability barrier. Oil and gas accumulations are thus formed.

Discussion: The described flow of petroleum "within and through the capillaries and narrow pores of a fine-grained source bed" is not backed by direct observations. The possibility of petroleum formation in a highly permeable rock to begin with seems more likely, as in this case oil migration into traps is local.

12.1.4 Petroleum Migration is Suggested to Have Happened by Active Water Flow, by Displacement, or by Diffusion

The previous citation goes on: "Since practically all pores in the subsurface are water-saturated, movement of petroleum compounds within the network of capillaries and pores has to take place in the *presence of the aqueous pore fluid.* Such movement may be due to active water flow or occur independently of the aquas phase, either by displacement or by diffusion."

Discussion: Active flow of groundwater at the several-thousand-meters depth of petroleum formation does not exist, as it is deep in the zone of zero hydraulic potential. What is this mechanism, "diffusion of petroleum"? How is it suggested to work? Over more than a few meters? And what is meant by "displacement"? Displacement of what? Where? And how far?

In any case, any petroleum migration mechanism related to formation water is limited to the hydraulically interconnected systems of individual rock-compartments.

12.1.5 Caution Regarding Primary and Secondary Migration

Reservations were raised by Tissot and Welte (page 294):

> Processes of primary and secondary migration are not yet understood in detail. Basic data on pore geometry, porosity and permeability relationship, and distribution of pore water in buried dense source rocks are rare. Likewise, there is little information on movement and distribution of petroleum compounds inside the pores of source rocks. Therefore, it should be realized that the...discussion on petroleum migration is largely theoretical and should not be considered definitive. The theoretical concepts should be quantified to determine their effects on the migration mechanisms for a better understanding of the migration phenomena.

Discussion: This scientifically responsible statement opens the road to different models, e.g., the formation of oil and gas within permeable rocks that united the functions of "source rock" and "reservoir rock," thus making "primary migration" and "secondary migration" irrelevant.

12.1.6 Suggested Scales of Petroleum Migration

On page 354, Tissot and Welte state:

> Widely differing opinions have been expressed about distances of secondary migration. In some instances, a short-range migration has been inferred, such as in the isolated sand lenses in the Tertiary of the Rhine Graben and the pinnacle reefs in the Devonian of Western Canada. Long-distance migration has been suggested for the Anthabasca heavy oils in Canada and for many Middle East oil fields. *In many instances, however, there is limited evidence and virtually no*

proof. An oil deposit extending over several hundred square kilometers, such as the Anthabasca tar sand in Canada, requires secondary migration over distances of *the order of 100 km and even more*. Such distances are not unreasonable, provided privileged migration avenues are available such as the major Cretaceous–Paleozoic unconformity in Western Canada.

Discussion: The statement that in numerous cases the deduced migration distance is short is in agreement with the hydraulically isolated rock-compartment structure deduced from the study of formation waters and petroleum deposits. In contrast, the long migration distances that have been suggested are in disagreement with a series of boundary conditions addressed at the beginning of the present chapter.

Migration of the order of 100 km and even more along the major Cretaceous-Paleozoic unconformity in Western Canada sounds entirely nonrealistic. Lateral migration for such distances? By what mechanism? By what energy? Along what paths?

The model of oil and gas formation within closed rock-compartments overcomes such axioms.

12.1.7 Petroleum Could Migrate Only Along Hydraulically Interconnected Spaces

Oil poured over water floats on top of it. If we shake water and petroleum, upon settling the oil will rise and spread on top of the water. Similarly, if we mix gas with water or oil, the gas settles on top of the petroleum or water. These features are the result of (1) the immiscibility of these different fluids and (2) the difference in their specific gravity. These basic laws of physics are applied to understand the formation of oil and gas reservoirs, and they point out very clearly that petroleum could migrate only along hydraulically open avenues, i.e., within the rock-compartment of origin.

Petroleum could migrate only upward and not laterally. Migration by flotation is intrinsically upward. There is no mechanism that could drive oil or gas laterally.

12.2 Coal—A Fossil Fuel Formed with No Migration Being Involved

Coal has a lot in common with petroleum, and in a way it is the more inland counterpart. Its occurrence demonstrates formation of fuel within closed rock-compartments that subsided deep enough to undergo pressure- and temperature-induced processes.

12.2.1 Key Observations and Conclusions

The following observations and conclusions lead to boundary conditions of the formation of coal:

Coal is affiliated to gas and petroleum formation environments. Observations: Bates and Jackson (1987) provide the following definition: "Gas coal—bituminous coal that is suitable for the manufacture of flammable gas because it contains 33 to 38% volatile matter." Similarly, there are cases of coal-related oil and gas occurrences. *Conclusion*: Coal is a specific facies of fossil fuel that was formed from plant raw materials.

Coal, like petroleum, was formed in sedimentary basins. Observation: In Bates and Jackson (1987) the following definition is given: "Coal basin—a coal field located within a basinal structure, e.g., the Carboniferous Coal Measures of England." Another example of a famous coal-related system is the Upper Silesian Coal Basin, Poland (Carboniferous to Miocene). *Conclusion*: Coal and petroleum were formed in similar paleo-landscapes.

Coal originated in low paleo-flatlands frequented by sea invasions and retreats. Observations: In Bates and Jackson (1987) the following description is given: "Coal measures—a succession of sedimentary rocks (or measures) ranging in thickness from a meter or so to a few thousand meters, and consisting of claystones, shales, sandstones, conglomerates, and limestones, with interstratified beds of coal." *Conclusion*: Coal was formed in evaporitic lowlands frequented by sea invasions and retreats. This is the same paleo-environment deduced in the last chapters for the origin of the meteoric formation waters, tagged by brine-spray.

The continental vegetation materials were confined by rapid sea transgressions. Discussion: Coal was formed only from well-preserved organic raw material. Thus the terrestrial vegetation had to be efficiently preserved from the oxidation that is common in the exposed landscape. The preserving mechanism is disclosed by the structure of the coal measures, which reflect frequent sea transgressions. The latter covered the vegetation and confined it by marine sediments.

Coal is the inland counterpart of petroleum. Observation: Coal was formed from continental organic raw material alone, in contrast to petroleum, which reveals formation from marine organic raw material as well. *Conclusion*: Lush coal-forming vegetation grew in the more inland periphery of the evaporitic paleo-lowlands that were more rainy; whereas the petroleum-forming vegetation grew in the more arid and evaporitic zones that bordered the retreated shallow sea. Thus coal is the inland counterpart of oil and gas.

Coal deposits disclose that permeable rocks, rich in organic raw material, thrived in the evaporitic paleo-flatlands. Observation: In Bates and Jackson (1987) the following description is given: "Coal—a readily combustible rock containing more than 50% by weight and more than 70% by volume of carbonaceous material, including inherent moisture, formed from compaction and induration of variously altered plant remains similar to those in peat. Differences in the kinds of plant materials (type), in degree of metamorphism (rank), and in the range of impurity (grade) are characteristic of coal and are used in classification." *Discussion*: Continental rock-units containing intensive accumulations of organic plant materials were common and constituted the required raw material for the formation of coal within its source rocks. The same was relevant to petroleum formation—porous rocks were rather often rich in plant material, of both continental and marine origins, and the formed oil had enough room to reside in the same rock-compartment.

12.2.2 Coal Formation in Rock-Compartments—the Pressure-Cooker Model. Working Hypothesis

The listed conclusions and boundary conditions lead to the following stages of coal formation:

Lush vegetation developed in flat and rainy lowlands of sand, conglomerate, and mud landscapes.

Sea transgression confined the vegetation by newly sedimented marginal marine rocks, forming hydraulically isolated rock-compartments, some of them with a high content of organic material.

The hydraulically isolated rock-compartments subsided in the sedimentary basins; at some point the pressure and temperature became high enough for coal formation.

Associated formation water was probably involved in the coal-formation process.

12.2.3 Further Insight into Coal Formation

The following comments are from Tissot and Welte (1984):

"Most important coal occurrences were originally laid down in basins of long-lasting subsidence in coastal or peralic (coastal swamp) environments."

"Important coal occurrences also developed around huge freshwater lacustrine basins, such as the Carboniferous coals of Bohemia, the Saar district in Germany, Massif Central in France, and Spain.... As a consequence of their deltaic and coastal depositional environments, recent peat bogs, as well as their ancient counterparts, are found associated with a variety of sedimentary types.... These associated sediments are generally fine-grained clastics (sandstone, mudstone, shale), but may develop coarser lithologies (conglomerates) locally. Carbonates are generally minor components. Repetitive sequences with several lithologies, typically sandstone, shale, coal and some limestone, are a feature of most coal-bearing strata.... Study of these rhythmic or cyclic sedimentation sequences has indicated that tectonic or climate events or both produce a relative rise and fall of sea level which is an important geological factor controlling the process for coal formation."

"The sedimentary cycles mentioned above represent deposition under subaerial, fluviatile, intertidal and shallow marine conditions."

"Hence it is well established that coals are able to generate and release sufficient gas to form large commercial gas accumulations."

"However, evidence of commercial oil accumulations derived from coal is not so abundant. It is known that some liquid hydrocarbons are generated in coal during catagenesis. There are numerous reports of oil shows, small oil seeps and oil-impregnated sand lenses etc. being closely associated with coals from all over the world."

"Crude oils associated with coals often belong to the high-wax type, as the 'Frankenholz' oil from the Saar region in Germany; 'La Machine' oil from Central France, and oils from the midland coal measures in Great Britain... Crude oils of the Mahakam Delta area of Indonesia... also have to be mentioned here. The close association of oil and coal in the Officina area of Venezuela... must be ranked as another indication of a liquid hydrocarbon potential of coals."

Discussion: The above report and discussion by Tissot and Welte (1984) provides a suitable background for the pressure-cooker model of coal formation, presented above (Section 12.2.2).

12.3 Boundary Conditions Set by Formation Waters and Petroleum and Coal Deposits

Study of the formation water systems reveals a number of key observations and deduced conclusions that lead to the formulation of a set of intrinsic boundary conditions. The following boundary conditions have to be considered by every model dealing with the formation of petroleum deposits:

Boundary conditions set by formation water studies:

> Formation waters are fossil connate waters.
> Formation waters are by and large meteoric waters, disclosing formation and entrapment during sea regression phases.
> Formation waters are common within sedimentary basins. The paleolandscapes of the forming basins were flat lowlands frequented by sea invasions and retreats.
> Formation waters are mostly saline and tagged by brine-spray (and not sea spray), disclosing formation at evaporitic environments.

Many formation waters have an ^{18}O signature that discloses subsidence into the temperature window of petroleum formation.

The formation waters are entrapped in hydraulically isolated rock-compartments.

Compartmentalization of the rock-units started from the initial subsidence stage.

The hydraulic barriers that separate the rock-compartments were improved by compaction and temperature rise during subsidence.

Boundary conditions set by petroleum studies:

Petroleum was formed from organic matter that thrived on the surface in marine, aquatic, or subaerial environments.

Petroleum was formed in thermal processes, disclosing subsidence to the required depth.

Petroleum deposits are as a rule closely associated with formation waters.

Petroleum deposits are quite often associated with halite, anhydrite, and other evaporitic sediments.

Exploitable oil and gas deposits often occur in structural traps.

Petroleum deposits were preserved in their rock-compartments for geologically long periods.

Boundary conditions derived from coal deposits:

Coal has been formed from continental plants.

Coal discloses that rocks of evaporitic flatlands included porous rocks, like sandstone and conglomerate, that were from the beginning rich in organic material.

12.4 The Pressure-Cooker Model of Petroleum Formation and Concentration Within Closed Compartments

Putting the listed boundary conditions together, the pressure-cooker model is reached. Accordingly, petroleum deposits were formed within hydraulically isolated rock-compartments that contained all the necessary ingredients, subsided deep enough so that temperature-induced processes formed the petroleum, and the latter was concentrated in traps within these original compartments, the source rocks often serving also as reservoir rocks. The following stages are inferred by the model:

1. *Sediment accumulation.* Petroleum deposits are common within sedimentary basins, and these had a paleo-topography of vast flatlands frequented by sea transgressions and regressions. Slight changes in the sea level and/or the continent level caused frequent

sea transgressions and regressions that affected vast areas of the flatlands.

In sea covered regions marginal marine sediments accumulated, including clay, mudstone, limestone, dolomite, sandstone, and clastic material that was washed from the exposed land. All these sediments had the potential to be rich in organic material—marine or washed from the land.

During regressions the subaerially exposed landscape was to a large extend an evaporitic environment dotted with lagoons, sabkhas, wetlands, sand bars, and sand dunes. The formed sediments were terrigenous muds, sandstones, halite, anhydrite (or gypsum), and conglomerates.

Similar sediments were accumulated also in certain inland basins and rift valleys, in which lakes substituted the sea in providing large aquatic environments.

2. *Organic material accumulation.* The marginal marine sediments were enriched in marine organic material as well as mangrove like flora and also washed-in terrestrial material.

The terrestrial climate was often rather arid, as evidenced from the associated evaporites, but sediments were enriched in a variety of organic products, mainly connected to the water bodies, e.g., marshes, wetlands, lagoons, sabkhas, and lakes. Farther inland and at higher landscapes even forests thrived, evidenced by coal deposits.

3. *Efficient preservation of the organic material.* The flatness of the landscape provided a tranquil environment in which the organic materials could efficiently accumulate in the mud and sands and in the shallow sea and lagoons. The frequent transgressions aided preservation and confinement of the organic materials.

4. *Filling by meteoric brine-tagged groundwater.* During rather short phases of subaerial exposure the lowland relief acted as a groundwater through-flow system. All the rocks within this low relief through-flow zone were filled with the meteoric water, and interstitial water of the originally marine sediments was flushed.

5. *Rapid confinement and compartmentalization.* Once the sea invaded again, the meteoric groundwater, contained in the subaerially exposed sediments, was not replaced by seawater, as manifested by the results of the Deep Sea Drilling Project (Chapter 5).

The new marine sediments included impermeable clays that confined the underlying sediments along with their contained meteoric water. Due to lateral facies changes, e.g., as a result of the continually changing distances from the seawater line, confinement of separated rock-compartments was also from their flanks. In this mode stacks of rock-compartments built of sandstone, carbonates, and clastic rocks were engulfed by clays and mudstone.

In the large sedimentary basins and rift valleys accumulated thousands of meters thick sequences of such rock-compartments with a structure reminiscent of a fruitcake

6. *The rock-compartments were hydraulically isolated*, as is reflected by different formation water compositions frequently observed in adjacent wells or at different depths in the same well. It is important to note that the compartmentalization is a lithological feature and does not necessarily follow stratigraphic units. The latter are commonly composed of successions of permeable and impermeable rock units that vary laterally.

7. *Certain rock-compartments contained all the ingredients needed for the formation of petroleum deposits.* Rock-compartments formed in the described environments contained the whole list of needed ingredients: *organic raw material, rocks with high enough conductivity*, and *water*. In many cases the source rocks could also be the reservoir rocks, e.g., sandstones that were rich in organic materials or limestone rich in cavities (original or secondarily induced by the CO_2 formed as a byproduct of the petroleum formation).

8. *Temperature-dependent processes produced petroleum and coal upon subsidence* to the proper depth of several kilometers. The associated formation water underwent an oxygen isotope exchange. In places containing more inland sediments, e.g., with incorporated tree remains, coal was formed.

9. *Liquid oil and gas fractions floated above the associated formation water and accumulated in traps.* This process depended on hydraulic connectivity, and thus was confined within the individual rock-compartments, as portrayed in Fig. 12.1. The floating process could not cause migration through compartment seals, as it is conditional on hydraulic connectivity.

10. *Tectonic petroleum traps.* The oil and/or gas accumulation was in traps shaped by tectonics. In nontilted rock strata relatively thin horizontal traps were available on top of the conductive rocks (Fig. 12.1a), whereas in tilted and folded structures the petroleum could accumulate into more efficient traps (Fig. 12.1b,c). The folding and tilting could occur prior to the deep subsidence, so the formed petroleum migrated right away into the efficient trap, or the migration was first to the horizontal top of the rock-compartment and later on it further migrated into the tectonic structures once they were formed. The same holds true for salt beds—they could act as seals as long as they were horizontal, and the oil could remigrate a little once they were compressed and formed diapirs.

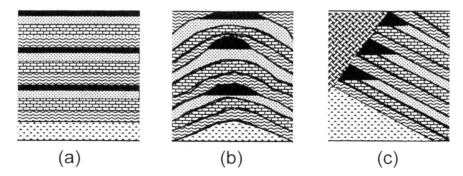

Fig. 12.1 Oil and/or gas accumulation was in traps located within the original rock-compartments: (a) In nontilted rock strata relatively thin horizontal traps were available on top of the conductive rocks. (b and c) In folded and tilted structures the petroleum could accumulate within the individual rock-compartments into more efficient traps. The folding and faulting could happen before the petroleum formation depth was attained, and in this case the newly formed petroleum could migrate right away into the tectonic traps; or the structuring was later and the petroleum emigrated to its present location.

11. *The dimensions of rock-compartments and traps vary over several orders of magnitude.* The size of individual systems can be mapped via the composition of the entrapped petroleum, the pressure, and the composition of the associated formation waters. Initial results reveal rock-compartments in the range of hundreds of meters to several kilometers across. The thickness seems to vary from tens of meters to hundreds of meters.
12. *The giant petroleum fields are clusters of separated compartments.* The distribution of petroleum-containing compartments is uneven—the sedimentary basins have zones rich with the latter and zones without them. In other words, even the giant petroleum fields are separated into hydraulically isolated compartments, a structure revealed by the variability of the composition of the petroleum phase, differences in fluid pressure, and the variable composition of the associated formation waters.

The large-scale oil and gas fields have their origin in suitable facies zones of the respective sedimentary complexes.

12.4.1 The Bottom Line: Connate Petroleum

We have followed the history of the term of connate water (section 7.5), which was coined to describe fossil groundwaters that remained within the host rocks since they were buried. Along the same line, petroleum is connate. Its organic raw material and associated formation water were entrapped in the rock-compartment that subsided until the temperature-induced processes manufactured the petroleum compounds, which then concentrated in traps within the individual compartments.

12.5 Another Case Study Supporting the Pressure-Cooker Model

In a way, it is possible to find occasional descriptions of the fruitcake structure of petroleum fields also in the traditional literature, e.g., the following citation from Tissot and Welte (1984), who described the giant Wilmington field in the Los Angeles basin of southern California as follows (page 361):

> The Wilmington structure is a broad, asymmetric anticline about 12 miles long and 4 miles wide. It is dissected by a series of transverse normal faults which separate the producing sand and sandstone reservoirs into *many different pools....* Seals are provided by claystone and shales. *Oil properties in the various fault blocks differ considerably, illustrating that the accumulation of hydrocarbons apparently occurred independently in each of the blocks.* The different oil properties also suggest that the faults provide effective barriers to communication between adjacent fault blocks. *There are seven major producing zone ranging in age from late Miocene to early Pliocene.* The upper part of the Wilmington structure was eroded during early Pliocene, and subsequent submergence resulted in the deposition of an additional 1800–2000 ft of almost horizontal Pliocene and Pleistocene sediments on top of the anticline. The entire Wilmington oil field is estimated to contain about 3 billion barrels of producible oil.

In the context of the fruitcake structure it is of interest to note that petroleum has been encountered in a list of different rock types, as reflected in the following remarks from Tissot and Welte (page 358):

> The majority of petroleum accumulations are found in clastic or detrital rocks, including sandstones and siltstones (highly siliceous, comprised mainly of quartz, feldspars and mica), and graywackes (sandstones with particles of dark, fine-grained igneous rocks). Clastic rocks probably comprise more than 60% of all oil occurrences, and an *additional 30% is found in carbonate reservoirs.* The remaining 10% of all petroleum occurrences are found in fractured shales, igneous and metamorphic rocks, etc.
> It is significant, however, that more than *40% of the so-called giant gas fields are found in carbonates.*

And, on page 360: "Well-sorted sands with nearly spherical grains, such as certain beach or dune sands, make excellent reservoirs with high primary porosity. The Rotliegend dune sands of the giant gas field of Groningen in the Netherlands are an example for the outstanding quality of such reservoir rocks."

12.6 Pressure-Regulating Mechanisms Within Rock Sequences Discussed in Light of the Fruitcake Structure of Isolated Rock-Compartments

Proponents of the theory of large-scale migration of petroleum evoked the mechanism of rock-compartment rupturing as a result of built-up overpressures. On the other hand, the observed fruitcake structure of a large variety of formation water types, as well as the long-preserved petroleum reservoirs, discloses that rupturing due to overpressure was uncommon. Let us have a qualitative look at processes evoked by subsidence, but first let us have a glimpse at the phenomenon of magmatic intrusive sills.

12.6.1 Rock Strata Lifted by Magmatic Sills

Observation: Beds built of magmatic rocks occasionally occur within sequences of sedimentary rock strata. *Discussion*: These are magmatic intrusions termed sills, defined in the geological dictionary as "A tabular igneous intrusion that parallels the planar structure of the surrounding rock." The sill has been produced by plastic magma that ascended through a neck and supplied the material to the laterally spreading sill. *Conclusion*: A sill lifted the overlying rock strata, making room for itself.

The mechanism by which sills were injected reminds us of the familiar jack with which each of us can lift a car. The jack operates by our movements that push a pump that pressurizes oil and lifts a stand, which lifts the car. In case studies it has been concluded that sills lifted overlying rock strata of a thickness of many hundreds to thousands of meters.

This process has direct bearing to the question of how the rocks in sedimentary basins reacted to increases of volume and fluid pressure.

12.6.2 As Rock Strata Subsided They Were Heated and Expanded

What happened with the added volume? Most probably the overlying rocks were simply uplifted. This is an example of a pressure-regulating mechanism. The included formation water was heated as well, but the rock-compartments were not ruptured, and the water did not burst out. The overlying rock strata were most probably lifted as by a jack, so the pressure within the burried rock-strata remained lithostatic and not more.

12.6.3 The Rock Voids Containing Formation Water Did Not Collapse Under the Pressure of the Accumulating Overlying Strata

Observations reveal that there are ample deep wells of artesian fossil water or of oil production that have high yields and operate for many years. This implies that there was a mechanism that countermanded the lithostatic pressure and preserved the rock voids and the included fluids. In the more competent rocks, like the carbonates, most probably the rock resisted the pressure, voids did not collapse, and the contained fluids were preserved. In other rocks the contained fluids helped to support the rock structure, and as a result the fluids were partly pressurized, but their pressure did not surpass the lithostatic pressure, and hence no rupturing happened. This is an example of a pressure-absorbing process.

12.6.4 The Combined Volume of Produced Petroleum and the Byproduct Compounds (Water, CO_2, Nitrogen Compounds, etc.) Was Larger Than the Volume of the Organic Raw Materials

Did this cause rupturing and expulsion of fluids outside the individual rock-compartments? Probably not. The gases, for example, could be dissolved in the fluid phases and did not consume additional volume, and the second process was most probably, again, uplifting of the overlying rocks.

12.7 Summary Exercises

Exercise 12.1: Diffusion has been suggested as a possible mechanism of petroleum migration (section 12.1). Could this be a mechanism by which noticeable petroleum migration occurred?
Exercise 12.2: Why are the 100 km and greater migration tracks (section 12.1) implausible?
Exercise 12.3: Are both the formation waters and the organic raw material from which the petroleum was formed connate? Did they stay together throughout? Discuss.
Exercise 12.4: How can we estimate the age of a petroleum reservoir?
Exercise 12.5: Petroleum is often encountered at depths that are significantly shallower than the depth required for the needed formation temperature. What can the explanation be?

PART V
HYDROLOGY OF WARM GROUNDWATER AND SUPERHEATED VOLCANIC SYSTEMS

~~~~~~~~~~~~~~~~~~~~~~~~~~~~

You want to understand the thermal processes?
Study the involved water!
It is deeply involved and records every step.

~~~~~~~~~~~~~~~~~~~~~~~~~~~~

13
MINERAL AND WARM WATERS: GENESIS, RECREATION FACILITIES, AND BOTTLING

The term *warm water* is here applied to all kinds of groundwater that are heated by the normal terrestrial heat gradient. A separate group is the boiling and superheated waters that are as a rule heated by recent shallow magmatic intrusions, a topic addressed in the next chapter.

13.1 The Anatomy of Warm Springs

Warm springs have attracted attention since ancient times, and relics of ancient bath houses are found near them in many countries. At warm springs spas developed, providing the grounds for a therapeutic culture and turning the warm groundwater into a resource for tourism. Also some countries pump hot groundwater for space heating.

13.1.1 Hydrochemical Profile of Warm Springs at Swaziland

About ten warm springs are known in Swaziland (Fig. 13.1), a few of them with several eyes. Seven of these have been sampled for detailed analyses of their dissolved noble gases, stable hydrogen and oxygen isotopic composition, dissolved ions, and tritium; on three springs ^{14}C dating was carried out as well. In addition, four adjacent rivers have been sampled for comparison. Liter-

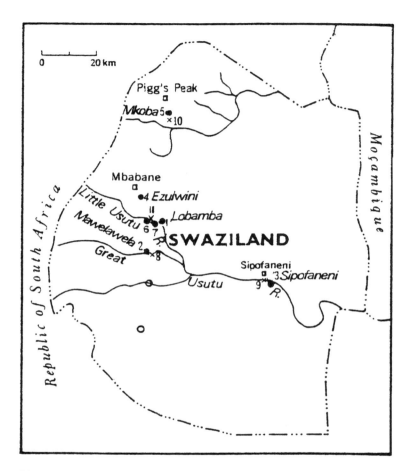

Fig. 13.1 Map of the studied warm springs (dark dots) and adjacent river samples (x) in Swaziland. Other warm springs that were not studied are marked by open circles. (Following Mazor et al., 1974.)

ature on the geology and hydrological studies has been covered by Mazor et al. (1974). The following observations and conclusions were obtained:

Temperatures and deduced depth of water storage. The springs issue at a temperature range of 35°C (Mawelawele) to 51°C (Mkoba). The average ambient temperature is around 25°C. Thus the warmest spring is 26°C warmer than the average ambient temperature. As the region has no active volcanism going on, the warming of the water is concluded to be by the terrestrial heat gradient. Taking the common value of 25°C/km, we reach a storage depth of slightly more than 1 km for the 51°C spring and a depth of 0.4 km for the 35°C spring. These are minimal depths as some cooling during ascent may occur.

Atmospheric noble gases disclose meteoric water. Figure 13.2a depicts the finger-print diagram of Ne, Ar, Kr, and Xe in the spring samples. The similarity to air-saturated water at around 25°C is conspicuous. This is taken as clear evidence that we deal with meteoric water that was entrapped at the above-concluded depth and is now ascending.

Supersaturation of atmospheric noble gases at the time of sample collection indicates that closed system conditions prevail at depth. The solubility of the noble gases, especially Ar, Kr, and Xe, is temperature dependent (Fig. 13.2b). Hence the question comes up whether the concentration of the observed atmospheric noble gases in the warm springs indicates re-equilibration at the observed elevated temperature. The measured concentration of each of the noble gases in the spring waters was observed to be higher than the expected concentration in water equilibrated with air at each spring temperature (read from the solubility graphs, Fig. 13.2b). The warm waters were found to be oversaturated. The percentage of noble gas is plotted in Fig. 13.3. All the samples were found to have maintained their original noble gas concentrations, indicating the respective groundwater systems are closed underground with respect to the atmospheric noble gases.

Radiogenic ^4He enrichment discloses high water age. The ^{4}He concentration in common air-saturated groundwater is around 4.5×10^{-8} cc STP/cc water, and in the studied Swaziland warm springs significantly higher He contents were found—up to $17,000 \times 10^{-8}$ cc STP/cc water. As the region is non-volcanic, it is concluded that the excess He is radiogenic, flushed from the host rocks in which the He is formed from the decay of the contained U and Th (section 3.3.5). The radiogenic He indicates a long storage time in the host rocks. Thus we deal with buried fossil formation water that is now pushed up through preferred flow paths.

Radiogenic ^{40}Ar enrichment confirms a high water age. The ^{40}Ar/^{36}Ar ratio of atmospheric argon is 293.6 and in the studied springs the ratio was significantly higher—by 5 to 23%. The ^{40}Ar excess is attributed to flushing out from the host rocks that contain a few parts per million of ^{40}K (section 3.3.5).

A positive correlation has been observed between the ^{4}He and ^{40}Ar excesses; the range of radiogenic He/Ar varied between 2.6 and 5.0. This is similar to the ratio range of 2 to 4 observed in New Zealand hot springs (Wasserburg et al., 1963) and equal to a value of 4 found at Yellowstone superheated waters (Mazor and Fournier, 1973). Thus the observed ^{40}Ar excess supports that the warm waters are fossil formation water.

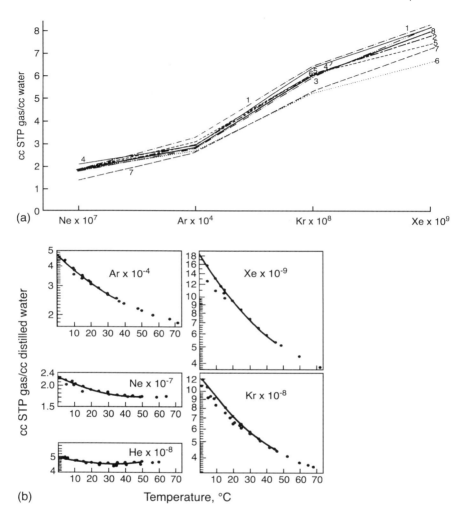

Fig. 13.2 (a) Atmospheric noble gases in warm springs in Swaziland. The isotopic compositions were found to be atmospheric, and the abundance patterns, similar to ASW 25°C, confirm this. The variations in concentrations are caused by different recharge altitudes. (Mazor et al., 1974.) (b) Solubility of atmospheric noble gases in fresh water at sea level (1 atm) as a function of the ambient temperature. (From Mazor, 1979.)

Mineral and Warm Waters

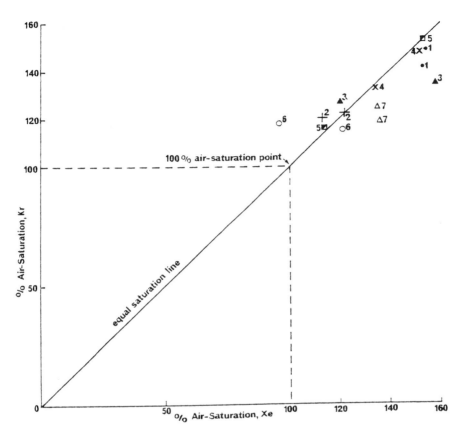

Fig. 13.3 Percent air saturation of krypton and xenon for warm springs in Swaziland. All samples were oversaturated at the temperature of emergence (35 to 52°C), indicating the warm groundwater systems were closed and the initial gases were preserved. (Following Mazor et al., 1974.)

The δD and $\delta^{18}O$ values reveal a nonevaporated meteoric origin of the water. The results from the warm springs group around $\delta D = -17‰$ and $\delta^{18}O = -4.2‰$, values that are slightly lighter than the composition of the rivers, indicating recharge by nonevaporated meteoric water. In addition, these values rule out any partial loss of a steam phase.

Absence of tritium rules out intermixing with shallow water. Most of the springs issue at, or very close to, rivers, and the question came up whether the collected samples are mixed with river water. Tritium in four river samples revealed values of 46 to 53 TU. The values in the warm springs were only 0.0 to 1.3 TU. This suggests that during ascent the spring waters are virtually unmixed with any river water or other recent shallow groundwater.

Carbon-14 content suggests a water age on the order of 5000 years. Three springs were analyzed for their ^{14}C content, and the values obtained are 43 to 48 pmc (section 3.3.5). This leads to a calculated age on the order of 5000 years.

Chemical composition: low salinity and $NaHCO_3$ dominance. The warm springs are surprisingly low in their salt contents, the total ions ranging between 123 to 400 mg/L. The dominant salt species is $NaHCO_3$, formed by CO_2-induced water–igneous rock interaction; and the Cl content is between 11 and 120 mg/L. The low salt content is in remarkable contrast to most warm springs around the world. The explanation of the low salt content is twofold: (1) igneous host rocks, in contrast to the more saline hot waters that issue in sedimentary rocks; and (2) the lack of deep-seated magmatic gases. In spite of the very low salinity, some of the waters are not potable because of a sulfur taste and an occasional H_2S smell.

13.1.2 Hot Springs of Zimbabwe

Zimbabwe has tens of thermal springs, out of which the four hottest complexes (Fig. 13.4) were closely studied for their atmospheric and radiogenic noble gases, isotopic composition, tritium, and chemical composition (Mazor and Verhagen, 1976). The following is a brief description of the studied springs:

Binga complex, situated on the shores of Lake Kariba. The group includes a jet of steaming water, 70 cm high, with a temperature of 100°C (March 1972), and earlier researchers reported a temperature of 97°C, approximately the boiling point at the 600-m altitude of the spring. One series of samples was taken from the gusher, a second was taken from a seepage 10 m southward at 98°C, and a third batch of samples was collected at a group of springs situated 600 m north of the gusher, revealing a temperature of 97°C.

The terrain around the springs is coated with a white silica sinter that covers an area of over 1 km^2, indicating that the thermal activity has been present for a long period and that the eyes shifted location. The springs issue in a Karroo terrain, and in adjacent trenches the sandstone is exposed.

Lubimbi. A 63°C spring issuing in an alluvial valley within a Karroo terrain. A small pool was built around the spring and plentiful bubbles ascend through the water.

Mineral and Warm Waters

Fig. 13.4 Location of hot spring complexes studied at Zimbabwe (dark dots) and river samples (x). (Following Mazor and Verhagen, 1976.)

Rupisi. A 62°C spring in an alluvial valley, located in an igneous terrain.

Hot springs. A 54°C spring complex, issuing in an alluvial valley located in an igneous terrain.

Absence of tritium rules out mixing with fresh shallow water. The tritium content was 0.2 ± 0.4 TU, as compared to 6 to 38 TU found in the adjacent rivers. This reveals that the samples were of indigenous deep-seated waters, and addition of recent shallow water is negligible.

Atmospheric noble gases indicate a meteoric origin and storage within closed systems. In the Lubimbi, Rupisi, and Hot Springs groups Ne, Ar, Kr, and

Xe were found in concentrations similar to air-saturated water at common surface temperature (Fig. 13.5). This indicates (1) the water had a meteoric origin, and (2) the water was stored as a closed system—no gas (or steam) phase was lost.

Oversaturation with noble gas explains variations between duplicate samples. The noble gases found in the named three spring groups were oversaturated with respect to their elevated temperature, explaining differences in the results obtained from duplicate samples; some gas phase had apparently been lost in part of the collected samples. Hence the highest noble gas concentration in repeated samples is closest to the true value (and not the average value).

Low atmospheric noble gas concentrations disclose steam separation in two of the Binga springs. The 100°C steaming gusher and the nearby 98°C spring were found to be substantially depleted in their atmospheric noble

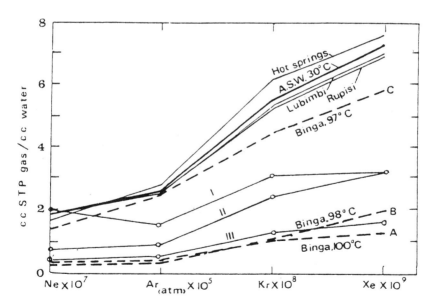

Fig. 13.5 Dissolved atmospheric noble gases in Zimbabwe hot springs. The lines of Lubimbi, Rupisi, and Hot Springs follow closely the line of air-saturated water (A.S.W.) at 30°C. This discloses the meteoric origin, as well as the closed system conditions (see text). The Binga 97°C spring has retained most of the atmospheric noble gases, but the 98°C spring and the gusher of 100°C are substantially depleted, indicating a gas phase escaped. (From Mazor and Verhagen, 1976.)

gases (Fig. 13.5), disclosing that a gas and steam phase has been separated, most probably during ascent. In contrast, the 600-m-away 97°C spring retained most of its atmospheric noble gases, as is seen in Fig. 13.5.

Radiogenic ^{4}He and ^{40}Ar indicate these are fossil formation waters. The concentration of He in meteoric water, saturated with air at ambient surface temperature, is around 4.5×10^{-8} cc STP/cc water (Fig. 13.2b). The He concentration in the studied springs exceeded this value substantially, reaching a concentration of up to $10,400 \times 10^{-8}$ cc STP/cc water. Similarly, an ^{40}Ar excess was observed, the excess ^{4}He/^{40}Ar ratio being in the range of 1.8 to 5.0, disclosing these are radiogenic gases released from common rocks, as

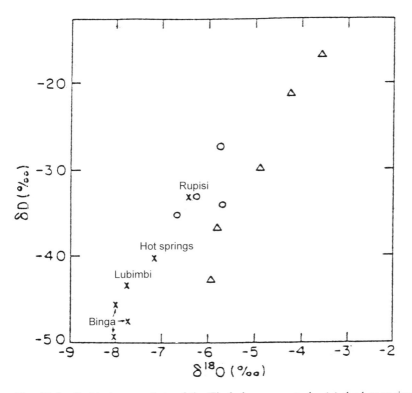

Fig. 13.6 Stable isotope data of the Zimbabwe case study: (x) the hot springs; (△) Zambezi River; and (○) Sabi River systems. The superheated springs of Binga have a distinctly light composition disclosing (1) recharge during a different climate with isotopically lighter rain, (2) recharge by nonevaporated rain, and (3) negligible steam loss during ascent. (Following Mazor and Verhagen, 1976.)

discussed above for the Swaziland case study. The high radiogenic noble gas concentrations disclose these are fossil waters.

Light δD and $\delta^{18}O$ values disclose an origin by nonevaporated rainwater during a different paleoclimate. Figure 13.6 depicts isotopic data for the studied Zimbabwe springs and related rivers. The light isotopic values indicate (1) recharge was during a different (colder?) paleoclimate, when the local rains were isotopically lighter, (2) recharge was by nonevaporated rainwater; and (3) steam loss during ascent is negligible.

Chemical composition and low salinity. The three Binga springs have practically the same low salinity and ion abundance, revealing a common deep fluid. Chlorine is around 40 mg/L in the springs, except Lubimbi, which contains 333 mg/L. The total salinity is low: 700 mg/L at Binga, 1340 mg/L at Lubimbi, and even lower, around 250 mg/L, at Rupisi and Hot Springs. This difference is readily explainable by the difference in the country rocks: Karoo sediments in the first case and igneous rocks in the second.

13.1.3 South Africa Thermal Springs

Sixteen thermal springs of South Africa were included in a special study, reported by Mazor and Verhagen (1983).

Temperature indicates depth of storage. The springs (Fig. 13.7) range in temperature from 26 to 61 °C (Table 13.1). No magmatic activity takes place, hence the warming is caused by the thermal gradient and indicates depths of storage of several hundreds of meters to over 1000 m.

Chemical composition reflects the type of host rocks. The chemical composition of the studied springs outlines two groups: (1) relatively saline waters of 936 to 2364 mg/L total dissolved ions, with NaCl as the major component, and (2) fresher waters, 90 to 432 mg/L total dissolved ions, with no dominant ion. The first group is associated with sedimentary rocks and the second with igneous rocks.

The δD and $\delta^{18}O$ values are significantly lighter than the nearby rivers. This is well seen in Fig. 13.8. This leads to a number of boundary conditions:

No intermixing with nearby river waters
No loss of a vapor phase
Origin by isotopically light rains during a different climate (colder?)

No measurable tritium. This indicates we do not deal with recently recharged groundwater, and no intermixing with such water has occurred.

Mineral and Warm Waters

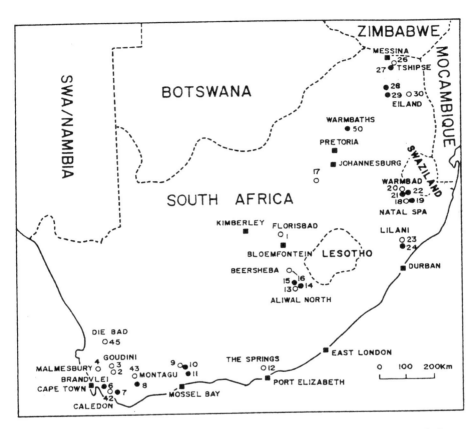

Fig. 13.7 Location and sample numbers of studied thermal springs (o) and river samples (•) for the South Africa case study. (Following Mazor and Verhagen, 1983.)

Range of ^{14}C indicates water ages of almost recent to 20,000 years. Only some of the springs have been measured. The ^{14}C values range between 8 and 78 pmc, indicating water ages in the range of around 20,000 years to almost recent. No other parameter is seen to covary with the ^{14}C values.

Atmospheric noble gases indicate meteoric origin. Figure 13.9 portrays the noble gas concentrations revealing that the Ne-Ar-Kr-Xe lines are parallel to the air-saturated lines, i.e., they have relative abundances as of meteoric groundwater.

Radiogenic helium indicates water ages from almost recent to several thousands of years. A radiogenic He excess is seen in some of the springs (Fig. 13.9).

Table 13.1 South Africa Hot Springs Chemical Composition

Spring	Temperature (°C)	Total ions (mg/L)	K (mg/L)	Na (mg/L)	Mg (mg/L)	Ca (mg/L)	Cl (mg/L)	SO_4 (mg/L)	Alk. (mg/L)
Goudini	37	91	1	6	8	10	14	25	27
Brandvlei	61	100	2	7	9	8	21	19	34
The Springs	24	105	1	18	10	4	43	15	14
Olifants	48	170	9	17	14	13	43	22	62
Natal Spa	37	196	2	52	7	8	36	37	54
Warmbad	36	223	2	60	5	8	21	31	96
Lilani	40	268	2	90	3	4	43	87	39
Tshipise	58	397	4	143	1	2	136	23	88
Beerseba	26	432	1	85	24	17	64	77	164
Eiland	39	936	14	305	2	32	473	41	69
Malmesbury	31	1096	11	340	13	51	571	56	54
Aliwal N.	34	1205	2	360	7	85	678	46	27
Florisbad	29	2364	8	800	7	93	1350	65	41

Note: The table is arranged in increasing order of the total ion content.
Source: Data from Mazor and Verhagen (1983).

Mineral and Warm Waters

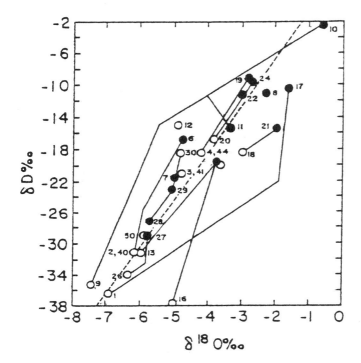

Fig. 13.8 Isotopic composition of the studied South Africa thermal springs (○) and adjacent rivers (●). The sample numbers are as in Fig. 13.5. The lines connect each spring value with the data of the nearest river, and it is clear that the springs are isotopically significantly lighter than the respective rivers, revealing the springs do not mix with river water. The dashed central line is the local meteoric water line, revealing that the springs originate from light meteoric water. (Following Mazor and Verhagen, 1983.)

Even some ^{40}Ar excess has been observed. The age range of the water of the different springs has been estimated to be between several thousands of years to almost recent.

13.1.4 Mixing of Cold and Warm Waters: Combioula Springs, Southern Switzerland

Samples were repeatedly collected and analyzed and the data are plotted in composition diagrams in Fig. 13.10. The conclusion reached by the researcher (Vuataz, 1982) was that the correlation lines of the concentrations of Li, Na, K, Mg, SiO$_2$, TDS, and temperature, as a function of Cl concentration, are

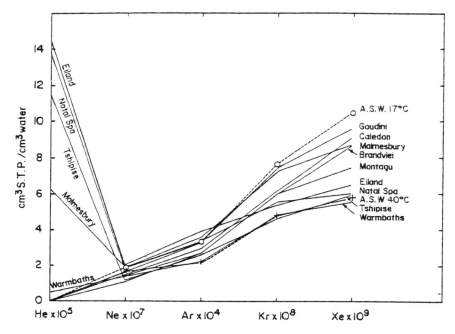

Fig. 13.9 Noble gases in thermal springs of South Africa. The Ne-Ar-Kr-Xe lines resemble air-saturated water (A.S.W.), indicating a meteoric origin. An excess of He is seen in some of the samples, indicating an age in the range of nearly recent to several thousands of years. (Data from Mazor and Verhagen, 1983.)

mixing lines of two types of water. Thus, at Combioula, a warm mineral water ascends and is mixed with shallow cold and fresh groundwater.

13.1.5 Mineral Springs Tinted by Dead Sea Brine: The Kanneh-Samar Springs

These springs issue on the western shore of the Dead Sea, Israel. They form a 400-m-long green oasis of tamarisk trees, with tens of small springs, conspicuous in the rocky desert environment and the blue-grey background of the Dead Sea. The taste varies from one spring to another; a few are fresh and others are saline. The Cl concentration varies, for example, from 3.3 to 213 meq/L, which is equivalent to 117 to 7650 mg/L.

The question came up as to the cause of such a variability. Are the springs outlets of different water systems, or do we see various degrees of dilution of a single type of saline water? Figure 13.11 provides the answer—the data of different ion pairs plot along mixing lines that extrapolate to the

Mineral and Warm Waters

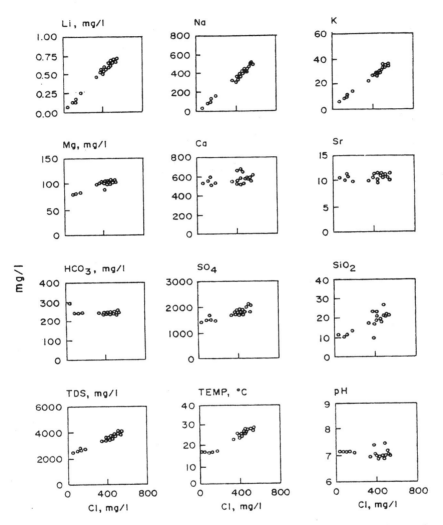

Fig. 13.10 Repeated measurements in a group of springs at Combioula, southern Switzerland. Positive correlation lines are seen for Li, Na, K, Mg, TDS, and temperature, plotted as a function of Cl. Mixing of cold fresh water with ascending warm saline water is indicated. (Data from Vuataz, 1982.)

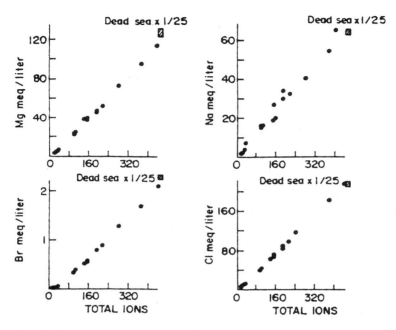

Fig. 13.11 The Kaneh-Samar springs on the western shore of the Dead Sea, Israel. A well-defined mixing line is seen, and the Dead Sea values plot on it. Fresh water that is drained from the Judean Mountains is drained to the surface of the saline lake and is mixed with it in the range of 0 to 3%. (Following Mazor et al., 1973b.)

zero points on the one end and meet the Dead Sea water composition on the other end. The conclusion is clear: we have fresh water, recharged at the Judean Mountains and drained to the Dead Sea surface, mixed with up to 3% of the brine.

13.2 Medicinal and Healing Aspects of Warm and Mineral Waters

Hot springs have the reputation of being healthy and remedial. The writer has had several opportunities to lecture to meetings of medical societies interested in spas. The presentation covered the large assortment of mineral and thermal waters occurring, for example, at the Dead Sea shore. The question would come up: for which health problems has each water type the potential to be curative. The answer I received, repeatedly, was that it is difficult to match in strict terms water composition and curable diseases or health problems.

Instead, the medical experts recommended that each warm or mineral spring and well will be matched, on the basis of its properties, to one of the well-known classic spas. The doctor can then tell us what is the respective curing capacity.

Maybe this is one of the ways to develop a new spa—to have the properties of every source well established and compared to existing famous spas. This challenge is well in the frame of the hydrochemist's tasks.

Many spas have at reachable distances several types of groundwaters. All natural water types are suitable for bathing, and many waters have also been found to be good for controlled drinking, as a sort of medicine. The latter were traditionally highly mineralized waters, each patient had a personal cup that would be filled at the spring or well head and drank three times a day during a stay at the spa. Thus a respectable spa provides access to a variety of waters. This may be solved by piping waters to a central "drinking hall" or having trails leading to the different water sources.

Natural heating of water is often preferred over artificial heating. Some people proclaim that bathing in naturally heated water is "healthier" than bathing in water heated by burning fuel or warmed electrically. This has no proven basis, but in the medicinal recreation industry we have to listen to the customers.

Water produced from a source has to be lead into the bathing facility through well-isolated pipelines in order to preserve the heat. In certain cases it is found that the warm end-member is cooled by intermixing with shallow cold water. In such cases drilling a well may be profitable if the hot end-member can be reached directly. Once a well is put into operation, it has to be seen whether additional warm water has been reached or the pumping reduces the output of the original spring. In the latter case an optimal operation has to be worked out.

Certain warm and hot mineral springs have a heavy smell due to a few parts per million of hydrogen sulfide (H_2S). To some, this smell is widely accepted as a sign of natural mineral warm water. Absence of a heavy sulfuric smell is good news to other customers. The therapeutic value seems, also in this case, to be little proven. Hydrochemically, presence of H_2S indicates a reducing environment, suggesting the water is not part of the through-flow zone, but entrapped fossil formation water.

Certain warm or mineral springs and wells are famous for their content of the radioactive gas radon and smaller concentrations of radium. There are those who believe that this is an additional therapeutic asset, but this has not been proven. We are warned from radioactive waters for use as our regular drinking water, but the rare exposure of spa visitors does not seem dangerous. If there is elevated radon or radium content, it may be added to the list of qualities promoting the local water: "Unique content of radon." If, on the other hand, the water of another spa is poor in the radioactive isotopes, this might be stated as an advantage as well: "Not radioactive."

The rule of the game is to study the water of each spa well and provide this knowledge to the public.

Traditionally hot springs were somehow more appreciated than hot wells, but this is changing. Springs are sometimes sensitive to pollution from nearby industry, including food manufacturing; and springs occasionally disappear altogether because of nearby intensive pumping.

In contrast, deep wells, tapping entrapped formation water, have the advantage of being pollution immune, and they better preserve some properties such as temperature and H_2S content.

13.3 Developing the Resource—The Hydrochemist's Tasks

All over the world, in a wide range of cultural communities, there is a constant demand for a variety of recreation and healing facilities related to natural hot and mineral water. The hydrochemist has a key role in these projects. At each site the groundwater system has to be carefully studied, including observations and measurements to be carried out at all accessible springs and wells in the selected site and around it. The issues to be addressed are as follows:

1. How many different types of groundwater are at practical reach, including hot, warm and cold waters, and mineralized and fresh waters?
2. What are the properties of each type of water—salinity and chemical composition, isotopic composition, concentration of age indicators, temperature, and pressure.
3. It is imperative to understand the hydrological system in terms of through-flowing water, entrapped fossil water, hydraulic interconnections or isolation, and, most important, degree of mixing of different kinds of water.
4. How much of each water type can be exploited over the long run, i.e., with steady supply (including drilling for end-members)?
5. Make sure the client gets the water with its advertised properties. The need for proper heat insulation from the spring or well to the bathing installations has already been mentioned. If gases like radon or H_2S are typifying the water, the baths have to be filled in such a way that the gases are retained.
6. Is the water corrosive? This has to be found out early enough to secure usage of resistant materials.

7. Careful preparation of the overview for the medical experts is essential in terms of similarity to famous spas and recreational facilities known throughout the world, as discussed in the previous section. The report has to be based on detailed measurements, thorough observations, and profound understanding of the system of each source.

13.4 Local Exhibitions Disclosing the Anatomy of Warm and Mineral Water Sources and Their Properties

The near-universal interest in nature provides an excellent educational opportunity. It is a sure bet that every visitor will be interested in the natural story of each spring and well on the grounds of a spa or a thermal water park. Explanations should be given at several levels—for kids, average adults, and some extras for specialists.

Label of specificity. Water installations open to visitors can be promoted by a label describing the local specifity. The aim in this respect is to identify the special characteristics and formulate it in a few words. Examples follow:

Dead Sea aroma springs (section 13.1.5)
King of the Radon Springs
Water boiled by nature
Steaming spring
Sulfur champion
Springs assortment
Refreshing spring
The Thousand Meters Well
The self-rising well
The Artesian Giant
Jurassic water
Karstic springs

Spring and well signposts are highly recommended. They should include the name, composition, temperature, depth, an understandable geological cross section, history, etc.

Visible measuring instruments. At each water source it is always interesting to place large measuring instruments that provide information in real time, for example, thermometer, conductivity measuring device (indicating the total salinity), scintillometer (for radon), and H_2S sensor. The instruments should be placed in an accessible location and be easy to read. Beside each one can be placed a graph with the parametric results over the last year or for a daily cycle. The parameters can be briefly explained.

Drinking hall and history of the place. As stated, spas and recreational facilities often include a drinking ceremony from one or several local springs or wells. It is recommended that this access is located in a specially designed building that, besides the respective source signboards, also hosts an exhibition of the history of the place, along with welcoming seating areas.

Remains of ancient bathing facilities or other modes of water usage are always of interest. For example, heating facilities, bathrooms, irrigation, or technical installations.

Trails leading to points of interest. At each water-related facility open to the public there is room for signposted trails to springs, wells, lagoons, travertine sediments, geology phenomena, or plants of interest.

13.5 Bottled "Mineral Water"

At restaurants it is common to order bottles of "mineral water" that is tasty because it contains very little minerals—it is fresh water. Why is it called so?

At the spas of Europe it was common that people came for several weeks in order to be medically treated. Besides bathing they also drank local water that was extra-rich in dissolved salts; some of the waters had special composition, like high content of sulfur, magnesium, and even iron. The taste was strong and unpleasant, but a cup twice or three times a day was part of the curing ritual. These were really "mineral waters" and they were later also sold in bottles. From this the term was transfered to the fresh and tasty bottled drinking water. The demand for the latter is rapidly increasing because of pollution that threatens the tap water in many of our cities.

On many of the drinking water bottles it is emphasized that they are "spring water." Is spring water automatically better than well water? Not at all. Let us remember that a main reason for buying bottled water is to ensure drinking of nonpolluted water. Wells, especially if they are deep, are often less vulnerable to contamination than springs. In any case, what really should be written on the bottles is that they contain clean water that has been well inspected.

13.6 Summary Exercises

Exercise 13.1: Saline water produced from a deep well on the grounds of an operating spa is found to be devoid of measurable tritium and ^{14}C but contains He in excess over the concentration in air-saturated water at 10°C. What label of specifity can you suggest?

Exercise 13.2: In a new well the following age indicators were found: measurable tritium and 70 pmc ^{14}C, along with a pronounced concentration

Mineral and Warm Waters 291

of radiogenic helium. Are these logical results? What can be deduced from these results about the hydrological anatomy?

Exercise 13.3: What is a likely mechanism that causes this mixing in the well? Should something be done? What?

Exercise 13.4: New drilling to a depth of 300 m is planned, 100 m east of a warm spring, in order to catch warmer water. What should be observed and measured during the drilling operation? For what purpose?

Exercise 13.5: Suppose it is decided to heat the water piped to the bathing facility of a spa. Is it advisable to do so by passing heated steam through the water pool or bath? What other heating method may be preferable?

14
WATER IN HYDROTHERMAL AND VOLCANIC SYSTEMS

Boiling springs, fumaroles (steaming from volcanoes), and wells that produce superheated steam (used to manufacture electricity)—all these water-related features have an aura of mystery. Let us have a closer look at these phenomena.

14.1 Hydrothermal Systems

The term *hydrothermal system* pertains to complexes of hot water, steam, and gases existing underground connected in some way to a magma body. In a narrow sense, *hydrothermal* is used for hot water and steam deposits exploitable for electricity production. In the broader sense *hydrothermal* relates to all occurrences of warm and hot water and gases, including boiling springs, fumaroles, and superheated water encountered in specially sited boreholes.

There is room for a long list of scientific research questions in each case study, including the following:

What is the depth and structure of the magma chamber that causes the heating?
Is the water recharged at present, or did it predate the magma intrusion?
What is the age of the water?
What is the source of the dissolved salts?
Which gases are dissolved in these waters? What are their sources?
Can we reach a general model explaining the large variety of fluid waters and steam?

We have at hand a long list of measurable parameters, each leading to definable boundary conditions that together disclose the anatomy and origin of hydrothermal systems. The parameters applied in hydrothermal investigations include the following:

Temperature
Pressure
Gas/water ratio
Chemical composition of dissolved salts and gases
Isotopic composition of deuterium and ^{18}O in the water
Carbon-13 in CO_2 and hydrocarbons
Age indicators like tritium, ^{14}C, ^{36}Cl, ^{4}He, and ^{40}Ar
The spatial distribution of the hydrothermal phases and their dynamics in natural conditions and in response to exploitation.

14.2 Yellowstone National Park, Western United States

The terrain including the geysers and boiling springs at Yellowstone, Wyoming, has been declared as a protected National Park since 1872—the first in the world. The region has been volcanically active since the Eocene, and this activity culminated some 600,000 years ago when a huge collapse caldera was formed, which was soon filled with young volcanic products. The six geyser basins are inside the caldera, whereas the Mammoth boiling springs complex is outside, north of it. Due to the local altitude high above sea level the springs boil at around 93°C. In research wells 102 to 210 m deep, temperatures of 160 to 195°C have been reached.

14.2.1 Chemical Composition of the Hot Waters

An intensive chemical composition study was conducted for years at the following subsystems (Fig. 14.1): Heart, Shoshone, Upper Basin, Middle Basin, Lower Basin, Norris, and Mammoth. At first glance one is highly impressed by the large variety of water chemistries and the large range of salinities. However, when plotted on composition diagrams (Fig. 14.2) some order and regularity is seen (Mazor and Thompson, 1982).

14.2.2 The Chemistry Discloses that the Mammoth Springs Are a Separate System

Its waters cluster, as is seen in Fig. 14.2, around a fixed composition, with the following pattern (meq/L):

$Ca \gg Mg > Na > K$ and $HCO_3 \geq SO_4 \gg Cl$

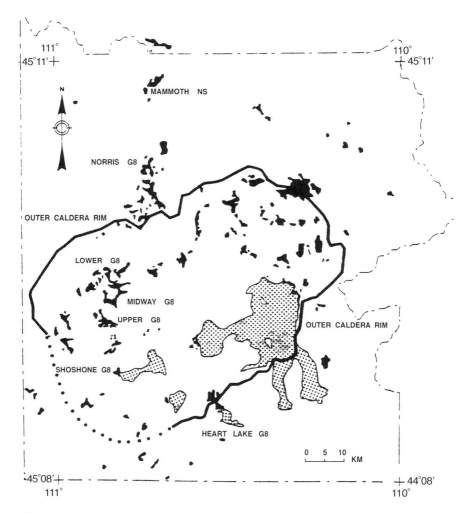

Fig. 14.1 Yellowstone National Park: the caldera rim and regions of intense heat flow (black) and main groups of hot spring. (Following Mazor and Thompson, 1982.)

Water in Hydrothermal and Volcanic Systems

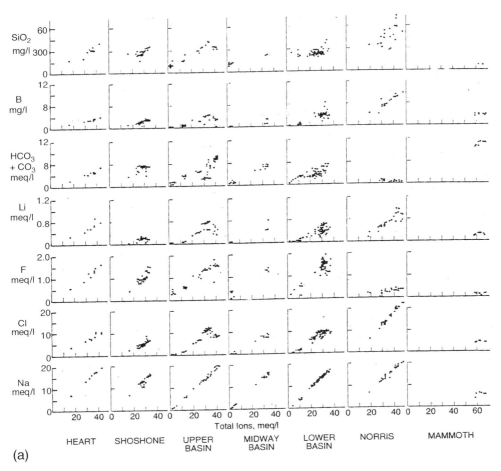

Fig. 14.2 Composition diagrams of seven major hot spring groups over the large terrain of the Yellowstone National Park (see Table 14.1). (a) Chemical compounds positively correlated to total ion content. (b) Compounds of low content and not correlated to the total ion content. (Following Mazor and Thompson, 1982.)

The elevated $HCO_3 + CO_3$ concentration (Table 14.1) indicates a significant contribution of CO_2, which is a common observation at active magmatic and volcanic systems. The elevated SO_4 concentration stems, most likely, from the oxidation of H_2S—another gas observed to emanate in systems of magmatic and volcanic activity. These gases induced water–rock interactions, and the high Ca concentration indicates that the interaction was with limestone. This

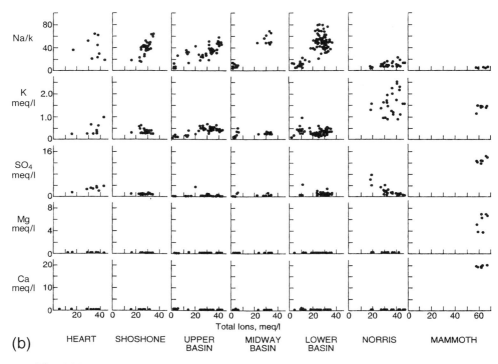

Fig. 14.2 Continued.

is one of the major geological differences between Mammoth and the spring groups of Yellowstone proper. The latter are situated within the caldera, which is built of volcanic rocks (Fig. 14.1), whereas the former are situated outside the caldera, in a carbonate rock region.

Terraces of travertine are the trademark of the Mammoth springs, formed as a result of CO_2 loss from the hot water as it reaches the land surface.

14.2.3 Yellowstone Springs: Intercorrelated Ions

The six spring groups, or geyser basins as they are locally called, that are situated within the caldera (Fig. 14.1) reveal the following common characteristics:

1. *Positive correlations of several ions* as a function of the total ion concentration (Fig. 14.2a). The list of these correlated ions includes Na, Cl, F, Li, $HCO_3 + CO_3$, B, and SiO_2. The best correlation is seen for Na, followed by Cl. The best correlations for the different

Table 14.1 Composition of the Most Saline Spring in each of the Major Groups at Yellowstone

Region	Number of samples	Na	K	Li	Ca	Mg	Cl	HCO$_3$ + CO$_3$	SO$_4$	F	B	SiO$_2$	Total ions
I mg/L													
Heart	10	450	40	5.5	10	1.2	355	300	187	30	3.9	390	
Shoshone	35	380	27	1.4	4	0.5	250	430	48	13	0.6	350	
Upper Basin	46	460	~20	4.9	12	3.6	~370	630	24	28	4.3	180	
Midway Basin	19	370	~16	4.9	12	0.7	320	460	24	25	2.8	220	
Lower Basin	46	390	~16	3.5	14	0.1	355	~300	34	30	4.5	350	
Norris	28	470	59	7.0	6	1.2	760	30	24	H	11.4	650	
Mammoth	7	125	55	1.4	400	79.0	160	880	650	3	3.8	58	
II meq/L													
Heart	10	19.7	1.0	0.8	0.5	0.1	10.0	5.0	3.9	1.6	0.36		43
Shoshone	35	16.5	0.7	0.2	0.2	0.04	7.0	7.0	1.0	0.7	0.06		34
Upper Basin	46	20.0	~0.5	0.7	0.6	0.3	~10.5	10.4	0.5	1.5	0.40		42
Midway Basin	19	16.0	~0.4	0.7	0.6	0.06	9.0	7.5	0.5	1.3	0.26		34
Lower Basin	46	17.0	~0.4	0.5	0.7	0.01	10.0	~5.0	0.7	1.6	0.42		36
Norris	28	20.5	1.5	1.0	0.3	0.1	21.5	0.5	0.5	0.4	1.06		47
Mammoth	7	5.4	1.4	0.2	20.0	6.5	4.5	14.5	13.5	0.14	0.35		67

Source: Mazor and Thompson (1982).

ions are seen at Heart Lake and Norris. These correlations indicate an external origin.

2. *The correlation lines extrapolate to the zero values*, indicating dilution of a relatively saline water with fresh water, the latter possibly being condensed steam and/or shallow groundwater. Some concentration by steam loss is possible as well.

3. *The composition (meq/L) is*

$$Na \gg K \simeq Ca\ Mg \quad \text{and} \quad Cl > HCO_3 \geq SO_4 \gg F$$

indicating addition of CO_2 and induced water interaction with Na-rich magmatic rocks. Added H_2S was oxidized and formed an acid that caused water–rock interaction as well.

4. *The concentrations in each site vary over a wide range of more than two orders of magnitude*, in contrast to the Mammoth springs that all have the same salinity. The most concentrated springs in each group are still relatively low in their salt content (Table 14.1).

5. *The concentration of SiO_2 is also correlated to the total ion concentration*, as is seen in Fig. 14.2a. This is of special interest, as the solubility of this ion is temperature sensitive (Fournier and Rowe, 1966; Truesdell an Fournier, 1976). This indicates that the water with the external ions was heated at depth where the elevated temperature increased the SiO_2 dissolution. The ascent of the water is relatively rapid so that no SiO_2 is precipitated as a result of cooling on the way up.

6. *Concentrations of Ca, Mg, and SO_4 are remarkably low* in all springs of the six Yellowstone groups (Fig. 14.2b), in noticeable contrast to Mammoth, which has high contents of these ions. Potassium is low as well, except at Norris, possibly reflecting the higher temperature of the water of this group of springs.

7. *Relative abundances of Ne, Ar, Kr, and Xe are similar to air-saturated water at surface recharge temperature*. Noble gases in samples collected at research boreholes and springs at Yellowstone National Park (Mazor and Fournier, 1973) revealed relative abundance patterns similar to air-saturated water at 10 to 20°C, as can be seen in Figs. 14.3 and 14.4. This indicates that the hot groundwater originated as shallow meteoric groundwater that was introduced to a significant depth.

8. *Radiogenic ^4He and ^{40}Ar indicate nonatmospheric contribution* (Fig. 14.5). Radiogenic He has been observed in significant amounts in all the Yellowstone samples, indicating prolonged entrapment in the host rocks (Mazor and Wasserburg, 1965). In samples retrieved at depth from special research wells radiogenic ^{40}Ar was also detected (Mazor and Fournier, 1973). The ratio of the two gases was

Water in Hydrothermal and Volcanic Systems

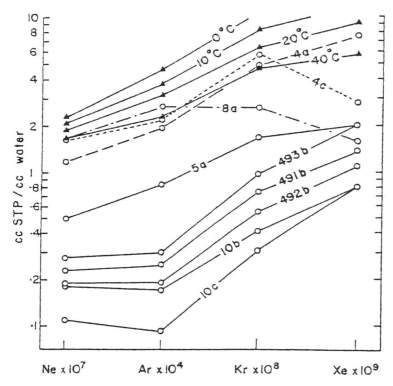

Fig. 14.3 Noble gases in water samples, Yellowstone National Park (Table 14.2). Solid symbols denote air-saturated water at sea level at given temperatures. The abundance lines of the samples resemble the abundance lines of air-saturated water at 10 to 20°C, disclosing an origin as meteoric waters and partial gas loss during ascent to the land surface. (Following Mazor and Fournier, 1973.)

observed to be in the range of 1 to 6, with an average around 4; this is within the order of magnitude expected from the U, Th, and K distribution in common magmatic rocks (Mazor, 1977).

14.2.4 Boundary Conditions to the Formation of the Thermal Waters at Yellowstone National Park Deduced from the Chemistry and Noble Gases

The following boundary conditions and deduced conclusions transpire from the reported data:

1. *There is one type of deep formation water, best represented at Norris.* The overall similarity of the water composition of the six Yellow-

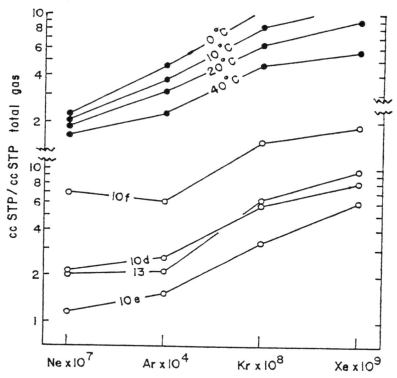

Fig. 14.4 Noble gas in gas samples, Yellowstone National Park (Table 14.2). Solid symbol denote-air-saturated water at sea level at given temperatures. The abundance lines of the samples resemble the lines of air-saturated water at 10 to 20°C. (Following Wasserburg et al., 1963.)

stone spring groups reveals that one type of deep water seems to prevail in this rather large region of boiling water sources, which are up to 40 km apart. A number of secondary processes acted on the original deep hot water, as discussed below. The Norris group best represents the deep, least-modified hot formation water. This is supported by the high SiO_2 and K contents, reflecting highest temperature and least dilution or cooling.

2. *Additions of magmatic CO_2 are inferred at the other five Yellowstone groups.* The boiling spring groups of Heart, Shoshone, Upper Basin, Midway Basin, and Lower Basin are distinctly enriched in $HCO_3 + CO_3$ relative to Norris (Fig. 14.2a). This reflects additions of magmatic CO_2 gas that induced interactions with the magmatic

Water in Hydrothermal and Volcanic Systems

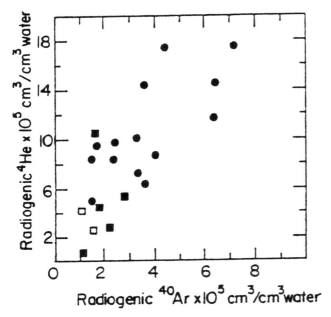

Fig. 14.5 Radiogenic ^{4}He and ^{40}Ar measured in Yellowstone boiling waters. The positive correlation is taken as an indication that these isotopes are flushed from common magmatic rocks. (From Mazor, 1977.)

rocks. This is well supported by another observation: at Norris both the Cl and Na are high and balance each other (Fig. 14.2a and Table 14.1, the meq/L section), whereas at the other five groups the Na concentration is distinctly higher than the Cl concentration, disclosing that Na was added by the CO_2-induced interaction with the magmatic rocks.

3. *Dilution occurs by condensed steam and/or shallow groundwater.* The large range of ion concentrations in the springs of each group and the observation that the correlation lines extrapolate to the zero values in the composition diagrams disclose that dilution is effected. This is probably by condensed steam as well as by the local shallow groundwater.

4. *Additions of magmatic CO_2 and H_2S can be deduced in the carbonate region of Mammoth.* The Mammoth boiling waters, as compared to the Yellowstone boiling waters, have remarkably low concentrations of Na, Cl, F, Li, and B, but are high in Ca and to a lesser degree

Table 14.2 Noble Gases in Samples from Yellowstone National Park (10^{-8} cc STP/mL water and 10^{-8} cc STP/cc STP total gas)

Source	Sample no.	Phase	Sampling date	He*	Ne	Ar	Kr	Xe	^{40}Ar/^{36}Ar	^{36}Ar/^{38}Ar
Well Y4, Lower Basin	4a	Water	May 24, 1970		11.9	19,500	4.87	0.75	303.9	5.42
Well Y4, Lower Basin	4b	Water	May 24, 1970		9.6	19,800			305.1	5.42
Well Y4, Lower Basin	4a	Water	June 1, 1971	642	16.2	22,000	5.69	0.28	289.2	
Well Y5, Lower Basin	5a	Water	May 26, 1970		5.0	8,300	1.70	0.20	293.9	5.37
Well Y5, Lower Basin	5b	Water	May 26, 1970		5.5	7,300	1.59	0.26	207.8	5.35
Well Y5, Lower Basin	5c	Water	May 29, 1971	122	11.8	16,000	3.40	0.41	291.2	
Well Y8	8a	Water	May 28, 1971		16.3	26,800	2.63	0.16	285.6	
Well Y8	8b	Water	May 28, 1971	43	18.9	27,600	2.79	0.19	291.7	
Well Y10, Mammoth	10a	Water	May 25, 1970		2.5	1,900	0.29	0.06	296.5	5.29
Well Y10, Mammoth	10b	Water	May 25, 1970		1.8	1,700	0.41	0.06	206.5	5.40
Well Y10, Mammoth	10c	Water	May 27, 1971	9	1.1	930	0.31	0.08	298.8	
Spring 491, Upper Basin	491a	Water	May 30, 1970		2.7	3,100	0.90	0.18	293.8	5.35
Spring 491, Upper Basin	491b	Water	May 30, 1970		2.3	2,500	0.75	0.14	293.8	5.39
Spring 492, Upper Basin	492a	Water	May 30, 1970		2.2	1,900	0.63	0.13		
Spring 492, Upper Basin	492b	Water	May 30, 1970		1.9	1,900	0.55	0.11	286.6	5.41
Spring 493, Norris Basin	493a	Water	May 31, 1970		3.1	3,000	0.86	0.17	292.5	5.44
Spring 493, Norris Basin	493b	Water	May 31, 1970		2.8	3,000	0.98	0.20	294.5	5.39
Well Y10, Mammoth	10d	Gas	May 25, 1970		21.9	26,700	5.02	0.81	811.9	6.40
Well Y10, Mammoth	10e	Gas	May 25, 1970		11.7	15,300	3.32	0.61	316.1	5.43
Well Y10, Mammoth	10f	Gas	May 27, 1971	12700	71.5	62,000	15.0	1.87	314.6	
Well Y13, Lower Basin		Gas	May 27, 1970		20.3	21,500	6.36	0.97	307.9	5.45
Air-saturation measurement with Y4 well water at 32°C	a			6.2	19.4	24,500	4.60	0.70		
	b			11.2	25.3	27,100	5.21	0.72		
	c			0.9	23.2	27,100	5.39	0.65		
Distilled water, 20°C				4.5	19.0	31,000	6.6	0.90	295.6	5.35
1 cm^3 air				500	1820	934,000	114	8.7	295.6	5.35

Source: Data from Mazor and Fournier (1973).

also Mg, both balanced by $HCO_3 + CO_3$ and SO_4. This discloses quantitative additions of CO_2 and H_2S, as discussed above—two gases common in many regions of magmatic activity.
5. *Atmospheric noble gases disclose an origin as meteoric waters*, recharged on the continent (Table 14.2).
6. *Radiogenic ^4He and ^{40}Ar indicate prolonged deep storage*, i.e., the nondiluted end-member is fossil water.
7. *Efficient mixing of the various components and heating prior to ascent and steam separation* are revealed by the correlation between the meteoric components like the allochthonous dissolved ions and atmospheric noble gases on the one hand and the CO_2 and H_2S-induced compounds and the radiogenic noble gases on the other.

14.2.5 Deep Formation Water Predated the Magmatic Intrusion

Water cannot flow into an active pressure cooker. The water in the hot zone of Yellowstone is pressurized, as is disclosed by the ascent of superheated water and steam along open flow paths feeding the boiling springs and geysers or the water rising in the drill holes. No gravity-flow water can enter and recharge such pressurized systems.

The deep fluid must have been in the host rocks when the last magmatic intrusion arrived. The deep fluid of the Yellowstone hydrothermal system must have been within the host rocks before the arrival of the last large-scale magmatic intrusion that caused the ongoing heating. This is borne out by the following arguments:

1. The meteoric water, allochthonous ions, and atmospheric noble gases are well mixed with the products of the temperature-induced water–rock interactions, the radiogenic noble gases, and the volcanic CO_2.
2. Thus the meteoric water was stored at below-boiling depths, i.e., depths at which compaction (or hydrostatic pressure) prevented boiling.
3. The high contents of radiogenic ^4He and ^{40}Ar indicate prolonged storage times prior to and after the heating.
4. Recent recharge cannot penetrate the pressurized systems; it can only dilute the ascending fluids as they reach a shallow depth.

The hydrothermal system could occasionally be recharged during stages of collapse and fracturing. Volcanic activity ejects material to the land surface, and as a result collapse structures are formed, e.g., calderas. At collapse phases, even minute ones, shallow groundwater can flow downward. Similarly, fracturing during dilatation phases may grant the shallow groundwater occasional access to a greater depth. The water stays down until a new volcanic activity stage evolves.

14.3 Cerro Prieto, Northern Mexico

The Cerro Prieto hydrothermal field is located near the northern border of Mexico, at the southern end of the Salton Sea–Mexicali Trough, which hosts several hydrothermal manifestations (Fig. 14.6). The depth of the superheated nonboiled fluid is 2500 m or more. A detailed study of 12 regularly monitored hydrothermal wells, located within 1 km^2, revealed the following set of observations and deducible conclusions (Mazor and Mañon, 1979; Mazor and Truesdell, 1984):

Operation-induced variations of the ion concentrations within individual wells. Repeated samples collected in a well reveled a wide range of ion concentration, as may be seen in Fig. 14.7. These variations were found to be related to details in the mode of operation of the well, diameter of pipes, and point of sample collection. These variations seem to reflect a combination of concentration increase by steam loss and dilution by condensed steam and shallow groundwater. The important lesson is that a single sample collected at a hydrothermal well may not reflect the concentration of the deep stored fluid.

External origin of Cl, Na, Li, and B. The data of the most concentrated sample observed at each well are plotted in Fig. 14.9, providing an overall picture of the field. The Cl, Na, Li, and B values are seen (1) to be positively correlated and (2) the correlation lines extrapolate to the zero values. This discloses that (1) these ions are external and (2) these ions are not involved in water–rock interactions.

Temperature-dependent concentrations of K, Ca, and SiO$_2$. Figure 14.8 reveals linear correlations between Cl, K, Ca, and SiO$_2$, but their correlation lines do not extrapolate to the zero values. This is in good accord with the work of Fournier and Rowe (1966) and Truesdell and Fournier (1976), which pointed out that the K, Ca, and SiO$_2$ concentrations increase with temperature in hydrothermal fluids. The temperature of the deep nonboiled fluid is above 260°C, explaining the pronounced SiO$_2$ and K concentrations.

One type of deep fluid is revealed by the similarity of the correlation lines seen in Figs. 14.7 to 14.9.

The relative abundances of Ne, Ar, and Kr resemble those in water saturated with air at 10°C, indicating a meteoric origin. Results of a noble gas study at Cerro Prieto (Mazor and Truesdell, 1984) are plotted in Fig. 14.10.

Water in Hydrothermal and Volcanic Systems   305

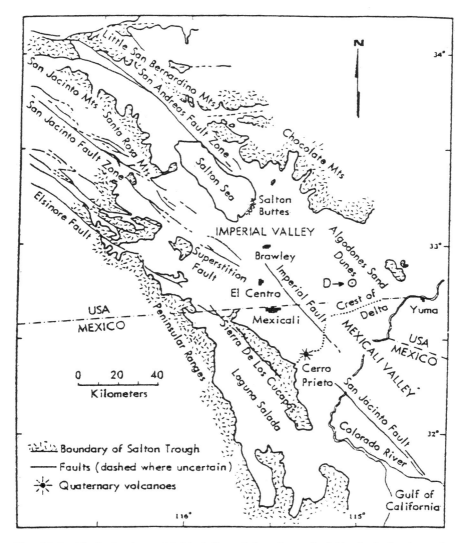

Fig. 14.6 The hydrothermal field of Cerro Prieto, located within the hydrothermal complex of the Salton Sea–Mexicalli Trough. (Following Mazor and Mañon, 1979.)

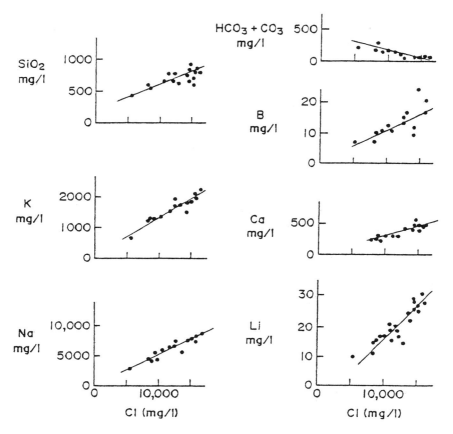

Fig. 14.7 Ion contents in repeatedly collected samples at well 11 at Cerro Prieto. The variations reflect different modes of well operation, changing the rate of concentration increase by steam separation as well as dilution by condensed steam and shallow groundwater. Thus samples collected at well heads sometimes do not represent in a straightforward mode the ion concentration of the deep nonboiled water. (Data from Mazor and Mañon, 1979.)

The lines of Ne-Ar-Kr are seen to vary in concentration, but they all resemble the line of water saturated with air at 10°C. This reveals that we deal with a fluid that originated as meteoric groundwater.

Pronounced radiogenic ^{4}He concentrations, correlated to the atmospheric noble gases, indicate one type of well mixed deep superheated fluid. The concentration of He is seen in Fig. 14.10 to covary with the Na, Ar, and Kr concentrations. This discloses that at depth there is one type of fluid in

Water in Hydrothermal and Volcanic Systems

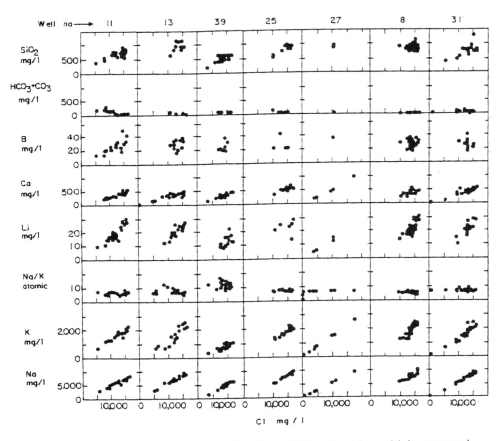

Fig. 14.8 Ion concentrations as a function of Cl content in multiple measured samples from wells at Cerro Prieto. The processes depicted are discussed in the text. (Adopted from Mazor and Mañon, 1979.)

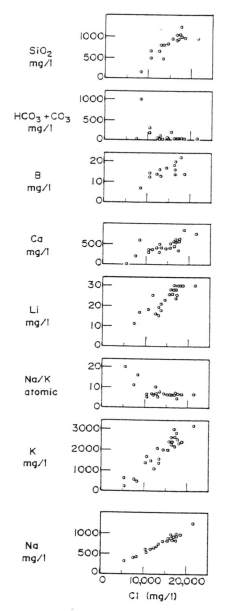

Fig. 14.9 Abundance of dissolved ions in the most concentrated sample obtained from the individual wells at Cerro Prieto. The correlation lines of Na, Li, and B as a function of Cl extrapolate to the zero values, revealing these ions are conservative. In contrast, K, Ca, and SiO_2 as a function of Cl plot on linear correlation lines that do not extrapolate to the zero values, disclosing the impact of temperature-induced reactions that took place. (Following Mazor and Mañon, 1979.)

Water in Hydrothermal and Volcanic Systems   309

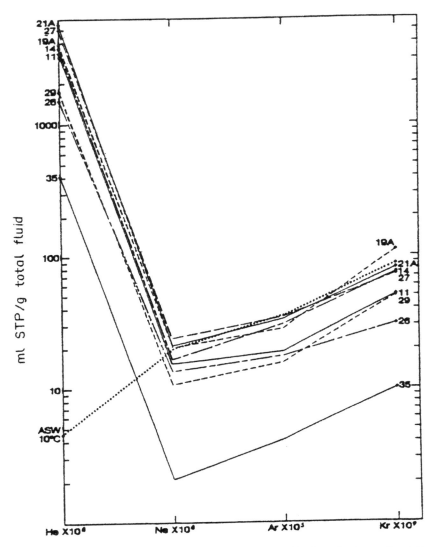

Fig. 14.10 Noble gas patterns in the grothermal fluid of producing wells at Cerro Prieto. Well numbers are indicated. The Ne-Ar-Kr lines resemble the 10°C air-saturated line (ASW 10°C), but the absolute concentrations reveal various degrees of depletion. Radiogenic He is in excess of the ASW value and is correlated to the atmospheric Ne, Ar, and Kr concentrations, disclosing long and deep burial of the original meteoric water. (From Mazor and Truesdell, 1984.)

which the atmospheric noble gases were well mixed with the radiogeic He expelled from the reservoir rocks. As a result of boiling, mainly during ascent, gases were lost in some of the wells, as reveled by the Ne, Ar, and Kr concentrations that are low compared to the implied origin from groundwater recharged at 10°C. The He concentration in the deep nonboiled fluid can be inferred from the samples of wells 21A and 27, which are seen in Fig. 14.10 to be nondepleted. The initial He concentration is 5500×10^{-8} mL STP/g total fluid (Mazor and Truesdell, 1984).

Steam was lost without fractionation of the noble gases. The pattern seen in Fig. 14.10 indicates that gas loss through steam separation was without fractionation.

Occurrence of measurable tritium in peripheral steam wells indicates that shallow groundwater penetrated the depressurized margins of the system. Carbon-14 determination may be useful to monitor such shallow water intrusions.

14.4 The Wairakei, Tauhara, and Mokai Hydrothermal Region, New Zealand

The study area, located in the North Island of New Zealand (Fig. 14.11), included: (1) at the Wairakei hydrothermal field 16 hydrothermal wells (8 to 300 m deep; aquifer temperatures 190 to 240°C; exploited since the 1960s), four fumaroles (95 to 105°C) and two drowned fumaroles issuing below a water body (98°C); (2) the Tauhara drowned fumaroles and hot springs (84 and 98°C); and (3) at Mokai two wells (606 and 1654 m deep; aquifer temperature 170 and 290°C), a drowned fumarole, and a hot spring were surveyed.

This large hydrothermal complex is related to the New Zealand island-arc tectonic system. The plumbing of this complex has been studied by many workers, applying geological and hydrological data, heat considerations, dissolved ions and gases, and stable isotopes (literature given in Mazor et al., 1990). In the present section let us have a look at the information gained from a study of the five noble gases, summed up in the named reference:

Atmospheric noble gas patterns indicate meteoric origin of the water. The isotopic composition of Ne, Ar (36 and 38), Kr, and Xe was found to be atmospheric. In addition, the relative abundance lines of Na-Ar-Kr-Xe are like air-saturated water, as may be seen in Figs. 14.12 and 14.13. Hence the original fluid was meteoric water. An ambient paleo-recharge temperature of around 10°C seems reasonable.

Separation into steam and residual water phases is observed in producing wells. Retention/enrichment factors were obtained for each sample by dividing the observed atmospheric noble gas (ANG) concentrations by the

Water in Hydrothermal and Volcanic Systems

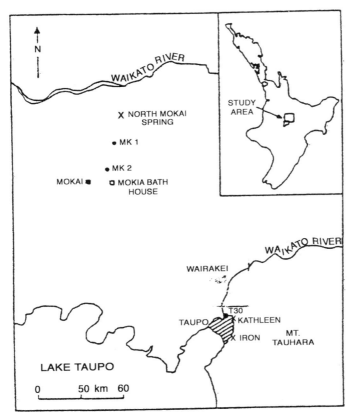

Fig. 14.11 Location of studied hydrothermal systems: Wairakey, Tauhara, and Mokai, on North Island, New Zealand. (Following Mazor et al., 1990.)

concentrations in air-saturated water at recharge conditions, which were assumed to have been 10°C. The data of hydrothermal wells vary from an enrichment by a factor as high as ~30 to a retention by a factor as low as 0.02.

Figure 14.12 discloses that there are wells that lost atmospheric noble gases by prolonged boiling, and other wells are enriched. The first group of wells produces a residual fluid that lost various portions of its steam phase, whereas the second group produces from a steam cap, enriched by the originally dissolved gases.

Enrichment of ANG discloses that fumaroles are fed by early boiling stages of the deep fluid. Figure 14.13 reveals the data obtained at the studied

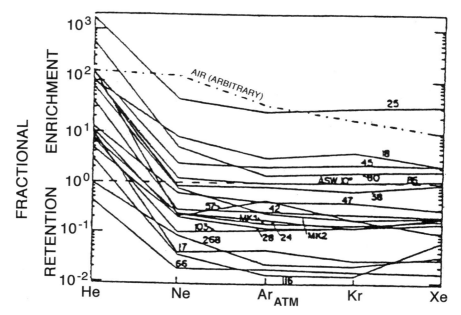

Fig. 14.12 Fractural retention/enrichment patterns of noble gases in the Wairakei and Mokai steam wells. The Ne-Ar-Kr-Xe lines are horizontal, i.e., they are parallel to the line of the assumed original meteoric water (air-saturated water at 10°C at sea level, horizontal dashed line), and they differ from the pattern of free air (upper dashed line). The data of hydrothermal wells vary from enrichment by a factor as high as ~30 to retention by a factor as low as 0.02. Excess of He is conspicuous, indicating prolonged storage time. (From Mazor et al., 1990.)

fumaroles. They are enriched in the atmospheric noble gases, indicating they are fed by gases expelled in the early boiling stage of the deep fluid. The same pattern is seen in Fig. 14.14 for the dry gas phase of the studied drowned fumaroles.

Depletion of ANG discloses that hot springs are fed by residual fluids. Most of the studied hot springs were found to be significantly depleted in their ANG, but they still contained some He excess. This observed noble gas depletion discloses that the springs are fed by a residual fluid that lost a substantial steam phase.

Radiogenic ^{4}He and ^{40}Ar disclose prolong storage of the hydrothermal liquid. The observed pronounced concentrations of nonatmospheric He and also nonatmospheric Ar-40 indicate that one does not deal with a recent

Water in Hydrothermal and Volcanic Systems 313

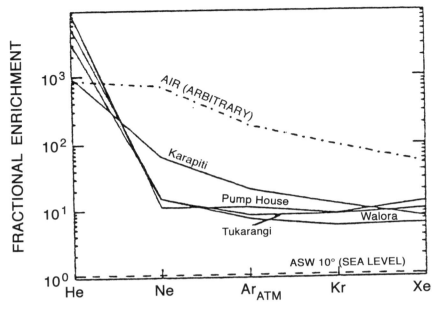

Fig. 14.13 Fractional enrichment of the noble gases in the New Zealand fumaroles compared to the assumed original water (ASW 10°C, dashed line). Excess He is well marked, indicating long storage time of the water phase. (From Mazor et al., 1990.)

through-flow system, but rather with water that was entrapped at depth and subsequently heated by the intruded magma.

Positive correlation between the atmospheric and radiogenic noble gases discloses the water penetrated deeper than the depth of boiling. The radiogenic and atmospheric noble gas concentrations are positively correlated, as is seen in Figs. 14.12 to 14.14. They are similarly enriched or depleted. This indicates that the deep nonboiled fluid is meteoric water that was stored below the depth of boiling, and the steam and gas separations take place during ascent.

Intermixing of shallow water and other secondary processes occur in some of the cases, as discussed by Mazor et al. (1990).

Exploitation effects. In the new wells the He excess was prominent—the He/Xe ratio was up to 1000 times the ratio in the original air-saturated recharge water. In contrast, the heavily exploited wells show a dramatic He drop (observed He/Xe ratios were in the range of only 29 to 68 times the ratio in the original air-saturated recharge water). This may reflect encroachment of recent (He-poor) groundwater as a result of intensive

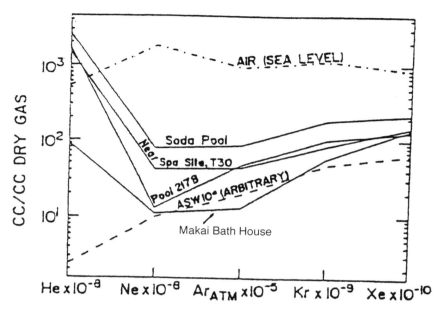

Fig. 14.14 Composition fingerprints of noble gases in the dry gas phase of drowned fumaroles of the Wairakey hydrothermal region. The Ne-Ar-Kr-Xe lines resemble ASW, reflecting the meteoric origin; and the high He content reveals a high age of the deep fluid. (From Mazor et al., 1990.)

exploitation and/or preferential escape of He into the separated steam phase.

The fluid of a new drill hole should be analyzed for its noble gases. If high depletion and a low He/Xe ratio are found, the borehole reached a niche of the field that has already been partially exploited by existing wells. In contrast, if the expected original composition is found, then the new borehole penetrated a section of the field that has not been exploited so far. In the latter case equipping for a production well is recommended, whereas in the first case equipping the well seems noneconomic.

14.5 Noble Gases in a Section Across the Hydrothermal Field of Larderello, Italy

Samples were collected from six wells along a section at the vapor-dominated hydrothermal field of Larderello (Fig. 14.15), 16 km southeast of Pisa.

Water in Hydrothermal and Volcanic Systems 315

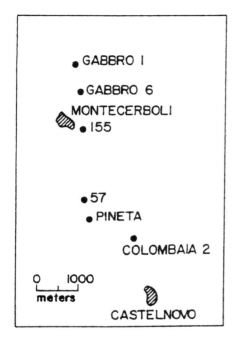

Fig. 14.15 Distribution of studied production wells at the Larderello hydrothermal field, Italy. (Following Mazor, 1979.)

From south to north the wells are (1) deeper, from less than 1000 m to over 1500 m; (2) relatively newer; and (3) closer to the center of the hydrothermal system.

The relative content of dry gas decreases with production. The oldest well of the study, Colombaia 2, contained less than 5% of the dry gas (mainly CO_2) that was contained by the newest well, Gabbro 6, and the dry gas concentration increased systematically along the section, as is seen in Fig. 14.16. This discloses that a gas and steam phase separates from the deep fluid as production goes on. Figure 14.17 is another look at the depletion of dry gas from the residual deep fluid as production continues.

Atmospheric noble gases provide a picture of the exploitation history of the hydrothermal field. Figure 14.18 depicts that Kr (representing the ANG) increases systematically along the section studied at Larderello (data from Mazor, 1979). This observation indicates independently that as

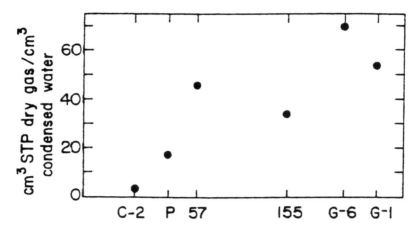

Fig. 14.16 Relative amount of dry gas in condensed steam along the traverse through the Larderello field (C-2) Colombaia 2; (P) Pineta; (57) well 57; (155) well 155; (G-6) Gabbro 6; (G-1) Gabbro 1. Wells are spaced according to projected distance. Gas content increases as the wells are newer, deeper, and closer to the center of the field. (After Mazor, 1979.)

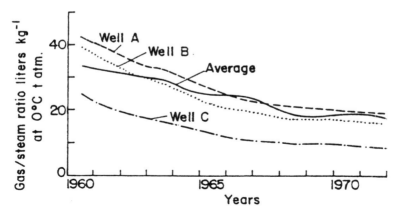

Fig. 14.17 Trend of gas/steam ratio as a function of production (time) in three wells in the intensively drilled area of Castelnuovo, Italy. The relative dry gas content decreases with production, an observation that may help to site new production wells (see text). (Data from Celati et al., 1973.)

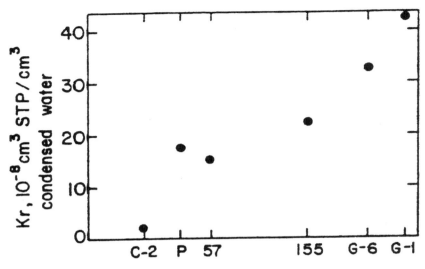

Fig. 14.18 Krypton in condensed steam along the transact through the Larderello hydrothermal field. (see Fig 14.16 for well abbreviation.) The atmospheric noble gas content increases as the wells are newer, deeper, and closer to the center of the field. (Following Mazor, 1979.)

production goes on a steam and gas phase leaves the system continuously.

Radiogenic He excess is positively correlated to the ANG, indicating long storage of the deep fluid. Figure 14.19 is a fingerprint diagram of the noble gases, revealing (1) a prominent He excess and (2) the He is positively correlated to the ANG. These two observations lead to the conclusion that the hydrothermal fluid was stored at depth for a long period, during which radiogenic He was added and well mixed with the ANG. No fractionation of the noble gases occurs as a result of the ongoing boiling process.

Radiogenic ^{40}Ar has been determined, and it is positively correlated to the radiogenic He, as is well seen in Fig. 14.20. This provides further evidence that the water phase was stored at depth for a long time.

Feedback to exploration and production. The long list of observable changes of composition of the extracted fluids, as production goes on, has a practical potential. Whenever a new drill hole encounters high-temperature hydrothermal water, measurement of a sample will disclose whether the new sample has the initial composition of the local deep fluid or a changed composition. In the first case, the new drilling has encountered

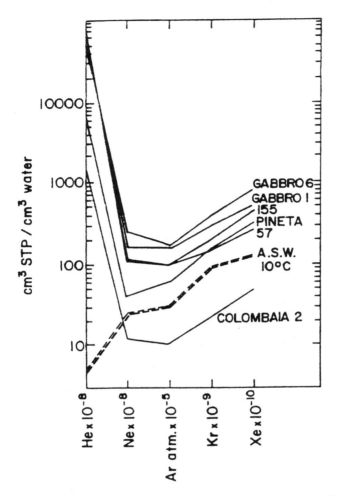

Fig. 14.19 Noble gases along a section in the Larderello hydrothermal field. Colombaia, the oldest well, was depleted in ANG relative to the suggested intake concentration as air-saturated water at 10°C. The other wells reveal ANG excess in concentrations that increase along the section. Radiogenic He is evident and proportional to the ANG. (Following Mazor, 1979.)

Water in Hydrothermal and Volcanic Systems

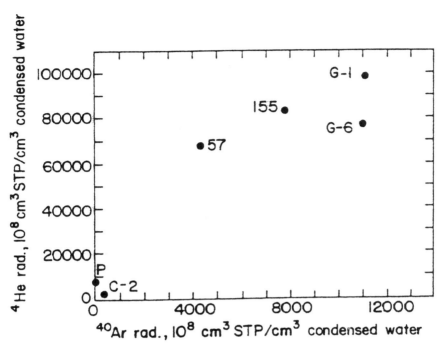

Fig. 14.20 Radiogenic He versus radiogenic Ar, Larderello hydrothermal field. A positive correlation is seen, indicating thorough mixing occurred prior to steam separation. (Following Mazor, 1979.)

a nonexploited niche of the system, and the drill hole may be equipped and operated. On the other hand, when the examined drill hole water sample resembles the changed fluid, this is an indication that water drawn from this specific point is already reached by one of the production wells, and there is no point in equipping it as a well, because the production will be competing with existing nearby wells.

14.6 Fumaroles of Vulcano, Aeolian Island, Italy

Fumaroles are vents occurring in semidormant volcanic systems from which vapors and gases are emitted. Various origins have been suggested for the water and gases of fumaroles, e.g., air entrapment, meteoric water, seawater intrusion, leaching of volatiles from country rocks, water–rock interaction, heat-induced decomposition of sedimentary rocks, outgasing of magma, and emanations from the mantle. In most cases a combination of several sources is concluded.

Sample collection at fumaroles is tricky, as one deals with superheated steam and gases that issue from a pressurized system. Upon ascent, the pressure drops, cooling takes place, and shallow groundwater might be added. In the aim to study these aspects, a field workshop was organized at the Island of Vulcano in September 1982, and the same fumaroles were sampled by all the participants and then analyzed at their laboratories (Corazza, 1986).

The chemistry of the Vulcano fumaroles has been studied by Martini et al. (1980), the noble gases were studied by Mazor (1985), and the data were further processed by Mazor et al. (1988).

Fumaroles occur inside the Vulcano crater, and drowned fumaroles, bubbling through surface water, occur at the Hypopotamus Pond and at the adjacent sea beach (Fig. 14.21). The temperature of the ejected fluids ranged at the Vulcano fumaroles from 149 up to 287°C. The following description addresses studies of the Crater fumaroles F-5 and F-1 (Mazor et al., 1988).

The Cl/Br weight ratio between HBr and HCl discloses seawater decomposition.
The Cl/Br weight ratio between the observed quantities of HBr and HCl was 346. This value is very close to the seawater value of 293 and very different from values reported for other volcanoes (up to 2700) (Martini et al., 1980). It seems that seawater was entrapped in the Vulcano deep system, and the salts are thermally decomposed to HBr and HCl, which are in this case not from a magmatic origin. The seawater entrapment may happen during stages of low internal pressure of the volcanic system, following intensive eruption phases.

The Cl concentration in the fumarole fluids indicates the relative amount of seawater. The F-5 fumarole was found in September 1982 to contain 14.7 g Cl/L. Mediterranean seawater contains around 20 g Cl/L, and the local fresh water contains around 0.2 g Cl/L. Thus, the fumarole fluid might well have been formed by 73% seawater that entered the Vulcano system at depth, along with 27% fresh water added at a shallow depth. Similar values were found for the adjacent F-1 fumarole. Such a mechanism implies a set of interactions between seawater and hot rocks, leading to the formation of HCl and HBr.

The δD values confirm the dual origin from heated seawater diluted by shallow groundwater. The δD value for fumarole F-5 was -1.0‰, whereas the value for Mediterranean Sea water is $+9.8$‰ and that of local fresh groundwater is -33‰, obtained from the Abbondanza well and cistern.

Water in Hydrothermal and Volcanic Systems 321

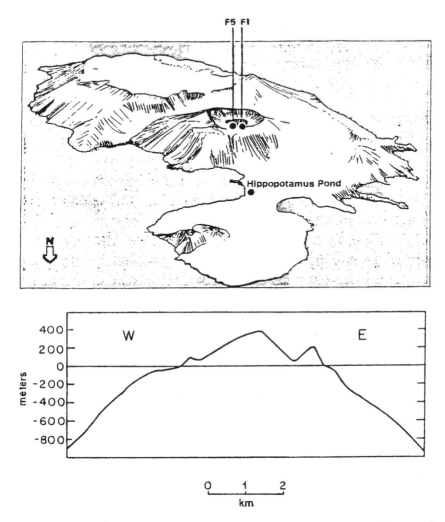

Fig. 14.21 Studied fumaroles at the Volcano crater and a cross section revealing Vulcano above and below sea level. (Following Mazor et al., 1988.)

Thus, the F-5 value lies between the values of the seawater and the local groundwater, suggesting formation of fumarole F-5 from 75% seawater and 25% fresh groundwater, in good agreement with the above derived figures that were based on the Cl concentration. This agreement is well seen in Fig. 14.22, at which the F-5 value lies on the seawater–fresh water mixing line.

The $\delta^{18}O$ value reveals temperature-induced isotopic exchange. A value of $\delta^{18}O$ = +4.76‰ was obtained for fumarole F-5, whereas local fresh water revealed a value of −5.07‰, and the Mediterranean Sea water showed +1.33‰. Thus the value of F-5 is heavier than seawater, disclosing an ^{18}O shift as a result of temperature-induced interaction with the silicate rock. This is direct evidene that the involved seawater has been entrapped at depth and heated by the intruding magma.

Low tritium content rules out direct recharge by rain. Tritium was measured in 1980, in many collected samples, and the content was from 0.0 ± 1.2 to 4.4 ± 3.0 TU. Local rain contained at that time around 40 TU. This rules out any significant contribution from rain directly recharged into

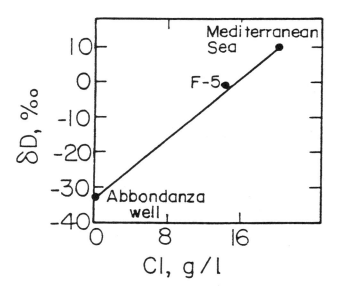

Fig. 14.22 Mixing line defined by the Cl and δD values of local seawater and local fresh groundwater, Vulcano region. The September 1982 value of fumarole F-5 plots on this line, revealing mixing of 75% seawater and 25% fresh groundwater. (Following Mazor et al., 1988.)

the fumarolic vents. Both the seawater and the groundwater components must have a pre-bomb age, i.e., of more than 40 years.

The CO_2, concentration discloses a source other than seawater or groundwater. A concentration of 4.3 mol CO_2/1000 g water was recorded in fumarole F-5, a value that is three orders of magnitude higher than the concentration in seawater or local groundwater. Hence the bulk of CO_2 is supplied by a special source.

The $\delta^{13}C$ value of CO_2 indicates thermal decomposition of underlying sediments as a possible source of the CO_2. A $\delta^{13}C$ value of $-2.0‰$ has been found in the fumarole CO_2. This value differs significantly from the range of -5 to $-8‰$ commonly accepted for mantle-derived CO_2 The observed fumarole value of $-2.0‰$ is close to values found in marine carbonate rocks. Thus thermal and chemical decomposition of underlying sediments seems to be a possible source of the observed CO_2 in the Vulcano fumaroles, rather than magmatic contribution.

High sulfur concentration indicates a source other than seawater sulfate. Fumarole F-5 was observed to contain 12.5 g S/1000 g water. Seawater contains 2.6 g SO_4/L, which equals 1.4 g S/L. Thus the fumarole contains nine times the S content in seawater. Hence it is concluded that the sulfur of the Vulcano fumaroles comes mainly from a source other than seawater sulfate.

Sulfur–chlorine negative correlation indicates that Cl and S originate from different sources. A distinct negative correlation between the Cl and S contents in collected samples of F-5 is seen in Fig. 14.23. Thus these two elements originate from different sources that intermix in different proportions.

Time covariation of SO_2, CO_2 and H_2O in different fumaroles indicates gas/ water variations occur at great depth. The H_2O, SO_4, and CO_2 concentrations were repeatedly measured in five fumaroles (Fig. 14.24). The concentrations covaried, indicating the change happened at the deep fluid, before its ascent in the individual vents.

The NH_4 concentrations indicate an origin from a source other than seawater. A value of 1.4 mmol NH_4/L fluid has been observed in F-5. For comparison, seawater contains only about 0.02 mmol/L. Hence the NH_4 in the Vulcano system originates from a source other than seawater ammonium.

Radiogenic ^4He indicates a long entrapment time of the deep fluid. In the 1982 samples of F-5 a He concentration of 1.3×10^{-3} cc STP/g water was found. For comparison, air-saturated water at 15°C contains only 4.8×10^{-8} cc STP/g water, and seawater contains even less. Thus the He content in the fumarole is five orders of magnitude higher than the

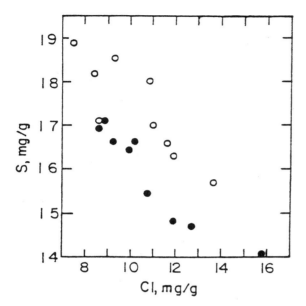

Fig. 14.23 Sulfur concentration as a function of Cl concentration in multiple collected samples of fumarole F-5 of Vulcano. The negative correlation discloses intermixing of two different sources of fluids. (Following Mazor et al., 1988.)

atmospheric He. This additional He is by and large radiogenic. The ^{3}He/^{4}He ratio was close to the atmospheric value and significantly lower than the magmatic value, supporting the radiogenic origin from the host rocks. This in turn discloses long entrapment time of the deep fluid.

Argon-40 excess discloses a radiogenic origin as well. The concentration of nonatmospheric ^{40}Ar in F-5 was 10.9×10^{-4} cc STP/g water. The ratio of He to nonatmospheric ^{40}Ar was 1.2, well in the range of radioactive production in common rocks. Thus long underground storage of the deep fluid is supported.

Modes of seawater intake. Water cannot flow into an active volcanic system because it is pressurized. Seawater and groundwater can be "pumped" into the system when it gets depressurized at the end of an eruptive phase—in extreme cases even rocks collapse in, forming a caldera.

The bottom line. The described study of the Vulcano system reveals how much can be learned about the anatomy of a volcanic system by measuring a

Water in Hydrothermal and Volcanic Systems

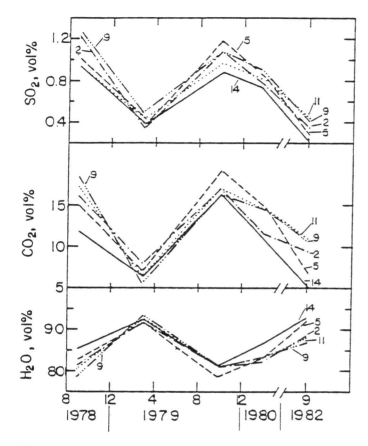

Fig. 14.24 Variations in the composition of five repeatedly sampled fumaroles at Vulcano. Substantial variations in the gas content happened at all the fumaroles at the same time, indicating the gas/water fluctuations happen at depth, prior to the ascent into the different vents. (From Mazor et al., 1988.)

list of parameters and applying the obtained results to define boundary conditions.

Results vary from one volcanic system to another, disclosing variations in the local constellation. For example, contribution of seawater is expected in some volcanic islands, but others may be found to be sealed and have no seawater involved. With all the observed variability, common threads come out—part of the fun of scientific investigations.

14.7 The Hydrology and Geochemistry of Superheated Water in Hydrothermal and Volcanic Systems

At the opening of the present chapter we posed a number of questions regarding the superheated water manifestations. Let us now try to provide answers in light of the described case studies:

1. *Magma is the heating agent.* Superheated waters, in the observed range of 120 to 340°C, are as a rule heated by shallow magma cells in cases active volcanically (e.g., Vulcano) and in other cases relatively calm (e.g., Yellowstone).
2. *Shallow depth of the magma body* is in certain cases definable via geophysical methods or hinted by periodic volcanic activity. It seems to be in the range of a thousand to several thousands of meters. The superheated fluids are encountered at depths that are significantly shallower than the heating depth of the normal terrestrial heat gradient.
3. *The superheated water is meteoric.* This is borne out by several indicators, e.g., the isotopic composition of the water, the allochthonous origin of Cl and correlated ions, and last, but not least, the isotopic composition and abundance pattern of the atmospheric noble gases.
4. *Entrapment of the superheated fluid within pressurized rock-compartments* is borne out by several observations: (1) high-pressure steam bursts out through penetrating drill holes; (2) the water has as a rule a high age, ruling out recent recharge; (3) radiogenic noble gases accumulated and were not flushed away by through-flowing groundwater.
5. *An old superheated deep fluid feeds the boiling springs or wells at every site.* The time period this fluid has been stored at depth is definable by low to zero content of ^{14}C and presence of high non-atmospheric ^{4}He and ^{40}Ar.
6. *Water may be sucked into volcanic systems when underpressurized at the end of eruptive stages*, observed in extreme cases to cause collapse calderas.
7. *Ongoing recharge water cannot penetrate into pressurized hot chambers; it can only intermix with ascending steam.* Such mixing is recognizable by the presence of tritium and ^{14}C along with elevated concentration of radiogenic helium and argon (sections 3.7.3 and 10.6).
8. *The water entrapment predates the hot magma intrusion*, as is borne out by the boundary conditions set by the above listed conclusions.

9. *Chlorine is external and so are correlated ions*, their list changing from one case to another.
10. *Carbon dioxide and H_2S are often added to magmatic and volcanic superheated waters*, as is reflected by the products of CO_2-induced water–rock interactions, by elevated SO_4 content, and by H_2S smell.
11. *The large variety of boiling water types is formed by local combinations of the environmental conditions*, as discussed above, and variability of the geological setups.

14.8 Summary Exercises

Exercise 14.1: How can we identify intermixing of shallow groundwater with ascending steam? Can it be caused by the drill hole?

Exercise 14.2: How are hydrothermal water systems changing due to intensive production? Base your answer on observations described in the discussed case studies.

Exercise 14.3: Which parameters would you recommend to be included in a regular monitoring scheme of a producing hydrothermal field?

Exercise 14.4: Why is such monitoring needed? Discuss.

Exercise 14.5: How can proper monitoring assist the exploration and exploitation management of a hydrothermal field?

Exercise 14.6: Energy production has to be friendly to the environment. What advantages can you see in the hydrothermal exploitation? Which environmental precautions seem necessary?

Exercise 14.7: Logically, can the residual water of hydrothermal electricity production facilities be automatically released into an adjacent river? Discuss.

PART VI
IMPLEMENTATION, RESEARCH, AND EDUCATION

~~~~~~~~~~~~~~~~~~~~~~

We depend on high quality water.
So let us
monitor,
understand what we are doing,
and last but not least
educate.

~~~~~~~~~~~~~~~~~~~~~~

15
DATA ACQUISITION, PROCESSING, MONITORING, AND BANKING

In the parade of case studies discussed in the previous chapters, we dealt with a large variety of field observations and measurements and a vast amount of laboratory-retrieved data. The wealth of data and observations can be overwhelming, and as researchers we might get lost. Let us devote some time to the organization and proper handling of data-related aspects of our research.

15.1 Sample Collection and In Situ Measurements

Our data have to be of first class quality. To achieve this, the in situ measurements and sample collections for laboratory analyses have to be done properly. The following are a few relevant points:

Guidance by the laboratories is highly recommended, and projects should be discussed in advance with the lab staff (discussed further below).

Mode of sample collections for the various groups of analyses, e.g., the needed volumes and relevant containers; chemicals that have to be added in order to stabilize the samples; auxiliary information needed; handling of samples, e.g., keeping in a cool and/or dark place; knowledge of the limits of detection and expected accuracy for each parameter.

Assistance of the lab experts is recommended in the collection of water samples. Some sample collections are tricky, e.g., samples for noble gas analyses (excluding absolutely entrapment of air; determining the right point at a spring pool or well head); H_2S (adding of a stabilizer) and other dissolved gases (avoiding losses); pH; and SiO_2 (distilled water has to be added as silica may precipitate due to cooling). At fumaroles and steaming wells, steam condensers have to applied and the dry gas content be established. It is highly recommended that the lab experts guide the field workers in these procedures and demonstrate them in the field.

Required in situ measurements include water pH, temperature, and conductivity (reflecting salinity). Such in situ data are part of the needed parametric values, and in addition they are needed to assess whether changes occured in the samples before they have been analyzed.

The exact point of sampling and in situ measurements has to be well selected. For example, at a spring pool the most representative spot is where rising water can be seen and where the temperature is maximally different from the ambient temperature (to minimize secondary effects). At a well head use an outlet at the straight pipe to minimize pipe bends that may cause local gas separation (enrichment or losses that may be reflected in a sample). Samples in drillings have to be collected after the drilling fluids have been pumped away (pump until water properties become stable).

Mode of sample collection and in situ measurements has to be described in the field notebook. It might be of importance in the data interpretation.

Spare samples. Questions pop up occasionally regarding obtained laboratory results. In such cases a duplicate sample has to be sent in for reanalysis (or, in extreme cases, to be sent to another laboratory). Hence collection of spare samples is highly recommended.

Sample library. As a project matures, the need for the analysis of additional compounds often comes up, yet collection of a proper sample may not be feasible, e.g., samples collected during a drilling operation, samples collected at a new well; water separated at the head of a new oil or gas well; samples collected at fumaroles prior to or after a volcanic eruption; or a sample collected at a salinized well before it was shut.

15.2 Checking the Laboratories' and Data Quality

A laboratory is a complex of activities that have to be checked constantly. Errors may happen due to instrumentation problems, introduction of new supplies of reagents or standards, mixing of sample numbers, or errors sneaking into the laboratory notes or data computations. For these reasons every laboratory has its checking procedures, and the field researcher has to do his or her own checking as well.

Check by duplication of one or two of the samples; this is a must with every batch of samples sent to one of the labs. The duplicates should have different numbers and names. If the results are satisfactory, the whole batch seems reliable. On the other hand, if the duplicates differ beyond the specified accuracy, a question mark has to be placed on the data of the entire batch; the problem has to be openly discussed with the laboratory staff, and repeated measurements will probably be done. The question whether something happened during the transport to the lab has to be discussed as well.

Have your own standard. It is recommended that in every study site the researcher collects and well preserves a large volume of water from a representative source. This large sample will be kept in a dark and cool place (preferentially refrigerated), and this will be the standard. A sample of it will be added to every batch of samples sent to the laboratory. This is another mode to survey the reliability of the lab results and detect any problem that may come up.

Discuss with the laboratory staff any problem that may come up with the results, e.g., illogical data or sudden and bizarre differences from previous measurements.

Is all this checking worthwhile? Yes! After all, the data have to be reliable. Applying incorrect data will lead to incorrect interpretation.

Check of the reliability of historical and published data. Historical data obtained from nonpublished archives and data tables included in professional publications often lack information regarding their analytical quality. How can we check? The answer is quite simple. Let us process them in a variety of diagrams—poor quality data will scatter around. In contrast, well-defined linear correlation lines, seen for data obtained from different wells, clearly confirm that the data are of high quality. The same is true for clear clusters seen for data from different wells.

Linear temperature–depth profiles confirm quality of samples. Linear positive correlation lines of ion concentrations, of the type shown in Figs. 10.2 or 10.5, may theoretically be dilution lines of a deep saline formation water that became intermixed with various amounts of shallow fresh water, as a result of short-circuiting caused by poor well casing. In the named examples this possibility could be ruled out in light of the perfect heat gradient line seen in the temperature-depth profiles of Figs. 10.6 and 10.16. The water samples were separated from the petroleum at the well heads, and hence these are well-collected samples, each representing a single rock-compartment.

15.3 Types of Wells

Data related to a well provide a general and useful picture. We can raise the level of our understanding of a studied system if we determine the type of each well and incorporate this knowledge in our interpretation.

Depth and range of perforation are essential information in order to understand to which specific aquifer unit, or rock-compartment, the obtained data are pertinent and/or whether several units have been short-circuited and intermixed.

A newly drilled borehole first has to be pumped in order to ensure getting rid of the drilling fluid. These data are important as we get information of a new point of the system.

Observation wells are wells that are left nonpumped. Their water reflects the water table, temperature, and composition that are relevant to the regional system. The farther away an observation well is placed from production wells, the better it reflects the behavior of the undisturbed natural system. Observation wells are of importance in all systems—unconfined, confined, and artesian—and placed in all the geosystems, including oil and gas fields as well as hydrothermal steam production fields.

A specific use is for seismic monitoring—observation of water table fluctuations prior to, during, and following seismic events.

Nests of observation wells are made up of a number of closely placed wells that each sample a water system at a different depth. Data obtained serve to get vertical insight into a studied system. The waters at the different depths may all have the same temperature and composition, providing an idea of the vertical extension of a studied groundwater body. In contrast, differences between the different wells of the same nest disclose that they tap different systems or separated compartments.

Production wells are continuously pumped. The water table is locally lowered, forming a *pumping depression cone*, and new water flows in from the aquifer. Thus the water table measured in an observation well is somewhat higher than the water table in a close-by production well that exploits the same water body. The first reveals the *static water table*, and the second reveals a *dynamic water table*. Differences between the two types of nearby wells may be seen also in the water temperature and composition, providing an insight into the dynamics created by the pumping operation, e.g., inflow from two or more water bodies.

The dynamic water table is rarely measured, but it is of prime importance for proper operation. The set of red lines of operation include the lowest dynamic water table tolerable.

15.4 Multiparameter Studies

Every additional parameter that is measured in a system defines specific boundary conditions and improves our understanding of studied systems (Mazor, 1976). The following is a list of the common parameters:

Water table depth and altitude measured in observation wells provide insight into whether we deal with a perched groundwater body (section 3.1.1), the regional unconfined system, or deeper water bodies, entrapped within the zone of zero hydraulic potential. Response of the water table in wells to pumping tests in adjacent boreholes discloses hydraulic interconnections. The water table measured in observation wells in an unconfined system discloses the inclination of the regional water table, which in turn disclosesthe location of the base of drainage.

Discharge and pumping yield. The discharge of a spring or yield of a well provides an idea of the size of the sustaining water body. Follow-up over time provides an insight into the dynamics of the respective system—its response to natural as well as manmade processes.

Temperature of a water source indicates the depth from which the water is coming. The temperature of the warm end-member has to be determined in the case of intermixing of different waters. In cases of intermixing of a deep warm water with a shallow cold water, a negative correlation between the temperature and other parameters, e.g., tritium or ^{14}C, can serve to find by extrapolation the temperature of the warm end-member (Mazor et al., 1985; 1992).

Common ions. Too often chemical data measured at a spring or well relate to Cl concentration alone, representing the water quality; occasionally nitrate is also analyzed, representing pollutants. Such measurements have a limited value. If Cl is low, does it mean high quality water? The SO_4 content may be high, so this information is needed as well. Increase in Cl in a coastal well may indicate seawater encroachment, but actually it might be fossil groundwater intrusion into overpumped regions. So, at least the Br, Ca, and Na concentrations have to be known. In brief, every analyzed ion adds to our understanding a studied system.

In the depth profiles of suboceanic interstitial waters and of formation waters in sedimentary basins the resolution of identifying isolated rock-compartments improves as more ions are analyzed. The common ions are, in general, Cl, Na, Ca, Mg, SO_4, and HCO_3, as well as Br and K.

Sea spray versus brine-spray tagging. The former facies can be identified by Na being the main cation that balances the Cl; and the Cl/Br weight ratio is around the seawater ratio of 293. The latter is identifiable by Ca being second to Na in balancing the Cl; and the Cl/Br weight ratio is in the range of 80 to 200 (Chapter 7). As we have seen in the discussed case studies, the two different modes of tagging are reliable identification criteria—the first identifies recent ongoing unconfined groundwaters, and the second typifies connate fossil waters, residing within the zone of zero hydraulic potential.

Rare elements include, in a way, the rest of the elements. Commonly measured are Sr, Li, B, Al, and SiO_2.

Water isotopes. The δD values help to identify meteoric groundwater formed on the continent, in contrast to seawater or residual brine. The $\delta^{18}O$ reflects in certain cases the meteoric origin as well, but in many formation waters and hydrothermal waters the $\delta^{18}O$ is heavy, revealing temperature-induced ^{18}O enrichment by water–rock interaction. This, in turn, discloses a deep subsidence phase.

Age indicators are crucial to the distinction between recent renewed water of the through-flowing water system and confined connate waters. The age indicators are (section 3.3.5) tritium (a few decades); ^{14}C (up to 2×10^4 years); ^{36}Cl (10^5 to 10^6 years), ^4He (10^4 to 10^8 years, and ^{40}Ar (10^5 to 10^6 years).

An entirely different age is the age of confinement (Chapter 8), which necessitates knowledge of the stratigraphic age of the host rocks of studied systems. Agreement between the isotopic age and the confinement age is an indication that deduced ages are valid.

The atmospheric noble gases (ANG) are helium, neon, argon, krypton, and xenon, which get dissolved in meteoric water in concentrations that depend on the ambient temperature (sections 3.2 and 3.3). They are indispensable in studies of the global water occurrences: (1) the ANG provide an independent marker to identify meteoric continental water, even when very old; (2) they are useful in the determination of paleo-temperatures; (3) the ANG are efficient markers of the physical processes going on, such as boiling and phase separations in hydrothermal and volcanic systems; (4) they provide means to determine the initial concentration of radiogenic ^4He and ^{40}Ar in water sytems that lost or gained gas and steam or were diluted by shallow recent water, thus paving the road to isotopic dating of ancient groundwaters (Chapter 8).

Petroleum compounds soluble in saline water are of prime interest in hydrological studies of formation waters in sedimentary basins and rift valleys, providing a promising petroleum exploration tool (Chapter 10).

Pollutants: a long list of polluting compounds are essential parameters in detailed studies of the unconfined groundwater system, and they may serve as useful tracers (Mazor, 2003). The absence of manmade pollutants supports the identification of fossil connate waters.

15.5 Multisampling

Study of a single water sample provides valuable information, but its relevance is very local in space and time. For the study of natural water systems samples from many sources, repeatedly collected, are needed in order to get the four-dimensional picture. This is true for a specific study area as well as for an entire geosystem—and the whole geohydroderm. When it comes to sample collection, the rule is "the more-the better" (Mazor, 1976).

Reconnaissance survey of maximum sources in an area. Study of a single well provides very local information. Multiple sources in a study area provide a wider perspective. Hence the planning of a study includes a survey of all accessible sources. The samples are best collected within a short time period in order to eliminate seasonal influence in the case of shallow groundwaters and exploitation influence in all systems.

Every study has budget and time limitations. However, sufficient sources have to be selected so that they cover all different occurrence, e.g., nearby wells that provide a detailed depth profile, wells tapping water at different depths all over a drainage basin, wells representing a whole basin, wells spread over an entire petroleum field or hydrothermal energy production field.

Wells of priority include those that have great yields, wells for which maximum information is available (the driller's notebook and historical data), or wells at locations where management problems have been detected or are anticipated.

Looking at a whole complex. To reach full understanding, even of a small study area, there is a need to explore and report a whole relevant unit, e.g., a drainage basin from the main water divide to the terminal base of drainage or the entire petroleum field or hydrothermal system. The specific study area may be small and intensively inspected, but its relation to the larger unit has to be established.

Measurements during a pumping test at a borehole reaching the unconfined groundwater system or a deeper confined system. The pumping is commonly rather intense and lasts for a day or two. The main purpose is in general to define the safe yield. An important task is to find out whether one type of water is involved or whether a second type of water intrudes into the system because of the pumping, possibly deteriorating the quality of the extracted water. Thus it is highly recommended that the pumped water is continually monitored for temperature and electric conductivity and that about ten samples be collected. If the properties remain unchanged the safe yield can be readily estimated, but if the quality deteriorated a relatively low yield may be concluded.

Measurements during a drilling operation. Every drilling operation has the potential to pass through several separated water bodies, and it is of high value to learn about them, as they may be potential future targets for exploitation of water and as part of our understanding of the studied system. In deep drilling operations bodies of formation waters may provide information on nearby petroleum occurrences. Thus the circulating drilling fluid may be continuously monitored for temperature and electrical conductivity in order to detect the appearance of a new water body. The drilling operation has to be stopped whenever a new water is encountered, and a representative sample of it has to be retrieved for detailed laboratory analyses.

Time series. Water in nature is dynamic in one way or another. In the unconfined cycling groundwater system there are seasonal variations, responding to rain and snowmelt; changes induced by exploitation, e.g., salination due to overpumping; and arrival of pollutants. Artesian well changes also do occur due to exploitation; and the same is true for formation waters in petroleum fields and the water in hydrothermal wells. Thus analysis of a single sample provides a limited glance at the studied system. Repeated analyses and repeated field measurements, conducted at different times of the year, provide the full dynamic picture, or a time series. Time series of different parameters shed light on the different aspects of water systems, e.g., time series of the water table, the temperature, or various dissolved ions.

15.6 Monitoring Networks

The topic of monitoring networks brings out the basic difference between the unconfined groundwater system and all the other large-scale confined systems.

15.6.1 Monitoring Networks of the Unconfined Through-Flow System

This necessitates both observation wells and producing wells. The observation wells monitor the static regional water table, whereas pumped wells are each characterized by a local lower water table—the dynamic water table. The water in these two types of wells often varies in salinity and ionic abundances, as saline water occasionally encroaches into overpumped wells. The combined data, from both types of wells, provide information on the renewability by seasonal recharge, water quality, and movement of pollutants. Both sets of data are essential for the understanding of the dynamics of the fresh groundwater resource and for its optimal management. The main management issue is to keep water production sustainable, e.g., by constraining production within the limits of the renewability of the resource.

The topic of monitoring networks for surface waters and unconfined groundwater systems is discussed in detail in the companion book of the present volume (Mazor, 2003).

15.6.2 Confined Formation Water Systems that Need Monitoring

The large group of confined formation waters includes the following subgroups listed along with their specificities:

Confined domestic water supply systems. These are growing in number in semiarid and arid areas in many countries, the wells being hundreds to a thousand meters deep. These are nonrenewed waters, stored in hydraulically separated rock-compartments. Supply is maintained by constantly drilling new wells, and this necessitates detailed knowledge of the three-dimensional local fruitcake structure.

The management needs are (1) to get in due time a warning that a certain well is running out of water and (2) locate the right site and proper depth of new wells.

Artesian basins supply water in many countries, perhaps the best known example being the Great Artesian Basin, northeast Australia, in which during the last century a few tens of thousands of deep wells have been drilled, each yielding for several years. Most of the water is used for agriculture and stock raising.

The principal management aim is here, too, to ensure steady water supply, ie., substitute the wells that lose yield with new ones. To properly site the new wells the detailed local fruitcake structure has to be known—from the data collected at the existing wells during their drilling stage and the monitoring that followed exploitation.

Two additional specific management targets are applicable in artesian well fields: (1) to construct the wells so that there will be no short-circuiting between rock-compartments of different water qualities and (2) to construct wells so that they can be easily closed and opened to allow use of only the needed quantities (for the time being excess water is allowed to flow away in many cases).

Oil fields. When the yield of an oil well gets too low and/or the amount of co-produced water becomes too high, a number of steps can be taken, e.g., slight change of the depth of casing perforation in order to reach less exploited niches of the rock-compartment trap, injection of water to wash more oil out of the rock pores, or drilling of a new well in a nearby less exploited niche. Another management operation is the reinjection of separated brines back into exploited parts of the operating field. A precise knowledge of the local compartmentalization and the induced changes is thus needed.

Therapeutic water installations. A spa builds up its reputation, and hence the long-range supply of the advertised type of water has to be secured. This necessitates continuous monitoring and strategies to follow the moment problems with existing springs and wells come up or when there is an increase in the demand.

Hydrothermal steam fields. The operators of a hydrothermal complex have to be alert to changes in the properties of the individual wells and the field as a whole. Continuous monitoring of the wells is aimed to provide an updated picture of the system and the ongoing changes in terms of water temperature, steam/water ratio, gas contents, encroachment of shallow groundwater, etc.

Nuclear waste repositories. A nuclear waste repository is planned to function as a hydraulically isolated compartment that will keep the waste enclosed for several thousands of years, until all the radioactive materials decay. Thus at a reasonable distance observation wells are needed in which relevant radioactive nuclei will be monitored.

15.6.3 Monitoring Networks of the Deep Confined Water Systems

A well-planned and well-defined monitoring network is a must for every operating field of any kind of formation water. The following are a few guidelines:

The networks have to be placed in three dimensions. To monitor setups of formation water occurring in fruitcake structures, a three-dimensional network is required. Therefore, in practice, all accessible wells are to be included in the monitoring networks.

One-time detailed analysis of every well or thermal spring is highly recommended along the list of parameters presented in section 15.4. This detailed analysis is needed (1) in order to fully understand the potential of the well and its position in the structure of the entire system and (2) to enable identification and evaluation of changes that may appear with time, as a result of corrosion and other modes of deterioration of the well or as a result of exploitation.

Defining minimal parameters to be regularly monitored at each well. The parameters that have to be routinely measured belong to two groups:

1. Basic parameters to be monitored at all wells. These include monthly measurements of temperature, water table or artesian pressure,

and electrical conductivity (indicating total salinity); and once a year the content of major ions.
2. Specific parameters, selected for each well. These will be defined on the basis of the one-time detailed analysis and the type of petroleum compounds in wells related to oil and gas fields, e.g., compounds poisonous to agriculture or stock raising in artesian water, therapeutic compounds at spas and thermal water parks; etc.

A computerized file for every monitored well is essential, including the following items:
1. The drillers report with all the observations and measurements conducted during the drilling operation
2. The results of the one-time detailed analyses
3. Discussion of the hydrological system and specific rock-compartment to which the well belongs
4. The reason the well is included in a monitoring network so the obtained data can readily be interpreted in management terms

An updated four-dimensional model for every operated system and location of all the monitored wells.

Operational "red lines" have to be defined in advance for every well, in terms of, for example, depth of water table, pressure, temperature, salinity, and other composition parameters. When such a critical value is observed in a well the information has to be immediately passed to the operator and relevant decisionmaker.

Monitoring networks are best operated via active data banks, as discussed in the next section.

15.7 Effective Data Banks

Data are valuable only if they are used. Hence efficient and active data banks are an essential part of any research setup—whether of a single researcher, an institute, or of an entire country. The following are a few guidelines of organization and data processing:

A portfolio for every operational or administrative system, e.g., a specific study area, a sedimentary basin or a defined section of it, an artesian system, an oil field, a hydrothermal system or energy production field, a therapeutic and recreational complex, etc. Each of these will have a portfolio in the data bank.

A file for every source, e.g., spring, observation well, operating water well, petroleum well (data of the continuously separated formation water), etc. The file will include the data obtained during the drilling operation, historical data, relevant publications and expert reports, data obtained from local authorities, and above all, reception of all the new obtained data of special operations and of the routine monitoring networks.

Direct reception of new data from the monitoring networks and from special studies by researchers and operating systems.

Issue of alarming observations in real time to the managing and directing authorities.

Modes of data processing and presentation that include, for example:

Time series
Composition diagrams
Fingerprints
Cross sections
Data maps
Search for outstanding cases

15.8 Summary Exercises

Exercise 15.1: You are in charge of the geohydrological section of a water-supplying organization operating in the northeast part of the Dream Geyser Basin. How will you organize the data bank?

Exercise 15.2: You moved and now work with the Black Gold oil field. Will you copy the data bank structure you suggested in the answer to the previous question?

Exercise 15.3: A set of parametric cross sections through certain wells is needed; how could it be easily retrieved in a user-friendly data bank?

Exercise 15.4: All of a sudden you remember that the water samples collected last week in the Bizarre Well were not yet delivered to the laboratory. The samples were kept in a car, and the last days were hot. The samples are by now really warm. What will you do?

Exercise 15.5: During a drilling operation a new water horizon is encountered. The drilling has to be stopped; a water sample has to be immediately collected; and drilling can then be continued. Is this the correct procedure?

16
CONCLUSIONS AND RESEARCH AVENUES

The global view of the occurrence of terrestrial fluid water is fascinating. In the previous chapters we became acquainted with water in its four dimensions, its direct relation to petroleum deposits, its involvement in hydrothermal processes and the key role it had in a long list of geological processes that shaped the unique landscape of our planet.

16.1 Criteria to Check Working Hypotheses Related to Global Water Occurrences

An integral part of every working hypothesis is identifying criteria to check its validity in new study areas. Hence the basic observations and deduced conclusions covered are summed up. The basic observations can be checked in every new study area.

16.1.1 Water Flow Occurs in Free Space in a Downward Direction

This is observable in the flow paths of runoff and rivers; they flow along the steepest paths available in free space above the landscape. The water table of the unconfined groundwater system is slightly inclined toward the base of drainage. This observation indicates that infiltration is at each point downward in free space of voids in the aerated zone until the water table is reached, where the flow direction is at once diverted to a nearly lateral base flow in the free spaces within the aerated zone on top of the saturated zone.

16.1.2 A Base of Drainage is the Lowest Free Space Available for Water Flow

This is well demonstrated by the flow of runoff and rivers that terminate at the local lowest points of the landscape, e.g., deep valleys or lakes, and the terminal base of drainage is the sea surface. The water table in the unconfined groundwater systems slopes to the terminal base of drainage, which is the sea surface. The surface waters and the unconfined base flow and water table, have never been observed below sea level.

16.1.3 "All the Rivers Lead to the Sea" (Ecclesiastics)

This is a two-millennia-old hydrological key observation that the sea surface is the ultimate base of drainage.

16.1.4 Identification of Meteoric Water Formed on the Continent

This is readily done by the following markers: (1) *negative (light) values of δD and $\delta^{18}O$, plotting on the GMWL* and (2) *the atmospheric noble gases (ANG) disclose original exposure to air.*

16.1.5 Identification of Recent Cycling Groundwater

This is made clear by the following characteristics:

Water table fluctuates seasonally
Tagging by sea spray
Meteoric δD and $\delta^{18}O$ light values
Occurrence within the landscape relief that is above sea level
Recent age by the isotopic age indicators
Frequent occurrence of manmade pollutants

16.1.6 Groundwater is a Closed System for the Dissolved Noble Gases

This conclusion has been based on the following observations:

ANG concentrations in unconfined groundwater reveal air saturation.
Fossil formation waters reveal reasonable ANG-intake paleotemperatures.
Radiogenic ^4He and ^{40}Ar accumulated in waters even 10^8 years old.

16.1.7 Radiogenic ^{4}He and ^{40}Ar Are Valuable Semiquantitative Groundwater Age Indicators

The concentration of radiogenic ^{4}He and ^{40}Ar, corrected by the ANG for occasional gas losses or gains during ascent, serve as reliable water age indicators, providing the order of magnitude of the age is in the range of 10^4 to 10^8 years. This is supported by the following observations: (1) *agreeing ^{4}He and ^{40}Ar ages* and (2) *agreement with confinement ages*.

16.1.8 Identification of Fossil Groundwater

This is made clear by the following characteristics:

Commonly saline
Ionic abundances significantly different from the local recent groundwater
Tagged by brine-spray
Old age by isotopic age indicators
Generally devoid of manmade pollutants

16.1.9 The Zone of Zero Hydraulic Potential

Water flows in a downward gradient in free space available in the zone of hydraulic potential differences that extends from the continental landscape down to the sea surface, which acts as the terminal base of drainage. The ocean surface is the plain of zero hydraulic potential. Below it extends the zone of zero hydraulic potential, at which the first principles consideration predicts that all the groundwater is static. This is validated by the following observations:

Suboceanic interstitial waters occur in isolated compartments
Formation waters are stored in isolated compartments (fruitcake structure)
The isotopically defined water ages are observed to increase with depth

16.1.10 Fossil Groundwater is Static—Stored in Isolated Rock-Compartments

This is revealed by the following observations:

Water is not recharged and is detached from the land surface.
Spatial diversity of formation water types.
Spatial differences of fluid pressure.
The isotopic ages increase with depth.

16.1.11 There Are Fully Impermeable Rock Barriers

This is revealed by the existence of very old formation waters—up to 10^8 years old—and very old traps of petroleum. The existence of even Paleozoic halite beds is another testimony of the existence of entirely impermeable rock barriers that engulfed each halite compartment and protected it from being dissolved and carried away.

16.1.12 A Fruitcake Structure Typifies the Zone of Zero Hydraulic Potential

This zone is built of rock-compartments that host different types of water and a variety of petroleum types and soluble evaporites. These various permeable rock-compartments are separated from one another by a network of impermeable rock barriers, altogether resembling the structure of a fruitcake.

16.1.13 Confinement Age and Isotopic Ages Are Mutual Checks

These two types of water ages are physically completely different, yet both have the potential to provide a semiquantitative age or order of magnitude of age. The application of both theses approaches to water dating is highly recommended in fossil groundwater studies, as they provide mutual checks.

16.1.14 Fossil Groundwater Stored in Sedimentary Rocks is Connate

The water is contained in its host rocks since the latter were confined by the overlying rocks, as is disclosed by the following observations and considerations:

The fossil waters reside statically in the zone of zero hydraulic potential.
The network of impermeable rock barriers blocked any through-flow.
The isotopic age indicators validate the deduced confinement ages.

Hence the only way the different types of fossil water could be introduced into their host rocks was a paleo-groundwater, which was confined in the rock-compartments the moment the latter were covered by a sea transgression and successively deposited sediments.

16.1.15 Brine Tagging Indicates Formation at Vast Evaporitic Flat Lowlands

The combination of meteoric water; external origin of most ions; and Cl/Br ratio and $CaCl_2$ as in evaporitic brines implies formation as continental groundwater in regions surrounded by evaporitic landscapes of sabkhas and lagoons. The large extension of many sedimentary rock basins that contain the brine-tagged formation waters reveals these paleo-flatlands were of a vast scale.

16.1.16 Formation Waters and the Host Rocks Disclose Frequent Alternations of Sea Regressions and Transgressions

The continental formation waters are entrapped in the vast sedimentary basins within rock beds that include sandstone, mudstone, clay, halite, and gypsum besides limestone and dolomite—a combination that reflects frequent alternations between shallow sea and low continental plains. The rock sequence recorded mainly the marine phases, and the meteoric formation water recorded the many phases of land exposure.

16.1.17 Deep Subsidence is Identifiable by Positive $\delta^{18}O$ Values of Formation Waters

Oxygen-18 enrichment is a rather common observation in formation waters, accompanying the light meteoric values of δD. This indicates temperature-induced exchange with the host rocks. This discloses, in turn, deep burial by subsidence. The last conclusion is supported by the frequent association with petroleum, which is a temperature-induced product in itself.

16.1.18 Petroleum and Coal Were Formed in Closed Rock-Compartments—The Pressure-Cooker Model

Petroleum and coal occurrences are an integral part of the fruitcake structure that typifies sedimentary basins. Petroleum deposits are encountered within traps that were well sealed over geological ages. These boundary conditions reveal that petroleum was formed within the rock-compartment that hosts it. The latter contained from the beginning the needed ingredients, e.g., organic raw material, permeable rocks, and water—all inside the compartment with impermeable rocks around it. The rock-compartment structure reveals formation on the Earth's surface and confinement by overlying sediments deposited by a subsequent sea transgression. Subsidence brought the compartment into the temperature–pressure zone that induced the formation of the petroleum and marked the associated water by the ^{18}O enrichment. Concentration of oil or gas happened within the domain of the original rock chamber. The size of the individual rock-compartments varies over orders of magnitude.

16.1.19 The Giant Petroleum Fields Are Clusters of Separate Compartments

The distribution of petroleum-containing compartments is uneven; the sedimentary basins have zones rich with petroleum and zones without them. In other words, even the giant petroleum fields are separated into numerous hydraulically isolated compartments—a structure revealed by the variability of the composition of the petroleum phase, differences in fluid pressure, and the variable composition of the associated formation waters. The large-scale oil and gas fields have their origin in suitable facies zones within the respective sedimentary complexes.

16.1.20 Mixed Water Samples—A Constant Threat to Our Research

Our in situ measurements and collected samples are always aimed to address a single water body, e.g., water from one rock-compartment in a sedimentary basin or from a single fracture-compartment in a crystalline shield. However, boreholes and wells are often drilled through several rock-compartments or fracture-compartments, enabling water from different chambers to come up as a mixture.

A conspicuous outcome is contradicting concentrations of isotopic age indicators, e.g., appreciable concentrations of tritium and post-bomb concentrations of ^{14}C, along with high concentrations of radiogenic ^4He and ^{40}Ar. The first parameters indicate an age of a few decades, and the latter define an age of tens of millions of years. In such a case the obtained ages are irrelevant and meaningless, and so are the measured water table or pressure and the measured temperature.

Hence intensive research is required in two avenues: (1) developing efficient ways to check for water mixing and (2) advanced methods to measure and sample single water bodies (preventing short-circuiting).

16.1.21 External Cl and Br

Practically as rule, Cl and Br dissolved in recent and fossil groundwaters are of an external origin; they are not a product of water–rock interactions. This is borne out by the following observations:

Undersaturation with respect to halite. Halite is the sole rock that can contribute Cl to groundwater. Water that is exposed to halite dissolves it readily, reaching a saturation concentration of around 180,000 mg/L of Cl, and the Cl in such a case is entirely balanced by Na. Formation waters commonly contain much less Cl, indicating they did not encounter halite, and hence the Cl is external.

The Cl/Br weight ratio is much different from halite. Halite has a ratio of >3000, but shallow groundwater has a ratio of around 300 (the marine value), and formation water has even lower ratios (as of evaporitic brines).

16.1.22 Other External Ions

These are identifiable by their linear positive correlations to the concentration of Cl, e.g., Na, Li, Sr, and often also Ca, Mg, or SO_4.

16.1.23 Internal Ions Are Rather Rare, Indicating that Water–Rock Interactions Are Limited

When the proven external ions are subtracted from the results of a chemical analysis of a formation water sample, in most cases just a small quantity of dissolved ions is left. These may still be external, like CO_2-induced interaction with the soil.

There is room for further research, based on case studies, to define the extent and nature of water–rock interactions in the domain of fossil formation waters.

16.1.24 Permeability and Hydraulic Connectivity Were Preserved in Spite of Compaction

As rock strata subsided they were buried under the gradually increasing weight of the overlying rocks. The latter exerted compaction and the question comes up: how did the different rock types respond? Did their porosity decrease due to collapse of the rock structure leading to denser repacking? A key observation is that high-yielding wells of deep formation water, including artesian wells, are common in practically all sedimentary basins. This is a direct indication that high porosities and hydraulic connectivity have been well preserved, even at a depth of several thousand meters. The same is revealed by high-yielding oil and gas wells. Thus the entrapped fluids supported the rock structure and resisted compaction.

16.1.25 Pressure Differences in the Zero Hydraulic Potential Zone Disclose Entrapment in Rock-Compartments Separated by Effective Impermeable Rock Barriers

In the unconfined groundwater systems we assume that a measured pressure difference indicates active water flow from the high pressure zone to the low pressure zone. The pressure difference is maintained in such cases by constant recharge.

The situation is intrinsically different in the deeper zone of zero hydraulic potential. A pressure difference observed between formation water encountered in adjacent wells or at different depths in the same borehole indicates that different rock-compartments exist, separated by effective impermeable barriers. Formation waters are fossil, and hence there is no recharge to maintain a dynamic pressure difference. This conclusion is supported by chemical differences that are commonly found between formation waters stored in separated compartments. The same is true for petroleum-containing rock-compartments; pressure differences indicate effective hydraulic isolation.

16.2 Geosystems that Host Fluid Water—Research Topics

We had a close look at case studies dealing with the different water-hosting geosystems that together make up the geohydroderm of planet Earth. The following key conclusions were reached, leading to management recommendations and inviting additional research.

16.2.1 The Oceans

The oceans cover two-thirds of Earth and serve as

The terminal base of water drainage (zero hydraulic potential at the surface)
The main terrestrial water reservoir
The main Cl and Br container
The main collector of erosion products

16.2.2 Surface Waters

These include lakes, wetlands, rivers, and runoff. Surface water

Makes up the terrestrial fresh water
Flows in availabe free space down to the terminal base of drainage
Recharges the unconfined groundwater system
Transports seaward the continental erosion products

16.2.3 The Unconfined Through-Flow Groundwater System

This system is

The main fresh groundwater reservoir
A renewed resource
Vulnerable to anthropogenic pollution

*Young, its age ranging from recent to a few thousand years
Sea spray tagged*

16.2.4 Suboceanic Interstitial Waters

In Chapter 5 we examined the achievements of the drilling operations below deep seawater and the analytical data obtained on the interstitial water, down to over a thousand meters of rock sequences. The following conclusions were reached:

*Seawater did not enter into the underlying soft sediments and hard rocks.
Water composition variability discloses rock-compartment structure.
In most cases a seawater origin discloses continued sea prevalence.
In certain cases a meteoric origin indicates a sea regression phase.*

Suboceanic interstitial water is an ideal research tool for mapping the marine–continental history of the present sea-covered regions.

16.2.5 Formation Water in Sedimentary Basins

This formation water resides in the zone of zero hydraulic potential, and it

*Occurs down to a depth of a few thousand meters and probably more
Resides in rock-compartments, disclosed by rock and water observations
Recorded phases of sea regressions
Occurs in association with all oil and gas deposits*

16.2.6 Rift Valleys

Rift valleys are elongated stretches of the crust that subsided along with the contained formation water. Open research topics include

*Did water also flow down in temporarily opened tectonic fractures?
Which evaporitic sources supplied in each case the brine tagging?
How are warm waters ascending? By temporary tectonic compression?*

16.2.7 Artesian Waters

Artesian waters, known from sedimentary basins and rift valleys, call for the following management steps:

*Installation of easy-to-close facilities to save water when not needed
Plug all abandoned drill holes to avoid short-circuiting damages
Develop as strategic reserves, usable at large-scale contamination accidents*

The last point warrants discussion. To be practical, there is a constant threat of large-scale pollution events, accidents or war acts that may happen anywhere and will put out of use the surface water reservoirs and place a question mark on the water of wells tapping the unconfined groundwater system. In such cases the artesian wells that tap fossil formation waters will be a safe water source, as they are hydraulically disconnected from the land surface. It is thus recommended that a certain number of artesian wells not be pumped, but kept ready as a strategic resource.

16.2.8 Crystalline Rock Terrains

These contain formation water within the zero hydraulic potential zone, raising a number of research topics:

How is hydraulic connectivity evolved in the shallow through-flow zone?
By which processes did fractures propagate to the observed great depths?
How can drillings be conducted without short-circuiting isolated fractures?

16.2.9 Hydrothermal Water Systems

These include steam jets, boiling springs, and exploitable steam fields. The following are open research questions to be addressed in new study areas:

Was the water already entrapped in the rocks when the magma intruded?
Did water enter into the hot systems during low pressure collapse phases?
From which depth is the heated water rising?

16.2.10 Volcanic Systems

Water occurrences in volcanic systems places fascinating research queries:

Did water in the rocks predate the magmatic intrusions?
Is water flowing in during calm and low-pressure phases?
Is arriving water triggering new eruptions?
In which cases is seawater introduced into the volcanic system and how?
Which gases are added and from what depth?

16.3 Geological Records—Research Avenues

Formation waters, along with the associated evaporites, petroleum and coal, are promising research subjects. A few examples follow:

16.3.1 Formation Water Applied as Paleo-Tectonic Records

Apply ^{18}O enrichment to deduce temperature and subduction depth.
Apply ^{18}O enriched shallow formation waters to deduce tectonic uplifts.

Apply the meteoric formation waters to identify phases of land exposure.
Apply formation water ages to deduce subduction velocity.
Determine low landscape periods that disclose orogenic tranquility phases.
Can pollen in formation water help to identify evaporitic facies?

16.3.2 Large Evaporitic Lowlands, Frequented by Sea Invasions and Retreats

These have been concluded to have prevailed at the time subsidence basins were filled with sedimentary rocks. A main boundary condition that leads to this conclusion was the brine tagging that typifies formation waters. Several questions may be addressed in the frame of research to be conducted in the rare coastal evaporitic lowlands currently existing on Earth:

Is brine tagging of meteoric water currently taking place in such terrains?
How long ago was the area last covered by sea?
What type of vegetation develops in the evaporitic lowlands?
What is the local height difference between seasonal high and low tides?

16.3.3 Halite and Gypsum Formations

Such formations of great thickness present fascinating research targets in light of the formation calculations presented in section 6.2, which concluded that for a 100-m-thick halite section, evaporation of at least 10,000 m of seawater were needed. Examples of research topics are

Search for locations at which formation of halite and gypsum occurs now.
Are these continental coastal flatlands, as suggested?
How are the residual bitterns replaced with new seawater supply?
How do halite and gypsum beds accumulate and how are they saved from erosion?
How are fresh halite and gypsum beds covered and confined?

16.3.4 Petroleum Systems

Petroleum systems call for a long list of research topics related to the deduced boundary conditions—that petroleum was formed in closed rock-compartments, which contained from the beginning all the needed ingredients. Examples are

Look for facies of sands rich in organic material.
Which organic materials can be wholly transformed into petroleum?
Which organic raw material may be supplied by black waters of wetlands?

From which source rocks could tar-rich sands be formed?
How can source rocks function also as reservoir rocks?
What is the fruitcake structure of giant oilfields like?
Did salts dissolved in formation water participate in petroleum formation?
How can formation waters serve to detect petroleum deposits?
How can brine injection help to map compartmentalization?
Calibrate petroleum to deduce temperature and subduction depth.
Apply shallow petroleum to identify phases of tectonic uplifting.

16.3.5 Coal Genesis

Coal raises, in a way, research topics that are similar to those listed for petroleum:

Did coal deposits evolve from inland vegetation of low paleo-flatlands?
Which types of vegetation were entirely transformed into coal?
What were the temperature of formation and depth of required subsidence?
Was the saline formation water a necessary catalyst in coal formation?

16.3.6 Hydroseismicity

This topic deals with the contribution of water to earthquakes. *Hydrofracturing* is a term that describes seismic tremors sensed following injection of water (industrial brines) into deep wells. Similar water-related seismic processes warrant research in light of former and future case studies, for example:

What is the mechanism of hydrofracturing?
At what depth interval is hydrofracturing effective?
Search for natural cases of hydrofracturing.
Are seismic tremors induced by water entering seismic dilatation fractures?
Can water table fluctuations serve to identify seismic pressure and extension zones?

16.4 Summary Exercises

Exercise 16.1: Petroleum exploration drill holes in the Petrohope Valley, finished 5 months ago, did not discover oil or gas, but in several wells artesian water rose to the land surface. The private exploration company abandoned the site, but three deserted wells keep flowing. What do you suggest the local authorities should do?

Exercise 16.2: Petroleum prospection drilling is aimed to find commercial deposits of oil and/or gas, so should one stop to collect samples only when oil or gas are encountered? Discuss.

Conclusions and Research Avenues

Exercise 16.3: Should these water samples be analyzed also for dissolved organic compounds? Why?

Exercise 16.4: Produced oil often contains some formation water. This water is separated and called "oil brines." Is it appropriate to let it flow in the adjacent river bed? Explain.

Exercise 16.5: A deep exploration driling project is planned in a new region. You are the consultant of petroleum-related hydrology. What will you suggest for the regular procedures of the drilling operations? Can you justify the cost?

17
EDUCATIONAL ASPECTS OF WATER, THE UNIQUE FLUID OF PLANET EARTH

Educational institutes are common in our societies, and they are devoted to a long list of variegated subjects. They are platforms at which the experts of a subject share their knowledge and enthusiasm with the wider public. It is high time that the very special topic of *water*, the unique fluid of planet Earth, is presented to our nations' people via the best educational facilities—our citizens deserve it. And who can drive this mission if not us, the experts who are familiar with the unique fluid and its story? In a way, the contents of the previous chapters serve as a beginning.

Why should this theme be publicized? For several reasons:

To contribute to the basket of general knowledge
To demonstrate the scientific approach of observation → conclusion → working hypothesis → checking, etc.
To increase understanding that will lead to appreciation of the resource
To recruit all of us to optimal management of water and related topics

In the companion book of the present volume, *Chemical and Isotopic Groundwater Hydrology: Third Edition* (Mazor, 2003), the educational aspect of water is dealt with as well. The emphasis there is on the active water cycle and man. The educational aspect of the vast topic of water is dealt with in the present chapter with respect to the four-dimensional occurrence of the terrestrial fluid that makes Earth such a unique place.

Educational Aspects of Water   355

References to sections from both books are given in the following sections as follows:

A (e.g., A6.16) refers to the *Chemical and Isotopic Groundwater* book
B (e.g., B3.3.5) refers to the present book.

17.1 List of Educational Topics

The list is general, aimed to serve as a basic set for use within a variety of educational activities discussed latter, for example, national "water and man museums"; local water-related exhibitions and information centers; water parks and spas; teaching opportunities at schools, and mini-research activities for students.

17.1.1 The Unique Fluid of Planet Earth

The blue planet. Earth is the only planet with fluid water, and plenty of it.
- Satellite photos; globes of water coverage; maps of water-rich landscapes

The solar system with respect to water availability
- Model of the solar system, including moons; photos of the planets and description of surface temperature, atmospheric pressure and composition, and lack of water (B1.2)

The unique terrestrial landscape shaped by water
- Earth's sculptor (B1.1)
- The accretion craters of the planets
- The water-shaped landscapes of Earth

The geohydroderm concept (B2.1.2)
- A globe showing the geohydroderm and the different water-containing geosystems

17.1.2 The Very Special Properties of Water

The list of 16 or more special properties (B1.3) can be highlighted with experiments and with discussions of how would the world be without this quality, etc.

Water has three phases—fluid, gas, and solid–and it readily passes from one phase to another, within the temperature range prevailing on the Earth's surface. On the other planets and the moon the prevailing temperature is in the range of freezing alone or vapor alone. *The crucial outcome*: The global water cycle is based on the three physical phases of water.

The solid phase of water is lighter than the fluid phase. This is anomalous—the solid phase of other materials is heavy and sinks in the fluid. *The crucial outcome*: If ice were heavier than water, we would have only tiny oceans! The water at the polar sections freezes seasonally, but the floating ice melts during the warm seasons or floats to warmer regions and melts there. If ice were heavy, it would sink to the bottom of the ocean and there it would be isolated from the seasonal sun. Thus huge parts of the oceans would be masses of ice and only at the upper few meters would there be water during the warm seasons. This, in turn, would severely limit evaporation and cloud formation, and the oceanic streams would be extremely limited in extent. The same holds true for lakes in cold climates.

Water can dissolve salts and gases but be readily separated as pure vapor. The sun evaporates vapor from the ocean, but the salts remain in the sea. *The crucial result*: Pure vapor clouds are driven by the wind (another process propelled by the sun), and the vapor condenses into rain and snow, which are the source of terrestrial fresh water that is eventually drained back to the ocean. In this mode the same water molecules have taken part uncountable times in the endless water cycle.

Water readily dissolves CO_2 and turns into an effective acid. The main source of CO_2 is provided to water on its passage through the root zone of soils. A very visual outcome of rocks attacked by this acid are the karstic features seen in many limestone terrains. But, actually, all rocks are dissolved by CO_2-induced water–rock interactions. This is a major mode of erosion by which material of the continental landscapes is transported into the sea.

Water has low viscosity and is not sticky. *The basic outcomes*: (1) Water flows efficiently as surface water, (2) water infiltrates into the soil cover, (3) it flows through permeable rocks, and (4) thanks to the low viscosity the oceans become well mixed.

Water is heavy enough to be moved by the gravitational field. *The crucial outcome*: Water on the land surface and in the unconfined groundwater system flows gravitationally.

Water can split into subparcels and reunite. *The fundamental result*: Infiltration through little voids between soil and rock particles is feasible, uniting to a larger-scale flow in the seaward base flow, and then joining the huge water body of the ocean.

Water has a specific gravity that is just right. *The crucial outcome*: Water vapor does not escape from Earth.

Educational Aspects of Water

The vapor and fluid phases are transparent. *The valuable result*: The sunlight can pass through the atmosphere and through the upper few hundreds of meters of seas and lakes—a prerequisite of aquatic life.

The water vapor phase is a greenhouse gas. *The crucial outcome*: The temperature on Earth is kept balanced.

The 4×10^9-year-old terrestrial catalyst. Water has existed on Earth for some 4 billion years and it is still around as it was on its first day. It is involved, for example, in photosynthesis; it is involved in the formation of sedimentary rocks; and it mobilizes and deposits ores; but it always comes back as pure water.

Oil is immiscible with water. This property is intrinsic in the formation of clean petroleum deposits.

Oil floats on water. This property is intrinsic to the migration of oil to traps (within the original rock-compartment; B Chapter 12).

Water has a very high dissolution capacity of chlorides. *The crucial outcome*: Efficient flushing of salts from the continent and their concentration in the ocean are operations that form a major part of the salt cycle.

Water is an effective lubricant. We know this from wet floors that are slippery, landslides that occur following rainy seasons, and hydrolubrication and hydrofracturing that are involved in seismic processes.

Last but not least, water is the base of life.

17.1.3 The Never-Resting Oceans

Key roles of the oceans in the dynamics of the global water cycle (B1.4)

- Base of drainage of runoff and of erosion
- Base of drainage of the groundwater base flow
- The source of fresh water, driven landward by clouds
- The saline ocean is the operational reservoir sustaining continental fresh water
- The major reservoir of chlorides and other soluble salts
- The receptor and storage place of continental erosion products
- The ocean frequently covered most continents

The story of sedimentary rocks

- Examples of marine fossils (macro and micro)
- Maps showing marine sediments covering the continents during various geological periods
- Plate tectonics drastically changed the location of the oceans; principles of reconstruction of continental drifting and, maps of the oceans and continents at various geological periods

17.1.4 Water Entrapped in Rocks Beneath the Vast Oceans

- The Deep Sea Drilling Project (B5.2)
- Water content in sub oceanic sediments (B5.3)
- The widespread marine facies of interstitial water (B5.4)
- Cases of continental water stored beneath the oceans—regressive stages (B5.5)
- Concluding that seawater did not penetrate into the underlying rocks (the no-flow zone)

17.1.5 Fresh Surface Waters

The miracle of fresh terrestrial water

- Fresh water can exist only on the continent.
- Modes by which continental water becomes saline.
- All is washed into the sea and new fresh water is delivered to the continents.
- Is the fresh water cutting the branch it sits on? (B1.5)

How old is a specific water? (A Chapters 10–12 and 14; B3.3.5 and B8.6)

- The isotopic dating principle
- The main water dating methods
- Examples of rock pockets that have contained water since the Paleozoic to recent times

The fresh cycling groundwater system

- The active fresh water cycle. (A Chapter 2; B2.2, B3.1.1, and B3.2. 4)
- Basic observations at wells disclose the anatomy of through-flowing groundwater.
- Coastal wells demonstrate that the sea surface is the terminal base of drainage.
- The physics of gravitational pull downward and the requirement of free space for flow.
- Tagging by sea spray all over the continents.
- The stories that the isotopic composition of water tells us.
- The efficient washing into the sea.
- Exploitation in the framework of renewability of the resource.
- Pollution hazards. (A Chapter 16)
- The need of nation-wide clean management.

17.1.6 The Vast Sedimentary Basins that Contain Fossil Formation Water

Examples of relevant case studies (B7.1)
The isotopic composition defines meteoric water formed on the continents
Tagging by evaporitic brine-spray
Fossil formation water trapped in deep rock-compartments (B1.6, B2.4, B3.1.2, B4.8.4, B7.1, B7.2, and B7.4)
The fruitcake structure of water-hosting rock-compartments separated by impermeable rock barriers, within the zone of no flow.
Artesian water (B3.2.5)

17.1.7 Warm, Boiling, and Steaming Groundwater (B2.9.1)

The anatomy of warm springs and relevant case studies (B13.1)
Spas—what is expected and classic examples (B13.2; 13.4)
Geysers, boiling springs, boiling mud pots, examples and anatomy
Black smokers on ocean floors
Volcanic fumaroles
Warm water parks, examples
Hydrothermal systems, examples (B Chapter 14)

17.1.8 Terrestrial Ice

Occurrences (a globe with polar ice, seasonal snow lands, freezing ocean zones, etc.)
The ice ages and their impact on sea-level

17.1.9 Water, the Landscaper

Former lowlands of sea transgressions and regressions (B2.4.6, B2.6.2, B3.2.6, and B6.3)
The geoquartet: water dancing between three giant energy sources (B Epilogue)
Fresh water erodes mountains but exists thanks to them (B1.5)

17.1.10 The Key Role of Water in the Formation of Fossil Fuels

The unique fuels of this planet are all products of water. (B1.9)
Petroleum hydrology. (B1.8, B2.6.6, B2.8, B3.2.7, B10.2–10.9, B11.5, and B12.4)
Entrapment within hydraulically isolated rock-compartments. The composition and pressure of petroleum-associated formation waters vary

between neighboring petroleum exploration drill holes and producing wells, revealing entrapment within separated rock-compartments. Hence the associated oil or gas is entrapped in hydraulically isolated compartments as well. The same picture is outlined by the composition of the petroleum in producing fields—different compositions and pressures disclose the existence of numerous neighboring separated traps.

The ingredients needed for petroleum formation were contained in the rock-compartments since their initial formation. The isolated rock-compartments were formed at the moment their rocks subsided and were confined by overlying rocks (B Chapter 11). The rock-compartments contained from the beginning all the ingredients needed for the formation of petroleum deposits: organic raw material, permeable rocks, and formation water.

Water prevented collapse of rocks as a result of compaction by the weight of the overlying rocks. This maintained the hydraulic connectivity that was essential in petroleum concentration in traps that make up exploitable deposits.

The petroleum-associated formation water subsided to depths of the oil-formation temperature window. We have seen that the isotopic composition of the hydrogen of formation waters reveals δD values of meteoric water, but the oxygen often has a relatively heavy $\delta^{18}O$ value, indicating temperature-induced isotopic exchange with the host rocks (B Chapter 10). This, in turn, indicates that the petroleum-associated formation water subsided to the depth of the petroleum-formation temperature window.

Petroleum concentration in traps within the original rock-compartments. Formation waters reveal confinement within rock-compartments that are separated by clay-rich rock barriers (B Chapters 7 and 10). This leads to the pressure-cooker model of oil formation and concentration (B Chapter 12).

Coal. (B12.2, B12.2.3)

Lush water-dependent vegetation developed in flat and rainy lowlands, of sand, conglomerate, and mud landscapes.

Sea transgression confined the vegetation by newly sedimented marginal marine rocks, forming hydraulically isolated rock-compartments, some of them with a high content of organic material.

The hydraulically isolated rock-compartments subsided in the sedimentary basins, until the pressure and temperature became high enough for the coal formation.

Associated formation water was probably involved in the coal formation process.

Educational Aspects of Water 361

17.1.11 The Key Role of Water in the Formation of the Ores Exploited by Humans

Salt: properties, precipitation, water-proof preservation over geological periods, its key role in human societies (B.5, B6.1, B6.2, B6.4, and B6.5)
Gypsum: properties, formation, uses since prehistoric times (B2.5 and B6.1)
Building rocks: main types and their aquatic formation
Clays: aquatic formation, impermeable barriers, usage
Phosphorite: its marine bioformation, composition, usage
Iron ore deposits: hydrothermal and weathering processes, production
Bauxite: paleosoil, composition, production of aluminum
Cooper: hydrothermal processes, types of ores, production
Soil: the current wet weathering product, properties, agricultural needs

17.1.12 The Dynamics of the Water Cycle Propelled Biological Evolution (B1.10)

Water is essential for life.
The cleaning up of a dense primordial atmosphere of CO_2. A large amount of sedimentary carbonate rocks cover the earth. A simple calculation shows that if we go back to the early history of Earth, before the marine carbonates were precipitated, all the involved CO_2 was in the atmosphere, which was as dense as the CO_2 atmosphere of our neighbor Venus. When our planet cooled sufficiently this CO_2 was gradually cleaned away by the water cycle and stored in the limestone and dolomite rocks. This was a prerequisite for the origin of life.
Provision of the initial free atmospheric oxygen. During the early terrestrial history water vapor was photochemically decomposed, the hydrogen escaped into space, and the oxygen was consumed by a list of terrestrial materials. Eventually free oxygen accumulated, the ozone layer could be formed, and the remaining water was shielded. Thus the initial dose of free oxygen was introduced into the ancient atmosphere, making room for advanced forms of life to evolve–thanks to the liquid water and the vapor.
Creation of a countless assortment of landscapes and ecological niches that life could select for its coming into being. Scientists are still wondering which environmental setup hosted the miracle of the origin of life, but it is sure that the activity of water prepared an enormous assortment of candidate locations.
Liquid water was an essential ingredient needed to sustain living forms, as it still is to this day.

17.1.13 Man and Water

The water at our tap—where is it coming from and how?
Water supply (A18.4)
Water quality and standards (A18.1, B3.2.1)
Bottled "mineral water" (B13.5)
Nation-wide clean management, needed to protect the water resource (A17.3)
Wise gardens (outdoor exhibitions)
Conserving water at our homes and households
Water monitoring networks, the aims, principles, and examples (A18.2)
Water data banks, aims, principles, and examples (A18.3)

17.2 National Water and Man Museums

The location can be in major cities, securing easy access, or at water-related sites. Let us imagine such a museum at a large waterfall, a picturesque lake, a location of geysers and boiling water, in the middle of a petroleum field, or at a dammed reservoir that supplies water to a major city. Each National Water and Man museum could emphasize the local water-related topics. It will be fascinating to see how different and original these museums will be.

The aims include

- To contribute to the general knowledge of the public
- To elucidate the fascinating story of terrestrial water, which is hardly known to the public
- To demonstrate how different fields of science can be focused and implemented on a common subject, in this case water
- To familiarize visitors with the water resource
- To build appreciation of water and provide the know-how needed to protect it

Some principles should be the following:

- Transfer of knowledge along the approach of applied science, namely, from observations and measurements (i.e., facts) to logical conclusions
- Conducting dialogues with visitors, who will use interactive exbihits to get feedback
- Use of case studies to illuminate concrete observations, problems, and solutions.
- Emphasis on simple experimental "hands on" displays
- Addressing all ages and levels of knowledge
- Maintaining an active information desk, supported by a library of short films and information sheets to provide answers to questions raised by visitors.

- Providing an outdoor exhibition space, hosting large water-related natural features and equipment elements

The list of topics presented in section 17.2 covers a wide range of subjects dealing with water, the unique fluid of planet Earth. In each group of topics there is plenty of room for the selection of items and issues that will be exhibited, preferably with regard to examples and case studies. Exposure to the activities of water-related volunteer organizations can provide further locally relevant subject matter.

17.3 Local Exhibitions and Water and Man Demonstration Centers

Local water-related exhibitions can be a valuable entity of every city and town. These exhibitions are small as compared to a full-sized museum—the length of a visit will be around one hour. Selected topics from the extended list in section 17.1 can be highlighted in connection with local natural assets or local human achievements. Emphasis in this case will be on the local features—those that every inhabitant and each visitor can see and sense.

The aims include

- To draw attention to local natural water-related assets
- To contribute to local consciousness and local pride
- To encourage local excursions and draw attention to points of interest
- To foster awareness of water-related issues
- To convince visitors to respect the resource and protect it

Some principles might be to have a dynamic exhibition at which displays are frequently changed and constantly updated so that repeat visits become a habit as well as to provide self-guided tours in the surroundings that expose water-related natural features or functioning installations.

Topics could include the following:

- An assortment of the topics listed in section 17.1 or additional ideas.
- Substantial information on local assets, e.g., springs, rivers, karstic features, lakes, wetlands, waterfalls, sea shores, wells, sections of the water-monitoring network, spas, hot springs or wells, ancient bath houses, water mills, ancient and modern irrigation systems, water towers, dammed water reservoirs, well fields, water-treatment stations, snow-related features, etc.
- For each of these assets there is high interest in providing geological/hydrological cross sections, the chemical and isotopic compositions, and what they tell us. Discuss the age of the water in these cases; are there any pollution problems?

- Local information of what has been done for the water resource and what was successful
- Visitors will be interested to contact the relevant water data bank.

17.4 Educational Water Recreation Parks

In every country there are national parks and other forms of public areas that include interesting and marvelous features related to water. A few examples are waterfalls, springs, rivers, lakes, wetlands, geysers, volcanos with fumaroles, desert oases, glaciers, and karstic caves. Any of these sites can be developed as an educational realm explaining the local features in the more general context of the global water story.

The idea may sound self-evident, but the reality is that only a score of natural water phenomena are really exhibited and adequately explained to visitors. The success might be great.

17.5 Spas

Section 13.4 presents a rather detailed discussion of local exhibitions disclosing the anatomy of warm and mineral water sources related to therapeutic and recreational installations.

17.6 Teaching at School and Student Mini-Research Projects

- At the kindergarten and the elementary level the active water cycle is a suitable topic, along with simple measures by which we can conserve water.
- At the middle school level the "man and water" topic may be developed, with visits to neighboring natural water assets, as well as coverage of the water supply issues.
- At the high school level the global water cycle can be tackled, and mini-research projects seem most suitable, a list of topics may be drawn from section 17.1.

Use of local case studies is only natural, and connection to the water data banks is most welcome. The above mentioned topics may be taught as part of various science courses. At the 11th and 12th grades there is room also for a full-sized course on water issues, to be based on selected topics from the list given in section 17.1.

Educational Aspects of Water

17.7 Teaching at Universities

An introductory course might deal with "water and man" or similar subjects. It would be a valuable addition to the introductory courses offered at colleges and universities. There are several aims:

- To familiarize students with the unique terrestrial fluid and the global water cycle.
- To expose students to the research methodology of observations and measurements that lead to conclusions, which provide the basis of our understanding of the studied natural systems
- To encourage students to be actively interested in the water-related issues that are on the decisionmakers' agenda

A graduate level course could be based on the present book. Such a course suits students of the following professions: hydrologist, hydrochemist, geologist, geomorphologist, petroleum expert, water-based recreation and therapeutic expert, ecologist, environmental scientist, urban and state-wide water supply operator, pollution control manager, and educators.

Field trips to relevant nature attributes and management installations are a must. Mini-research projects, on worldwide problems or local issues are most welcome. Acquaintance with monitoring networks and data banks is an intrinsic part of such a graduate water course.

EPILOGUE: THREE ENERGY SOURCES AND ONE TRANSPORTER—THE GEO-QUARTET UNIQUE TO PLANET EARTH

Fluid water has existed on Earth approximately four billion years, and it has not sat idle for a moment.

There is the regular earth life cycle of water being lifted from the ocean, by the sun's energy, and deposited on the top of the landscape relief, in the form of rain. Terrestrial gravity does the opposite—it pulls the water down and back into the oceans. The water does not become this turbulent for nothing—it washes down heavy loads of raw materials and transports these materials into the ocean, from which sedimentary rocks are formed.

Over time, this erosive activity lowers the landscape, and taking into account the observable water's efficiency, it turns out that in about 20 million years this earth life cycle could be over. Once the continents are eroded down to sea-level, the sun's energy finds no mountains to lift the water to, and the terrestrial gravity force finds no more water to pull down. This will practically be the end of terrestrial fresh water.

However, earth's life cycle has a third actor—the internal terrestrial energy. This energy source shifts the continents and lifts mountains, thus supplying new protruding landscapes to the industrious water, and more occupation to both the sun and master gravity. If water-based erosion did not operate, how would Earth's landscape look? Significantly higher and steeper than the present landscape!

In the middle of this turnover, life popped up! So many different environments constantly evolved so that eventually the right recipe turned up, but not before the young water tied up the primordial CO_2, making the atmosphere so friendly.

Thanks to the geo-quartet's tireless activity man has a very unique realm to live in: water to drink; oceans to sail; sedimentary rocks to build from; an assortment of fuels—coal, gas, and oil; a unique atmosphere that acts as Earth's air conditioner; and the entire domain of plants and animals to consume.

All these water-related niches are open to scientific research, following the old rule—the more the better. Thus, let us treasure the water and protect it!

ANSWERS AND DISCUSSION OF THE EXERCISE QUESTIONS

Answer 1.1: Erasing of the primordial accretion landscape of our planet and endless reshaping of the terrestrial landscape.

Answer 1.2: The oceans did not turn into huge ice bodies. Explain.

Answer 1.3: Water has existed on Earth for around 4×10^9 years, and its properties did not change in spite of its involvement in so many large-scale processes.

Answer 1.4: Yes, water erodes the landscape and lowers it, a process that would result in the vanishing of the landscape relief, enabling the sea to cover the entire Earth. At this moment no fresh water would be formed any more. (What has been left out in this account? What saves the destiny of terrestrial fresh water?)

Answer 1.5: Definitely yes! The present chapter is actually a continuous song of praise to water. The first step is to study the geohydroderm and its major groundwater-containing geosystems, which are addressed in the following chapter. Once we understand water, we can have a constructive look at our dealing with this resource.

Answer 2.1: *Observation*: Filling by gravel, deep beneath the sea level, has been found in the channels of major rivers draining into the Mediterranean Sea. *Conclusion*: In the geological past there was a stage at which the level of the Mediterranean Sea was significantly lower, and possibly it even dried up.

Latter on the sea level rose, and the river beds were filled with gravel. *Second observation*: At the rock sequence underlying the bottom of the Mediterranean Sea a thick bed of halite is encountered of a Messinian age. *Discussion and conclusion*: Salt starts to precipitate from seawater when the latter is evaporated so that only one-tenth of the initial volume is left. The present Mediterranean Sea is around 3000 m deep, but the salt discloses that when it was formed the depth was only 300 m or less. In other words, the sea level during the Messinian was at least 2700 m lower that the present level. *Third observation*: Groundwater encountered in Mediterranean coastal wells beneath the recent through-flowing groundwater system (i.e., deeper than sea level) is saline and brine tagged. *Conclusion*: this is folsil groundwater that was (1) formed when the base of the drainage (i.e., the sea surface) was substantially lower than at present; (2) this paleo-groundwater was formed in a large-scale evaporitic landscape—the land that was exposed after the sea retreat; and (3) when the sea level rose, the present groundwater system was developed at a higher elevation, confining the deeper fossil groundwater, and the new one is again tagged by sea spray.

A far-reaching geological conclusion: The drying up of the Mediterranean Sea could be caused only by a plate collision that temporarily closed the Strait of Gibraltar.

Answer 2.2: No. This zone has no base of drainage; in other words, it is the zone of zero hydraulic potential.

Answer 2.3: Sea spray has a Cl/Br weight ratio of around 293, and $CaCl_2$ is negligible. Brine-spray has a Cl/Br weight ratio in the range of 80 to 200, and $CaCl_2$ is second to the NaCl concentration. These differences are meaningful; they clearly distinguish between recently formed groundwater and the deeper domain of fossil formation waters.

Answer 2.4: The chlorine is internal if it originated from inside the rock system by water–rock interaction; and it is external if it was brought in by the recharge water. Let us find the answer in the following stages: There is only one mineral that contains Cl. (Which one?) Water in contact with this mineral reaches a saturation concentration. (Of how much?) Hence in all groundwater systems that contain less Cl, the source was external. (From where?)

Answer 2.5: Yes, observations reveal that at present the forming groundwater is fresh in regions with a wet climate (Cl concentration 5 to 10 mg/L), and it is saline in regions of an arid climate. (The Cl concentration even surpasses seawater salinity—which contains how much Cl?)

Answer 2.6: Meteoric is the term that specifies that water was formed on the continent. It can be identified by (1) negative δD and $\delta^{18}O$ values (of how much?) and by the concentration of the atmospheric noble gases (which ones?) that equals saturation with air at common surface temperatures (e.g., 10 to 20°C).

Answer 2.7: The pronounced variability of salinity, ionic ratios, and fluid pressure often seen between neighboring boreholes.

Answer 2.8: Groundwater encountered in rock formations beneath the ongoing through-flow groundwater system, i.e., at depths greater than sea level. (Why is the sea level important in this context?) It is fossil. (How can we find this out?)

Answer 2.9: Connate water is groundwater that stayed in its host rocks since the latter were confined by overlying rocks. Formation waters are connate at all the sites at which a fruitcake structure of rock-compartments, isolated by a network of impermeable barriers, is identified. This is found to be the prevailing case at depths below sea level.

Answer 2.10: The answer is given in section 2.3.2 and is well seen in the depth profiles of Chapter 5.

Answer 2.11: To identify meteoric water (how?), highly evaporated water and water that was enriched by ^{18}O (by which process?).

Answer 2.12: It is necessary to find out whether the new drilling reached a niche of the hydrothermal field that had already been exploited by an existing well, or whether a nonexploited niche has been reached. (Why is this information needed?) Hence the steam has to be analyzed for its concentration of CO_2, the atmospheric noble gases, and the radiogenic ones, as well as the concentration of tritium and other contaminants. (What can each of these parameters tell us?)

Answer 3.1: Permeability.

Answer 3.2: (b) and (d). Discuss.

Answer 3.3: Samples have to be sent to the laboratories for detailed analyses in order to find out whether we deal with recent seawater intrusion (which parameters will identify this source?); fossil and locally entrapped seawater (which parameters are informative?); or fossil, slightly saline groundwater (what kind of results will indicate this?).

Answer 3.4: This is recent groundwater formed at an arid region so that evaporation losses resulted in a high salinity, the tagging being by airborne sea spray.

Answer 3.5: (1) The properties of formation water provide a tool to map the spatial structure of the rock-compartments within oil and gas fields. (2) Enrichment of formation water in ^{18}O at relatively shallow depth indicates a former phase of deep subsidence of the hosting rock-compartment, where the conditions for petroleum formation did exist. (3) Formation water with relatively high contents of petroleum compounds reflects close association with a potential deposit.

Answer 3.6: Not at all. The always open possibility, that we have sampled a mixture of waters, has to be checked. Significant concentration of tritium along with low ^{14}C-14, or presence of ^{14}C along with significant concentration of ^{4}He, indicates that we sampled a mixture of old and young groundwater.

Answer 4.1: The phenomenon is of a continental scale—the river is 2500 km long. The sea spray is fed from north, east, and south (Fig. 4.1).

Answer 4.2: No fractionation is observable, as comes out of Figs. 4a and 4b. The fact that all over the globe sea spray is identified in flowing groundwater indicates no secondary interactions take place along the inland transport.

Answer 4.3: Check your answer with Table 4.1.

Answer 4.4: Jingellic is located near the head of the river, and Tailem Bend is far away downstream, 87 km from the outlet into the Southern Ocean (Fig. 4.1). At the first point the water is very fresh, and the dissolved ions are dominated by water–rock products. At the last point the water is saltier and dominated by sea-derived ions. The latter originate from sea spray transported by clouds and precipitated with the rain, and a large portion is airborne and deposited as dust. Runoff washes all these salts into the river so their concentration increases downstream.

Answer 4.5: The isotopic composition of the respective water phase provides the answer.

Answer 4.6: These are data from three shallow lakes at Australia. They are 10 km inland, and hence no seawater is washed in. The salinity did not reach the point of halite deposition. (What would be the Cl concentration at such a case?) Hence evaporation is partial, which indicates the lakes have an outlet—possibly seasonal underground drainage.

Answer 4.7: The Merredin shallow groundwaters are saline because, due to the semiarid climate, evaporation is intensive. Sea-derived airborne salts accumulate on the ground and are washed into the groundwater system by occasional strong rains. In other locations at the Wheatbelt concentrations of over three times those for seawater are observed (Fig. 4.7).

Answer 5.1: Not at all! Diagenetically changed seawater, e.g., depleted of its SO_4 or having δD and $\delta^{18}O$ values different from seawater, are preserved even

Answers and Discussion

at a depth of a few meters (find examples in the tables). So seawater does not flow down and does not replace the interstitial water.

Answer 5.2: The rocks beneath the ocean reside in the zone of zero hydraulic potential.

Answer 5.3: The term *connate* is discussed in section 5.7.

Answer 5.4: The Cl concentration is close to the seawater concentration of 19 g/L, indicating the observed interstitial water is basically entrapped seawater. The SO_4 concentration varies drastically along the depth profile revealing diagenetic changes.

Answer 5.5: The interstitial water in the deeper sections of the named depth profiles reveals an origin as groundwater formed on the continent at a flat arid lowland. (On what is this statement based?) Thus for a period of time the bottom of the Mediterranean Sea was a bare land.

Answer 5.6: The topic is discussed in section 5.5.1.

Answer 5.7: The data of Hole 374, reported in Table 6.5, reveal in the depth profile interstitial waters that are drastically different from seawater. (Which parameters reveal this? What type of water facies is revealed?)

Answer 6.1: Ten thousand meters of seawater. With an evaporation of about 1 m/year it turns out the process had to last at least 10,000 years. Observations at recent lagoons reveal that part of the formed halite is washed back into the sea, and hence the calaculated values are minimum figures.

Answer 6.2: Airborne sea spray.

Answer 6.3: Occurrence of widespread deposits of halite and other evaporitic sediments. (What is the aerial extent of the "saline giant" of the Salado? What is its age?)

Answer 6.4: Yes, the Zechstein complex, briefly described in section 6.5.

Answer 6.5: The salt beds buried beneath the Mediterranean Sea reach up to 2000 m. They were formed when the Strait of Gibraltar was closed, so that a closed basin was formed for about 1 million years.

Answer 6.6: Not at all. (What does this water reflect?) (Section 6.6).

Answer 6.7: This is a concrete demonstration that salt beds situated within the static groundwater zone (what is this?) are well preserved, whereas salt beds exposed to the ongoing groundwater base flow are quickly dissolved.

Answer 7.1: *Groundwater*—in general, any water stored in rocks underground; in daily use, water encountered in wells in the unconfined system. *Formation water*—the water first encountered when a new drilling enters a given formation; all the confined groundwater occurring in rocks of the continents, occasionally associated with petroleum. *Interstitial water*—a term mainly applied to water that is stored within loose or lithified sediments beneath the ocean floors. *Connate water*—water that has been stored in its host rock since the latter was buried and confined by newer overlying rocks.

Answer 7.2: In both formation water groups the following pattern is observable in the more saline members: Na > Ca ≥ Mg > K > Sr; and Cl is the dominant anion. To this may be added the information given in the text that the Cl/Br weight ratio is around 160 in the first system and 50 to 104 in the second. This pattern reveals a clear evaporitic brine imprint.

Answer 7.3: In both profiles the salinity and composition vary abruptly. This discloses that the waters are stored in hydraulically separated strata. Such variability has been observed also laterally, revealing water storage in isolated rock-compartments.

Answer 7.4: By entrapment in subsided rock-compartments. The different water compositions reflect changes that took place in the local evaporitic environment.

Answer 7.5: The values are discussed in Chapter 6.

Answer 7.6: Each sample has to be collected at a well-defined depth in order to avoid mixing of water from different depths.

Answer 7.7: Temperature; Cl, Br, Na, and Ca concentrations; and δD and $\delta^{18}O$ values are the minimal parameters to be measured and analyzed, but additional ions will help formulate a better picture, e.g., SO_4, HCO_3, Mg, K, Sr, Li, and age indicators like tritium, ^{14}C, ^{36}Cl, ^4He, and ^{40}Ar.

Answer 8.1: This is an example of a mixed water sample, collected at a borehole that is open to several aquifers—at least one that is recent and shallow and one that is old and deep. The recent one contains tritium and, hence, has an age on the order of a few decades, and the old one is older than the obtained ^4He age of 1.8×10^7 years.

Answer 8.2: The low atmospheric noble gas content reveals that the water was rich in gases and they partially escaped due to the pressure release in the well. The ^4He was lost at least in the same order of magnitude, and hence the measured ^4He concentration should be corrected by multiplying by a factor of 5, and the corrected value will be applied for the age calculation (i.e., the order of magnitude of the age—section 8.2).

Answer 8.3: The highest value is closest to the real value, as discussed in section 8.2.

Answers and Discussion

Answer 8.4: The limit of the ^{14}C water dating method is around 30,000 years. Hence both values are beyond the limit of detection, and all that can be concluded is that the age of the two waters is >30,000 years. (It may be any age greater than the limit of detection.)

Answer 8.5: Not at all! The Darcian type hydraulic flow equations were constructed for flowing water systems, but they are irrelevant for static groundwater.

Answer 8.6: Yes, as listed in point 11 of section 8.7.

Answer 9.1: The mode by which the formation waters entered the fracture compartments is demonstrated in Fig. 9.1. Borehole C short-circuits all the fracture compartments it passed, and in addition shallow through-flow water and even runoff may enter as well.

Answer 9.2: The answers are discussed in the opening paragraph of Chapter 9 and in section 9.1.6.

Answer 9.3: 19,200 mg/L of Cl. This is indeed very close to the Cl concentration in seawater. (However, is this a possible origin in light of the reported Cl/Br ratio? So what was the origin of this formation water? Can it be recent water?)

Answer 9.4: There are ample observations that at shallow depths crystalline rocks are (1) rich in interconnected fractures and (2) host fresh groundwater (section 9.2.1), whereas at greater depths fractures are much lower in number, contain a diversity of saline waters, and are thus hydraulically isolated from the land surface and from one another (sections 9.2.1 to 9.2.3).

Answer 9.5: (1) The Cl concentration is very high, namely, 168 g/L; hence this is a real brine, entirely different from the local recent groundwater, indicating this water was formed under completely different environmental conditions—this is fossil water. This conclusion is supported by the Cl/Br ratio (what is it?) and the dominance of Ca. (2) The very high salinity excludes any appreciable dilution by fresh water, indicating the reported light isotopic water composition is indigenous. The lack of traceable tritium supports this conclusion. The light isotopic composition discloses an origin as meteoric water, tagged by brine-spray and formed in an arid paleoclimate.

Answer 9.6: The answer is provided in section 9.6.

Answer 9.7: Yes, high ^{4}He and ^{40}Ar concentrations indicate presence of an old water end-member and a minimum age can be derived.

Answer 10.1: You have, most probably, arranged the table with an increasing order of Cl concentration. Which regularities can be seen in the concentration of other ions? Are your conclusions similar to the discussion in section 10.2?

Answer 10.2: The answers are included in the discussion of each of the five case studies (sections 10.2 to 10.8). The resemblance is impressive.

Answer 10.3: *External ions*—originated outside the hosting rocks and were brought in as salts dissolved in the recharged groundwater; common examples are Cl, Br, and Na. *Internal ions*—originated inside the hosting rock-compartments as a result of water–rock interactions. The topic is discussed in section 10.2.

Answer 10.4: The topic is addressed in section 6.1.

Answer 10.5: The host rocks were tectonically uplifted afterward and/or a substantial part of the overlying rocks was eroded away.

Answer 10.6: The most informative parameter in this respect is the isotopic composition of the water. It is clearly meteoric. This means the water originated as continental groundwater.

Answer 10.7: These are external salts, mainly brine-spray and evaporitic dust.

Answer 10.8: No, positive correlations between the concentrations of Cl, Br, Na, K, Li, and Sr (Fig. 10.8) disclose an external origin.

Answer 11.1: Your list probably includes terms such as delta, sabkha, marine, continental, evaporitic, lacustrine, deltaic, and others.

Answer 11.2: Yes, $MgCl_2$ is highly soluble and is enriched in seawater evaporation brines after the gypsum and halite precipitation stages. This Mg displaces Ca in carbonates in one of several dolomitization processes. Probably the involved limestone was often windborne and washed-in dust and detritus.

Answer 11.3: Chemical and physical changes that a rock underwent following the initial sedimentation. Examples: dolomitization; de-dolomitization (Ca introduced into a dolomite and Mg expelled into the interacting water); silicification of carbonate rocks, anhydrite nodules, or clays, and cementation of sandstone.

Answer 11.4: Frequent vertical, as well as horizontal, alternations of rocks with a high capacity to contain fluids and impermeable sealing rocks. (Give examples of each.)

Answer 11.5: Yes, whenever they contain petroleum and their formation waters have a pronounced ^{18}O shift.

Answers and Discussion

Answer 12.1: No, diffusion of oil in associated formation water would cause mixing of the different fluids, but solubility of petroleum in water is minute. Furthermore, diffusion is restricted by hydraulic connectivity and hence could not operate between compartments separated by clay, mudstone, or other impermeable rocks.

Answer 12.2: For several reasons: (1) a 100-km-scale route means mainly horizontal migration, and there is no process to drive it (within the zero hydraulic potential zone); (2) along such long routes the petroleum had to pass through an uncountable number of clays, shales, and other barriers—an impossible mission! (Can you think about additional reasons?)

Answer 12.3: Yes. The formation water was meteoric, and hence was introduced to the host rocks that were exposed on the paleo-land surface. Both the water and rocks were confined and subsided together, even into the depth of petroleum formation (as is disclosed by the isotopic exchange of ^{18}O). The organic raw material was enclosed in the subsided rocks as well.

Answer 12.4: The composition of petroleum deposits differs between adjacent rock reservoirs, as part of the fruitcake structure. The only way different types of petroleum could have entered into their host rocks was by in-compartment migration after formation, with formation occurring following subsidence to the required depth. This leads to the conclusion that petroleum ages are systematically increasing in each location with depth, so there is a relative age sequence. In addition, the subsidence age is often retrievable from tectonic information.

Answer 12.5: The oil-bearing strata were uplifted after petroleum was formed. This is a valuable information for tectonic studies.

Answer 13.1: "Fossil mineral water."

Answer 13.2: The well produces a mixture of deep fossil and shallow recent waters (section 8.5).

Answer 13.3: Short-circuiting, i.e., the well casing is open to both the deep and shallow groundwaters. If only the deep water is needed, e.g., to attain a higher temperature, the casing has to be repaired.

Answer 13.4: The temperature as well as the electrical conductivity (indicating salinity) of the circulated drilling fluid should be continuously monitored. When a change of properties is encountered, drilling has to be stopped, the temperature measured, and a proper water sample collected and sent for detailed analyses. The purpose is to find out which different types of water occur so that appropriate water horizons can be pumped later on.

Answer 13.5: Passing heated steam will flush away dissolved gases, e.g., H_2S and radon, which may be part of the local water specifity. It may be preferable to boil a small amount of water and add it carefully to the piped water in a mode that the gases will not be flushed out.

Answer 14.1: By a "contradicting" combination of recent and old age indicators. Yes, corroded casing may cause short-circuits.

Answer 14.2: The steam and boiling water properties change gradually, e.g., decreases in dry gas content; in the radiogenic noble gas concentrations, and eventually in the temperature of the produced fluid.

Answer 14.3: Temperature, dry gas content, dissolved ions, noble gas concentrations, concentrations of the various age indicators, isotopic composition of the water.

Answer 14.4: To better understand the dynamics between the natural components and to follow up exploitation-induced changes and processes. (What kinds of changes?)

Answer 14.5: A hydrothermal field sustains electricity-producing turbines and a constant steam supply has to be ensured, or even enlarged if the capacity of the installation is to be expanded. Hence the dynamics of the field has to be well known, e.g., in which wells production may soon drop, where to place new exploration drill holes, which drill holes encounter new sections of the geothermal system so they should be equipped as production wells, etc.

Answer 14.6: The advantage is that no burning of fossil fuel is involved and no disturbing gases and dust are released. But the heat extraction leaves behind large volumes of water that has to be released in a controlled mode.

Answer 14.7: No! Such water is often of low quality because of high salinity. In such cases it is best to reinject this water into depressurized parts of the same field. This has occasionally a profitable by-product, as the injected water is heated by the hot rocks and it revives the exploited systems.

Answer 15.1: By including the following sections: geological, hydrological, and other maps with all the wells marked on them; reports and articles dealing with the basin; a file for each well, including all the available data—historical and new.

Answer 15.2: Yes, in the general lines. Petroleum compounds dissolved in the associated waters are of special interest in the oilfield; and the mapping of separated rock-compartments is important management information as well.

Answers and Discussion

Answer 15.3: By marking on the map a line through the relevant wells and selecting from the menu "cross section" and marking the parameters for which sections are needed.

Answer 15.4: Collect new samples and bring them the same day to the laboratory.

Answer 15.5: No! After drilling has been stopped, a pump has to be lowered and enough water be removed so that a water sample can be collected with no intermixing with the drilling fluid. A sample of the drilling fluid has to be sent to the laboratory as well.

Answer 16.1: The water should in no case be allowed to continue to flow, as this is a waste of a nonrenewed resource. The water should be analyzed for its quality. If it is saline, the drill holes have to be carefully plugged at their entire length. However, if the water is of good quality, the drill holes should be cased and equipped as production wells for immediate use or, better, turned into a strategic reserve.

Answer 16.2: The drilling should always include collection of representative samples of formation water horizons that are encountered. These water samples warrant detailed laboratory analyses in order to gain information essential for the mapping of the existing rock-compartments. (Why is this important?)

Answer 16.3: Yes, presence of anomalous concentrations of certain organic compounds may indicate close contact with a petroleum deposit. The results may help to better select the site of future drillings.

Answer 16.4: Of course not! This saline water is rich in petroleum compounds, and if let free on the land surface, it will pollute the groundwater. The oil brines cannot be diverted to the sea as they contain petroleum compounds that are poisonous to the marine flora and fauna. These waters have to be carefully returned to the location they came from—the exploited petroleum-containing rock-compartments. (Is there need for any care in this operation?)

Answer 16.5: The first part of the question you have practically answered above (Answer 16.2), but detailed instructions have to be worked out for every project. Yes, it costs quite a lot, but the reward is worth it as it leads to the understanding of the studied system in the three dimensions, applying encountered formation waters to map out the rock-compartments and to prospect for petroleum.

REFERENCES

Acheche, M.A.; M'Rabat, A.; Ghariani; A.; Montgomery, S.L. Ghadames basin, southern Tunisia: A reappraisal of Triassic reservoirs and future prospectivity. AAPG Bull. 2001, *85* (5), 765–780.

Andrews, J.N.; Lee, D.J. Inert gases in groundwater from the Bunter sandstone of England as indicators of age and paleoclimate trends. J. Hydrology 1979, *41*, 233–252.

Andrews, J.N.; Goldbrunner, J.E.; Darling, W.G.; Hooker, P.J.; Wilson, G.B.; Youngman, M.J.; Eichinger, L.; Rauert, W.; Stichler, W. A radiochemical, hydrochemical and dissolved gas study of groundwaters in the Molasse basin of upper Austria. Earth and Planetary Sci. Lett. 1985, *73*, 317–332.

Andrews J.N.; Hussain N.; Youngman, M.J. Atmospheric and radiogenic gases in groundwaters from the Stripa granite. Geochim. Cosmochim. Acta. 1989, *53*, 1831–1841.

Arad, A.; Evens, R. The hydrology, hydrochemistry and environmental isotopes of the Campaspe River aquifer system, north-central Victoria, Australia. J. Hydrology 1987, *95*, 63–86.

Arkin, Y.; Gilat, A. Dead Sea sinkholes—an ever-developing hazard. J. Environ. Geol. 1999, *39*, 711–722.

Bates, R.L.; Jackson, J.A., Eds.; *Glossary of Geology*, 3rd Ed.; American Geological Institute: Alexandria, Virginia, 1987.

Bottomley, D.J.; Ross, J.D.; Clarke, W.B. Helium and neon isotope geochemistry of some groundwaters from the Canadian Precambrian Shield. Geochim. Cosmochim. Acta. 1984, *48*, 1973–1985.

Bottomley, D.J.; Gascoyne, M.; Kamineni, D.C. The geochemistry, age and origin of groundwater in a mafic pluton, East Bull Lake, Ontario, Canada. Geochim. Cosmochim. Acta. 1990, *54*, 933–1008.

Bottomley, D.J.; Gregoire, D.C.; Kenneth, G.R. Saline groundwaters in the Canadian Shield: geochemical and isotopic evidence for a residual evaporite brine component. Geochim. Cosmochim. Acta. 1994, 58, 1483–1498.

Breen, K.J.; Masters R.W.; Chemical and isotopic characteristics of brines from three oil- and gas-producing sandstones in eastern Ohio, with applications to the geochemical tracing of brine sources: U.S. Geological Survey. Water-Investigations Report. 1985; 84–4314.

Celati, R.; Noto, P.; Panichi, C.; Squarci, P.; Taffi, L. Interactions between the steam reservoir and surrounding aquifers in the Larderello Geothermal Field. Geothermics. 1973, 2, 174–185.

Connolly, C.A.; Walter, L.M.; Baadsgaard, H.; Longstaffe, F.J. Origin and evolution of formation waters, Alberta Basin, Western Canada Sedimentary Basin. I. Chemistry Appl. Geochem. 1990a, 5, 375–395.

Connolly, C.A.; Walter, L.M.; Baadsgaard, H.; Longstaffe, F.J. Origin and evolution of formation waters, Alberta Basin, Western Canada Sedimentary Basin. II. Isotope systematics and water mixing. Appl. Geochem. 1990b, 5, 397–413.

Cooper, M.; Weissenberger, J.; Knight, I.; Hostad, D.; Gillespie, D.; Williams, H.; Burden, E.; Porter-Chaudhry, J.; Rae, D.; Clark, E. Basin evolution in western Newfoundland: new insights from hydrocarbon exploration. AAPG Bull. 2001, 85 (3), 393–418.

Corazza, E. Field workshop on volcanic gases, Vulcano, Italy, 1982, general report. Geothermics. 1986, 15, 197–200.

Craig H. The isotope geochemistry of water and carbon in geothermal areas. In *Nuclear Geology of Geothermal Areas*; Tongiorgi, E., Ed.; Spoleto, CNR (Cent. Naz. rich.), Lab. Nucl. Geol., Pisa, 1963; 17–53.

Dickin, R.C.; Frape, S.K.; Fritz, P.; Leech, R.E.J.; Pearson, R. Groundwater chemistry to depths of 1000 m in low permeability granitic rocks of the Canadian Shield. *Groundwater Symposium on Groundwater Resources Utilization and Contaminant Hydrology*; IAH: Montreal, Canada, 1984; Vol. 2, 357–371.

Dollar, P.S; Frape, S.K.; McNutt, R.H. Geochemistry of formation waters, southwestern Ontario, Canada and southern Michigan. U.S.A.: implications for origin and evolution. Ontario geological Survey Open File Report, 1991; 5743.

Egeberg, P.K.; Aagaard, P. Origin and evolution of formation waters from oil fields on the Norwegian shelf. Appl. Geochem. 1989, 4, 131–142.

Emmermann, R.; Lauterjung J. The German Deep Drilling Program KTB: overview and major results. J. Geophys. Res. 1997, 102 (B8), 18179–18201.

England, W.A., Fleet, A.J., Eds.; *Petroleum Migration*; The Geological Society: London, 1991; 280 pp.

Fournier, R.O.; Rowe, J.J. Estimation of underground temperatures from the silica content of water from hot springs and wet-steam wells. Am J. Sci. 1966, 264, 685–697.

Frape, S.K.; Fritz, P.; McNutt, R.H. Water–rock interaction and chemistry of groundwaters from the Canadian Shield. Geochim. Cosmochim. Acta. 1984, 48, 1617–1627.

Gat, J.; Mazor, E.; Tzur, Y. The stable isotope composition of mineral waters in the Jordan Rift Valley, Israel. J. Hydrology 1969, 7, 334–352.

References

Gavrieli, I.; Yechieli, Y.; Halicz, L.; Foreign, D. Survey of mineral waters along the Dead Sea coast. Geological Survey of Israel Current Research 1997, *11*, 71–75.

Haak, V., et al. KTB and the electrical conductivity of the crust. J. Geophys. Res. 1997, *102* (1b), 18,289–18,305.

Hanson, A.D.; Riots Bad; Sinker, D; Meltdown, Lam; Buffet, U. Upper Oligocene lacustrine source rocks and petroleum systems of the northern Qaidam basin, northwest China. AAPG Bull. 2001, *85*, 601–619.

Hardie, L.A. Evaporites: marine or non-marine. Am J. Sci. 1984, *284*, 193–240.

Heaton, T.H.E. Rates and sources of ^{4}He accumulation in groundwater. Hydrology Sci. J. 1984, *29*, 29–47.

Herczeg, A.L.; Simpson, H.J.; Mazor, E. Transport of soluble salts in a large semiarid basin: River Murray, Australia. J. Hydrology. 1993, *144*, 59–84.

Herzberg, O.; Mazor, E. Hydrological applications of noble gases and temperature measurements in underground water systems: examples. Israel. J. Hydrology 1979, *41*, 217–231.

Hitchon, B.; Friedman, I. Geochemistry and origin of formation waters in the western Canada sedimentary basin. I. Stable isotopes of hydrogen and oxygen. Geochim. Cosmochim. Acta. 1969, *33*, 1321–1349.

Hitchon, B.; Billings, G.K.; Klovan, J.E. Geochemistry and origin of formation waters in the western Canada sedimentary basin. III. Factors controlling chemical composition. Geochim. Cosmochim. Acta. 1971, *35*, 567–598.

Hsü, K.J. When the Mediterranean dried up. Sci. Am. 1972a, *227* (6), 26–36.

Hsü, K.J. Origin of saline giants: a critical review after the discovery of the Mediterranean evaporites. Earth Sci. Rev. 1972b, *8*, 371–396.

Jenden, P.D.; Gieskes, J.M. Chemical and isotopic composition of interstitial water from Deep Sea Drilling Project sites 533 and 534. Initial Reports DSDP, 76; U.S. Government Printing Office: Washington; 1984; 453–461.

Kelly, W.C.; Rye, R.O.; Livnat, A. Saline mine waters of the Keweenaw Peninsula, northern Michigan: their nature, origin, and relation to similar deep waters in Precambrian crystalline rocks of the Canadian Shield. Am J. Sci. 1986, *286*, 281–308.

Kharaka, Y.K.; Berry, F.A.F. The influence of geologic membranes on the geochemistry of subsurface waters from Miocene sediments at Kettleman North Dome, California. Water Resources Res. 1974, *10*, 313–327.

Kharaka, Y.K.; Berry, F.A.F.; Friedman, I. Isotopic composition of oil-field brines from Kettleman North Dome Oil-field, California, and their geologic implications. Geochim. Cosmochim. Acta. 1973, *37*, 1899–1908.

Kharaka, Y.K.; Maest, A.S.; Carothers, L.M.; Law, L.M.; Lamothe, P.J.; Fries, T.L. Geochemistry of metal-rich brines from central Mississippi Dome Basin, U.S.A. Appl. Geochem. 1987, *2* (5/6), 543–561.

Lane A.C. Mine waters and their field assay. Bull. Geol. Soc. Am. 1908, *19*, 501–512.

Lowenstein, T.K. Origin of depositional cycles in a Permian "saline giant": The Salado (McNutt zone) evaporites of New Mexico and Texas. Geol. Soc. America Bull. 1988, *100*, 592–608.

Lubeking, G.A.; Longman, M.W.; Carlisle, W.J. Unconformity-related chert/dolomite production in the Pennsylvanian Amsden Formation, Wolf Springs fields, Bull Mountains basin of central Montana. AAPG Bull. 2001, *85* (1), 131–148.

Mackay, N.; Hillman, T.; Rolls, J. Water quality of the River Murray: review of monitoring 1978–1986. Murray-Darling Basin Commission, Canberra, A.C.T., Water Qual. Rep. No. 1, 1988.

Mahabubi, A.; Moussavi-Harami, R.; Brener, R.L. Sequence stratigraphy and sea level history of the upper Paleocene strata in the Kopet-Dagh basin, northeastern Iran. AAPG Bull. 2001, 85 (5), 839–859.

Manheim, F.T.; Sayles, F.L. Interstitial water studies on small core samples. Deep Sea Drilling Project, Leg 1. DSDP, 1969.

Manheim, F.T.; Chan, K.M.; Sayles, F.L. Interstitial water studies on small core samples. Deep Sea Drilling Project, Leg 5. DSDP, 1970.

Manheim, F.T.; Sayles, F.L. Interstitial water studies on small core samples. Deep Sea Drilling Project, Leg 6. DSDP, 1971.

Manheim, F.T.; Sayles, F.L. Interstitial water studies on small core samples. Deep Sea Drilling Project, Leg 10. DSDP, 1973.

Manheim, F.T.; Schug, D.M. Interstitial waters of Black Sea cores. Initial Reports, DSDP; U.S. Government Printing Office: Washington, 1978; Vol. 42, 637–651.

Martini, M.; Piccardi, G.; Cellini Legittimo, P. Geochemical surveillance of active volcanoes: data on the fumaroles of Vulcano (Aeolian Islands, Italy). Bull. Volcanology 1980, 43, 255–263.

Martinsen, R.S. Stratigraphic controls on the development and distribution of fluid-pressure compartments. In *Seals, Traps, and the Petroleum System*; Surdman, R. C., Ed.; AAPG Memoir 67, 1997; 223–241.

Marty, B.; Criaud, A.; Fouilac, C. Low enthalpy geothermal fluids from the Paris sedimentary basin. 1. Characteristics and origin of gases. Geothermics 1988, 17, 619–633.

Marty, B.; Torgersen, T.; Meynier, V. Helium isotope fluxes and groundwater ages in the Dodger Aquifer, Paris Basin. Water Resources Res. 1993, 29, 1025–1035.

Mazor, E. Paleotemperatures and other hydrological parameters deduced from noble gases dissolved in groundwaters: Jordan Rift Valley, Israel. Geochim. Cosmochim. Acta. 1972, 36, 1321–1336.

Mazor, E. Atmospheric and radiogenic noble gases in thermal waters: their potential application to prospecting and steam production studies. Proc. 2nd U.N. Symp on the Use of Geothermal Resources; San Francisco, May 1975 I,; 793–801.

Mazor, E. Multitracing and multisampling in hydrological studies. In *Interpretation of Environmental Isotope and Hydro-Chemical Data in Groundwater Hydrology*; IAEA: Vienna, 1976; 7–36.

Mazor, E. Geothermal tracing with atmospheric and radiogenic noble gases. Geothermics 1977, 5, 21–36.

Mazor, E. Noble gases in a section across the vapor dominated geothermal field in Larderello, Italy. Pure Appl. Geophys. (PAGEOPH) 1979, 117, 262–275.

Mazor, E. Sampling of volcanic gases—the role of noble-gas measurements: a case study of Vulcano, south Italy. Chem. Geol. 1985, 49, 329–338.

Mazor, E. Atmospheric and radiogenic noble gases as tracers for migration of fluids in the upper crust. Japan–U.S. Seminar on Terrestrial Rare Gases; Yellowstone N.P., Sept. 1986; Proc. 1986, 47–51.

Mazor E. Reinterpretation of Cl-36 data: physical processes, hydraulic interconnections, and age estimates in groundwater systems. Appl. Geochem. 1992a, *7*, 351–360.

Mazor, E. Interpretation of water–rock interactions in cases of mixing and undersaturation. In *Water–Rock Interaction*; Kharaka, Y.K., Maest, A.S., Eds., Balkema: Roterdam, 1992b; Vol. I, 233–23.

Mazor, E. Stagnant aquifer concept. I. Large scale artesian systems—Great Artesian Basin, Australia. J. Hydrology 1995, *174*, 219–240.

Mazor, E. *Chemical and Isotopic Groundwater Hydrology: The Applied Approach*; 3rd Ed.; Marcel Dekker: New York, 2003.

Mazor, E.; Bosch, A. Dynamics of groundwater in deep basins: He-4 dating, hydraulic discontinuities and rates of drainage. International Conference on Groundwater in Large Sedimentary Basins, Perth, Western Australia, July 9–13 1990, Department of Primary Industries and Energy; Proc., 1990; 380–389.

Mazor, E.; Bosch, A. He as a semi-quantitative tool for groundwater dating in the range of 10^4 to 10^8 years. In *Isotopes of Noble Gases as Tracers in Environmental Studies*; IAEA: Vienna, 1992; 163–178.

Mazor, E.; Fournier, R.O. More on noble gases in Yellowstone National Park hot waters. Geochim. Cosmochim. Acta. 1973, *37*, 515–525.

Mazor E.; George R. Marine airborne salts applied to trace evapotranspiration, local recharge and lateral groundwater flow in Western Australia. J. Hydrology 1992, *139*, 63–77.

Mazor, E.; Mañon, A. Geochemical tracing in producing geothermal fields: a case study at Cerro Prieto. Geothermics 1979, *8*, 231–240.

Mazor, E.; Mero, F. Geochemical tracing of mineral water sources in the Lake Tiberias basin, Israel. J. Hydrology 1969a, *7*, 276–317.

Mazor, E.; Mero, F. The origin of the Tiberias–Noit mineral water association in the Tiberias–Dead Sea Rift Valley, Israel. J. Hydrology 1969b, *7*, 318–333.

Mazor, E.; Molcho, M. Geochemical studies of the Feshcha springs, Dead Sea Basin. J. Hydrology 1972, *15*, 37–47.

Mazor E.; Nativ R. Hydraulic calculation of groundwater flow velocity and age: examination of the basic premises. J. Hydrology 1992, *138*, 211–222.

Mazor E.; Nativ R. Stagnant groundwaters stored in isolated aquifers: implications related to hydraulic calculations and isotopic dating—Reply. J. Hydrology 1994, *154*, 409–418.

Mazor, E.; Thompson, J.M. Evolution of geothermal fluids deduced from chemistry plots: Yellowstone National Park (U.S.A.). J. Volcanology and Geothermal Res. 1982, *12*, 351–360.

Mazor, E.; Truesdell, A.H. Dynamics of a geothermal field traced by noble gases: Cerro Prieto, Mexico. Geothermics 1984, *13*, 91–102.

Mazor, E.; Verhagen, B.T. Hot springs of Rhodesia: their noble gases, isotopic and chemical composition. J. Hydrology 1976, *28*, 29–43.

Mazor, E.; Verhagen, B.T. Dissolved ions, stable and radioactive isotopes and noble gases in thermal waters of South Africa. J. Hydrology 1983, *63*, 315–329.

Mazor, E.; Wasserburg, G.J. Helium, neon, argon, krypton and xenon in gas emanation in Yellowstone and Lassen Volcanic National Parks. Geochim. Cosmochim. Acta. 1965, *29*, 443–454.

Mazor, E.; Rosenthal, E.; Eckstein, J. Geochemical tracing of mineral water sources in the south-western Dead Sea Basin, Israel. J. Hydrology 1969, *7*, 246–275.

Mazor, E.; Kaufman, A.; Carmi, I. Hammat Gader (Israel): geochemistry of a mixed thermal spring complex. J. Hydrology 1973a, *18*, 289–303.

Mazor, E.; Nadler, A.; Harpaz, Y. Notes on the geochemical tracing of the Kaneh-Samar spring complex, Dead Sea basin. Israel J. Earth Sci. 1973b, *22*, 255–262.

Mazor, E.; Verhagen, B.Th.; Negreanu, E. Hot springs of the igneous terrain of Swaziland: their nobles gases, hydrogen, oxygen and carbon isotopes and dissolved ions. In *Isotope Techniques in Groundwater Hydrology*; IAEA: Vienna, 1974; Vol. 2, 29–47.

Mazor, E., Vuataz, F.D.; Jaffe, F.C. Tracing groundwater components by chemical, isotopic and physical parameters—example: Schinznach, Switzerland. J. Hydrology 1985, *76*, 233–246.

Mazor, E.; Jaffe, F.C., Flück, J.; Dubois, J.D. Tritium-corrected C-14 and atmospheric noble gases corrected He-4, applied to deduce ages of mixed groudwaters: examples from the Baden region, Switzerland. Geochim. Cosmochim. Acta. 1986, *50*, 1611–1618.

Mazor, E.; Cioni, R.; Corazza, E.; Fratta, M.; Magro, G.; Matsuo, S.; Hirabayashi, J.; Shinohara, H.; Martini, M.; Piccardi, G.; Cellini-Legittimo, P. Evolution of fumarolic gases—boundary conditions set by measured parameters: case study at Vulcano, Italy. J. Volcanology 1988, *50*, 71–85.

Mazor, E.; Bosch, A.; Stewart, M.K.; Hulston, R. The geothermal system of Wairakei, New Zealand: processes and age estimates inferred from noble gases. Appl. Geochem. 1990, *5*, 605–624.

Mazor, E.; Gilad, D.; Fridman, V. Stagnant aquifer concept: II. Small scale artesian systems—Hazeva, Dead Sea Rift Valley. Israel J. Hydrology 1995, *173*, 241–261.

McCaffrey, M.A.; Lazar, B.; Holland, H.D. The evaporation path of seawater and the coprecipitation of Br^- and K^+ with halite. J. Sedimentary Petrology 1987 *57*, 28–37.

McDuff, R.E.; Gieskes, J.M.; Lawrence, J.R. Interstitial water studies, Leg 42A. DSDP; 62 part 1, U.S. Government Printing Office: Washington, 1978, 561–568.

Möller P.; et al. Paleo-fluids and recent fluids in the upper continental crust: results from the German Continental Deep Drilling Program (KTB). J. Geophys. Res. 1997, *102* (B8), 18,233–18,254.

Noll, R.S. Geochemistry and hydrology of groundwater flow systems in the Lockport dolomite, near Niagara Falls, New York. M.Sc. thesis, Syracuse University, 1989.

Nordstrom, D.K.; Ball, J.W.; Donhaoe, R.J.; Whittemore, D. Groundwater chemistry and water–rock interactions at Stripa. Geochim. Cosmochim. Acta. 1989a, *53*, 1727–1740.

Nordstrom, D.K.; Lindblom, S.; Donahoe,R.J.; Barton, C.C. Fluid inclusions in the

Stripa granite and their possible influence on the groundwater chemistry. Geochim. Cosmochim. Acta. 1989b, *53*, 1741–1755.

Nordstrom, D.K.; Olsson, T.; Carlsson, L.; Fritz,. P. Introduction to the hydrogeochemical investigations within the International Stripa Project. Geochim. Cosmochim. Acta. 1989c, *53*, 1717–1726.

Nurmi, P.A.; Kukkonen, I.T.; Lahermo, P.W. Geochemistry and origin of saline groundwaters in the Fennoscandian Shield. Appl. Geochem. 1988, *3*, 185–203.

Ortoleva, P.J.; Basin compartmentation: definitions and mechanisms; In *Basin Compartments and Seals*; Ortoleva, P.J., Ed.; AAPG Memoir 61; American Association of Petroleum Geologists: Tulsa, Oklahoma, 1994a; 477.

Ortoleva, P.J., Ed. *Basin Compartments and Seals*; AAPG Memoir 61, 1994b.

Patterson, R.J.; Kinsman, J.J. Hydrologic framework of a sabkha along the Arabian Gulf. AAPG Bull. 1981, *65*, 1457–1475.

Pekdeger, A.; Sommer-von Jarmersted, C.; Thomas, L. Hydrochemical sampling of the formation water at the KTB-borehole and their chemical composition. Scientific Drilling 1994, *4*, 101–111.

Person M. Crustal-scale hydrogeology—an emerging paradigm. Hydrogeology J. 1996, *4*, 2–3.

Phillips, F.M.; Benrley, H.; Davis, S.N.; Elmoer, D.; Swanick, G.B. Chlorine-36 dating of very old groundwater 2. Milk River Aquifer, Alberta, Canada. Water Resources Res. 1986, *22*, 2003–2016.

Plummer, L.N.; Busby, J.F.; Lee, R.W.; Hanshaw, B.B. Geochemical Modeling of the Madison aquifer in parts of Montana, Wyoming and South Dakota. Water Resources Res. 1990, *26*, 1981–2014.

Pluta, I.; Zuber, A. Origin of brines in the Upper Silesian Coal Basin (Poland) inferred from stable isotope and chemical data. Appl. Geochem. 1995, *10*, 447–460.

Presley, B.J.; Pettrowski, C.; Kaplan, I.R. Interstitial water chemistry: Deep Sea Drilling Project, Leg 10. DSDP, 1973a.

Presley, B.J.; Pettrowski, C.; Kaplan, I.R. Interstitial water chemistry: Deep Sea Drilling Project, Leg 13. Init. Rep. DSDP, 1973b; 809–812.

Raven, K.G.; Bottomley, D.J.; Sweezry, R.A.; Smedley, J.A.; Ruttan, T.J. Hydrogeological characterization of the East Bull Lake research area. Environment Canada, IWD Sci. Series, No. 160; NHRI Paper No. 31., 1987.

Rogers, S.M. Deposition and diagenesis of Mississippian chat reservoirs, north-central Oklahoma. AAPG Bulletin 2002, *85* (1), 115–129.

Rosen, M.R.; Coshel, L.; Turner, J.V.; Woodbury, R.J. Hydrochemistry and nutrient cycling in Yalgroup National Park, Western Australia. J. of Hydrology 1996, *185*, 241–274.

Ruble, T.E.; Lewan, M.D.; Philp, R.P. New insights on the Green River petroleum system in the Uinta basin from hydrous pyrolysis experiments. AAPG Bull. 2001, *85* (8), 1333–1371.

Ruppel, S.C.; Barnbay, R.J. Contrasting styles of reservoir development in proximal and distal chert facies: Devonian Thrtyone Formation, Texas, AAPG Bull. 2001, *85* (1), 7–33.

Saller, A.; Ball, B.; Robertson, S.; McPherson, B.; Wene, C.; Nims, R.; Gogas, J.

Reservoir characteristics of Devonian cherts and their control on oil recovery: Dollarhide field, west Texas. AAPG Bull. 2001, *85* (1), 35–50.

Sayles, F.L.; Manheim, F.T. Interstitial water studies on small core samples the Mediterranean Sea. DSDP, 1973.

Sayles, F.L.; Manheim, F.T.; Chan, K.M. Interstitial water studies on small core samples, Leg 4. DSDP, 1970.

Schlosser, P.; Stute, M.; Dorr, H.; Sonntag, C.; Münich, K.O. Tritium/^{3}He dating of shallow groundwater. Earth and Planetary Sci. Lett. 1988, *89*, 353–362.

Stute, M.; Schlosser, P.; Clark, J.F.; Brocker, W.S. Paleotemperatures in the southwestern United States driven from noble gases in ground water. Science 1992, *256*, 1000–1003.

Taylor, J.C.M. Late Permian Zechstein. In *Introduction to the Petroleum Geology of the North Sea*; Glennied, K.W., Ed.; Oxford, Blackwell, 1985.

Tepper, D.H.; Goodman, W.M.; Gross, M.R.; Kappel, W.M.; Yager, R.M. Stratigraphy, structural geology, and hydrogeology of the Lockport Group: Niagara Falls area, New York. In *Field Trip Guidebook*, New York State Geological Association; 62nd Annual Meeting, Fredonia, NY; Lash, G.G., Ed.; 1990; B1–B25.

Tellam, J.H. Hydrochemistry of the saline groundwaters of the lower Mersey Basin Permo-Triassic sandstone aquifer, U.K. J. Hydrology 1995, *165*, 45–54.

Tissot, B.P.; Welte, D.H. *Petroleum Formation and Occurrence*; Springer-Verlag, 1984; 699pp.

Torgersen, T.; Clarke, W.B. Helium accumulation in groundwater. 1. An evaluation of sources and the continental flux of crustal ^{4}He in the Great Artesian Basin, Australia. Geochim. Cosmochim. Acta. 1985, *49*, 1211–1218.

Toth, J.; Sheng, G. Enhancing safety of nuclear waste disposal by exploiting regional groundwater flow: the recharge area concept. Hydrogeology J. 1996, *4*, 4–24.

Truesdell, A.H.; Fournier, R.O. Calculation of deep temperatures in geothermal systems from the chemistry of boiling spring waters of mixed origin. Proc. 2nd U.N. Symp. Development and Use of Geothermal Resources; San Francisco 1975; 1976; 837–844.

Vengosh, A.; Starinsky A. The origin of saline waters from the interface zone along the coastal plain of Israel. Israel Geol. Soc., Ann. Meeting, Askkelon 1992, 156–157 (Abst.).

Vuataz, F.D. Hydrologie, géochimie et géothermie de eaux thermales de Suisse et des regions Alpines limitrophes. Matériaux pour la géologie de la Suiss—hydrologie No. 29. Kummerly & Ffey Geographischer Verlag, Berne, 1982.

Vuataz, F.D.; Schneider, J.F.; Jaffe, F.C.; Mazor, E. Hydrogeochemistry and extrapolation of end members in a mixed thermal water system, Vals, Switzerland. Ecologae Geological Helvetiae 1983, *76*, 431–450.

Wasserburg, G.J.; Mazor, E.; Zartman, R.E. Isotopic and chemical composition of some terrestrial natural gases. In *Earth Sciences and Meteorites*; Amsterdam, 1963; 240–261.

Watney, W.; Guy, W.J.; Byrnes, A.P. Characterization of the Mississippian chat in south-central Kansas. AAPG Bull 2001, *85* (1), 85–1134.

References

Weaver, T.R.; Frape, S.K.; Cherry, J.A. Recent cross-formational fluid flow and mixing in the shallow Michigan Basin. GSA Bulletin 1995, *107*, 697–707.

Yechieli, Y.; Magaritz, M.; Levy, Y.; Weber, U.; Kafri, U; Woelfli, W.; Bonani, G. Late Quaternary geological history of the Dead Sea area, Israel. Quaternary Res. 1993, *39*, 58–67.

Zaikovsky, A.; Kosanke, B.J.; Hubbard, N.; Noble gas composition of deep brines from the Palo Duro Basin, Texas. Geochim. Cosmochim. Acta. 1987, *51*, 73–84.

Zartman, R.E.; Wasserburg, G.J.; Reynolds, J.H. Helium, argon and carbon in some natural gases. J. Geophys. Res. 1961, *66*, 277–306.

INDEX

Aerated zone, 83
Africa, North, oil fields, 241
Aquifer, 44
Argon-40 dating, 39, 59, 163
Artesian system, 48, 186
Australia
    Campaspe River Basin, 76
    Great Artesian Basin, 90
    Murray River, 68
    Wheatbelt, 78
    Yalgorup National Park, 75

Base-flow, 84
Biological evolution, 13, 361
Bottled water, 290
Brine tagging, 32, 33, 171

Canada
    Alberta Basin, 127
    Canadian Shield, 176, 179, 182, 184, 186, 188, 197
    Milk River, 168
    Ontario, 212
    Sudbury, 181, 185
    Western Canada Basin, 206, 207, 210
Chemical parameters, 55, 330

China, Qaidam oil fields, 250
Coal, 250, 261, 262
Compaction, 89, 108, 153
Compartmentalization, 238, 252, 253, 263
Confinement age, 59
Connate water, 30, 111, 122, 154
Crystalline shields, 33, 78, 171, 189

Data Banks, 339
Dating water, 57, 157, 160, 165, 196
Delta environments, 246
Diffusion, 109
Dissolved ions, 47

Earth from space, 8
Education, water related, 354
England, Mersey Basin, 136
Erosion, 11
Europe
    Fennoscandian Shield, 176, 177, 182, 184, 186
    Norwegian Shelf, 235
    Upper Silesian Coal Basin, Poland, 137
    Zechstein Formation, 119
Evaporitic landscapes, 117, 156

Exhibitions, water related, 289, 355, 362, 363
External ions, 22

Facies, groundwater, 66, 94, 98
Flow patterns, 20
Formation water, 27, 45, 126, 151, 190
Fossil fuels and water, 359
France
  Paris Basin, 167
Fracture
  compartments, 172, 183
  propagation, 173
  short-circuiting, 174, 189
Fresh water, 358
Fruitcake structure, 28, 153, 268

Geohydroderm, 16, 17, 66, 126
Geosystem, 17
Germany, KTB deep drill, 184, 194, 199
Gravitational flow, 19, 83, 86
Groundwater
  confined, 43
  static, 43
  unconfined, 18, 41
Gypsum, 31, 115

Halite, 31, 105, 111, 115, 118
^{4}He dating, 39, 59, 160, 162, 197
Heat gradient, terrestrial, 37
Hydraulic barriers, 46
Hydraulic continuity, 46
Hydraulic potential, 84, 88
Hydrothermal groundwater, 38, 50, 292

Impermeability, 41
Internal ions, 22
Interstitial sub-oceanic water, 24, 92, 108
Iran, oil fields, 243
Isotopic composition of water, 24, 27, 56, 74
Israel
  Coastal Plain, 150
  Dead Sea, 121

[Israel]
  Kanneh-Samar springs, 284
  Lake Tiberias Basin, 142
  Rift Valley, 141
Italy
  Larderello steam field, 314
  Vulcano, Aeolian Islands, 319

Landscaping, 31, 155
Low flatlands, 240

Management implications, 169, 329
Mars, 5
Medicinal aspects of water, 286
Mediterranean Sea, 98, 103, 111, 119, 140
Mercury, 3
Meteoric water, 28, 42, 193
Mexico, Cerro Prieto steam field, 304
Mixing, groundwaters, 60, 164, 175, 187, 197
Monitoring networks, 336
Moon, 6
Multiparameter studies, 332
Multisampling approach, 334
Museums, Water and Man, 362

New Zealand, hydrothermal fields, 310
Noble gases, 38, 57
Nuclear waste management, 35, 52, 175, 201

Ocean
  and water cycle, 10
^{18}O shift, 33, 36

Permeability, 30, 41, 87, 269
Petroleum
  connate, 266
  exploration, 49, 60
  formation, 35, 255, 263
  hydrology, 12, 205, 212
  migration, 255
  traps, 266
Physical parameters, 54

Index

Pollution, 46, 61
Pressure cooker model of oil and coal formation, 255, 263
Pressure regulating mechanisms, 268

Quality of data, 330

Rift valleys, 148, 249
Rocks, 13
Rock-compartments, 27

Sabkha, Abu Dhabi, 248
Sampling, water, 62, 63
Saturated zone, 83
Sea-spray, 21, 70, 75, 76, 78
Seawater encroachment, 26
Sedimentary basin, 27, 126, 151, 240
South Africa, thermal springs, 280
Spas, 286, 288
Subsidence phenomena, 238
Swaziland, hot springs, 271
Sweden, Stripa mine, 180, 198
Switzerland, Combioula Springs, 283

Teaching, water related, 364
Tectonic dynamics, 48, 200
Thermal fluids, 50
Through-flow, 84, 176

USA
    California, Los Angeles Basin, 267

[USA]
    Green River petroleum system, 261
    Kansas, oil fields, 240
    Kettleman oil field, California, 213
    Michigan Basin, 217
    Mississippi Salt Dome oil fields, 228
    Montana, Yellowstone Country, oil fields, 245
    Niagara Falls, borehole, 132
    Ohio oil fields, 222
    Oklahoma, petroleum fields, 247
    Ontario, 120
    Saline Giant, 118
    Texas, Palo Duro Basin, 166
    West Texas, 240, 252
    Yellowstone National Park, 293, 299, 302

Venus, 4
Volcanic water system, 50

Warm water, 37, 271
Water
    active cycle, 18, 19
    properties, 7, 355
    quality, 45
    superheated, 326
Water-rock interaction, 23, 82, 192
Well types, 331

Zimbabwe, hot springs, 278